STARS

Edited by Thomas A. Fleming

cognella™

First published in the United States of America in 2011 by Cognella, a division of University Readers, Inc.

15 14 13 12 11   1 2 3 4 5

Printed in the United States of America

ISBN: 978-1-60927-843-4

cognella™

www.cognella.com 800.200.3908

TABLE OF CONTENTS

Source Material

Stars by James B. Kaler, 1998, Scientific American Library, W.H. Freeman & Company, ISBN 0–7167–5033–3

The Cosmic Perspective, 5th Ed., by Bennett, Donahue, Schneider, & Voit, 2009, Pearson Education (Addison–Wesley), ISBN 0–321–55817–0 [6th Edition]

Cosmos by Carl Sagan, 1980, Random House Inc., ISBN 0–394–50294–9

Variable Star Astronomy by The American Association of Variable Stars Observers, 2008

Tales of the Constellations: The Myths and Legends of the Night Sky by Marianne McDonald, 1996, Michael Friedman Publishing Group, ISBN 0–8317–7277–8

PREFACE

In 1998, the University of Arizona adopted a new general education program that contained two tiers of course requirement: The standard introductory survey course (Tier 1) and a new type of specialized course, designated Tier 2. We in the Department of Astronomy were tasked to create a suite of new courses that were more narrowly defined than the introductory course, which covered the topic in much more depth than the introductory course did, and yet were still appropriate for the non-science major. In 2000, I created such a Tier 2 course, entitled *Stars*.

My idea for the course was to cover every aspect of the stars, from how they capture the human imagination, to how we obtain information about them, to how we develop an understanding of their true nature, namely, that they are nuclear fusion reactors that create the elements of which our very selves consist. I wanted my students to learn how and where the various elements are created by learning the details of the nuclear fusion reactions and neutron capture processes that occur in the stars, but at the same time keep the romance in the course by having them learn about the classical mythology behind the naming of the stars and constellations. Finally, I would tie all of these topics together within Carl Sagan's theme of "we are all made of star stuff."

The learning goals for this course are these: You will have an appreciation for the role that the stars have played in the history of humankind; you will be able to read and understand the information contained in a basic star catalog; you will know how the elements were created; and you will have exercised your critical thinking and problem solving skills.

The difficulty that I faced in implementing this course was finding an appropriate textbook. There are many introductory astronomy textbooks on the market, but they would not suit the purpose of my new course. We would not even cover ⅔ of the content in a typical introductory astronomy textbook, and the ⅓ of the material that we did cover did not cover the topic of stars in enough depth. I did find a couple of textbooks that treated the topic of stars in depth, but they were written with the presumption that the student had been exposed to calculus, which made them inappropriate for the non-science major.

I did find a wonderful book by Prof. James B. Kaler, which was entitled, appropriately, *Stars*. Published as part of the Scientific American Library, this inexpensive book was not written as a textbook, but it did cover most of the topics at the level that I wanted to cover them. So I made this the official "textbook" for my course. Unfortunately, the *Stars* book went out of print in 2002, and I have not used a textbook in my *Stars* course ever since.

This brings us to the current volume. This University Reader is an attempt to resurrect the Kaler *Stars* book and supplement it with readings on other topics that I've added to the course over the years. I've found that students better appreciate the popular subject of black holes after first being introduced to the topics of special and general relativity. So I have included two chapters on Einstein's Special and General Theories of Relativity from the excellent astronomy textbook *The Cosmic Perspective* by Jeff Bennett, Megan Donahue, Nick Schneider, & Mark Voit, as well as a chapter on celestial coordinates and navigation. On the topic of constellation mythology, I think it best to hear from an expert, classics and theater arts Professor Marianne MacDonald. I include her analysis of how and why different cultures created the constellations and their myths from her book *Tales of the Constellations: The Myths and Legends of the Night Sky*. On the topic of variable stars, I include a very nice primer on the topic, complete with light curves, from the publication *Variable Star Astronomy* by The American Association of Variable Star Observers. Finally, I always end my course by showing my students the episode *The Lives of the Stars* from Carl Sagan's *Cosmos* television series. No one can provide the sheer poetry to summarize what my students have learned like Carl Sagan. So, I include parts of the corresponding chapter from the *Cosmos* book.

The stars have always been our guideposts. Our ancestors used them to navigate and to measure time. They made up stories about them, incorporated them into their religions, and even lived their lives according to their dictates. As our science developed over the centuries, we began to comprehend the true nature of the stars. They are nuclear reactors spread throughout our Galaxy, which provided us with energy. They are element factories which, when they die, become recycling plants that enrich our Galaxy with the very elements needed to construct planets like our Earth and begin the process of life. We have come to the realization that we truly do have a deep connection to the stars, just as our ancestors believed!

CHAPTER 1
WHERE CAN WE FIND THE STARS?

Socrates: *Shall we make astronomy the next study? What do you say?*
Glaucon: *Certainly. A working knowledge of the seasons, months, and years is beneficial to everyone, to commanders as well as to farmers and sailors.*
Socrates: *You make me smile, Glaucon. You are so afraid that the public will accuse you of recommending unprofitable studies.*
—Plato, *Republic*

In ancient times, the practical needs for timekeeping and navigation were important reasons for the study of astronomy. The celestial origins of timekeeping and navigation are still evident. The time of day comes from the location of the Sun in the local sky, the month comes from the Moon's cycle of phases, and the year comes from the Sun's annual path along the ecliptic. The very name of the "North Star" tells us how it can be an aid to navigation.

We can now tell the time by glancing at an inexpensive electronic watch and navigate with handheld devices that receive signals from satellites of the global positioning system (GPS). But knowing the celestial basis of timekeeping and navigation can still be useful, particularly for understanding the rich history of astronomical discovery. In this chapter, we will explore the apparent motions of the Sun, Moon, and planets in greater detail, which will allow us to study the principles of celestial timekeeping and navigation.

1.1: ASTRONOMICAL TIME PERIODS

Today, our clocks and calendars are beautifully synchronized to the rhythms of the heavens. Precision measurements allow us to ensure that our clocks keep pace with the Sun's daily trek across our sky, while our calendar holds the dates of the equinoxes and solstices as steady as possible. In earlier chapters, we saw how this synchronicity took root in ancient observations of the sky. However, it is only much more recently that we have come to understand timekeeping in all its details. In this section, we will look in more depth at basic measures of time and our modern, international system of timekeeping.

How do we define the day, month, year, and planetary periods?

By now you know that the length of the day corresponds to Earth's rotation, the length of the month to the cycle of lunar phases, and the length of the year to our orbit around the Sun. However, when we look carefully at each case, we find that the correspondence is not quite as simple as we might at first guess. Instead, we are forced to define two different types of day, month, and year. We also define planetary periods in two different ways. Let's take a look at how and why we make these distinctions.

The Length of the Day

We usually think of a day as the time it takes for Earth to rotate once, but if you measure this time period you'll find that is *not* exactly 24 hours. Instead, Earth's rotation period

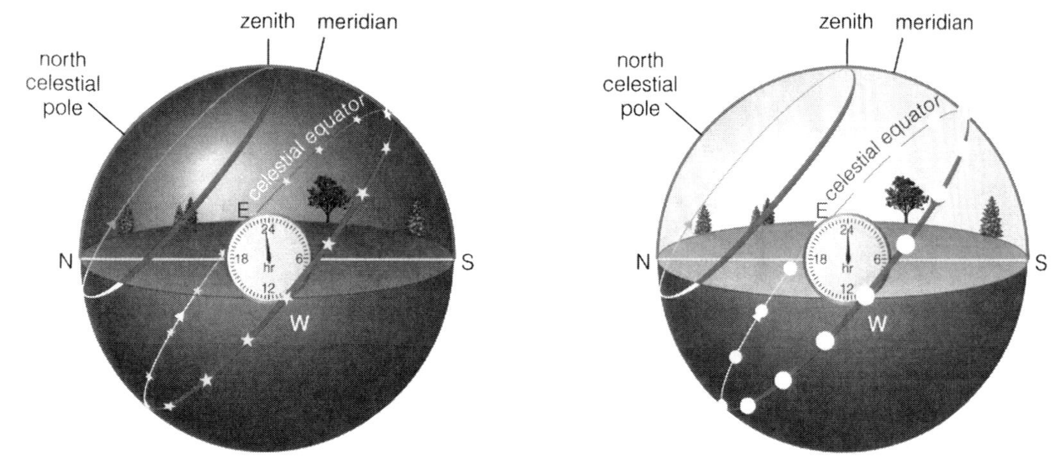

Figure 1.1 Using the sky to measure the length of a day.

a A sidereal day is the time it takes any star to make a circuit of the local sky. It is about 23 hours 56 minutes.

b A solar day is measured similarly. but by timing the Sun rather than a distant star. The length of the solar day varies over the course of the year but averages 24 hours.

is about 4 minutes short of 24 hours. What's going on?

We can understand the answer by thinking about the movement of the stars and Sun across our sky. Remember that the daily circling of the stars in our sky is an illusion created by Earth's rotation (see Figure 2.9). You can therefore measure Earth's rotation period by measuring how long it takes for any star to go from its highest point in the sky one day to its highest point the next day (Figure SU a). This time period, which we call a **sidereal day**, is about 23 hours 56 minutes (more precisely, $23^h56^m4.09^s$). *Sidereal* (pronounced *sy–DEAR–ee–al*) means "related to the stars"; note that you'll measure the same time no matter what star you choose. For practical purposes, the sidereal day is Earth's precise rotation period.

Our 24-hour day, which we call a solar day, is based on the time it takes for the Sun to make one circuit around the local sky. You can measure this time period by measuring how long it takes the Sun to go from its highest point in the sky one day to its highest point the next day (Figure 1.1b). The solar day is indeed 24 hours on average, although it varies slightly (up to 25 seconds longer or shorter than 24 hours) over the course of a year.

A simple demonstration shows why the solar day is about 4 minutes longer than the sidereal day. Set an object to represent the Sun on a table, and stand a few steps away to represent Earth. Point at the Sun and imagine that you also happen to be pointing toward a distant star that lies in the same direction. If you rotate (counterclockwise) while standing in place, you'll again be pointing at both the Sun and the star after one full rotation (Figure 1.2a). However, because Earth rotates at the same time it orbits the Sun, you can make the demonstration more realistic by taking a couple of steps around the Sun (counterclockwise) while you rotate (Figure 1.2b). After one full rotation, you will again be pointing in the direction of the distant star, so this represents a sidereal day. But it does not represent a solar day, because you will not yet be pointing back at the Sun. If you wish to point again at the Sun, you need to make up for your orbital motion by making slightly more than one full rotation. This "extra" bit of rotation makes a solar day longer than a sidereal day.

The only problem with this demonstration is that it exaggerates Earth's daily orbital motion. Because Earth takes about 365 days (1 year) to make a full 360° orbit around the Sun, Earth actually moves only about 1° per day around its orbit. Thus, a solar day represents about 361° of rotation, rather than the 360° for a sidereal day (Figure 1.2c). The extra 1° rotation takes about 1/360 of Earth's rotation period, which is about 4 minutes.

The Length of the Month

Our month comes from the Moon's 29-day cycle of phases (think "moonth"). More technically, the 29-day period required for each cycle of phases is called a synodic month. The word synodic comes from the Latin *synod*, which means "meeting." A synodic month gets its name because the Sun and the Moon "meet" in the sky with every new moon.

Just as a solar day is not Earth's true rotation period, a synodic month is not the Moon's true orbital period. Earth's motion around the Sun means that the Moon must complete more than one full orbit of Earth from one new moon to the next (Figure 1.3). The Moon's true orbital period, or sidereal month, is only about 27 days. Like the sidereal day,

Figure 1.2 A demonstration showing why a solar day is slightly longer than a sidereal day.

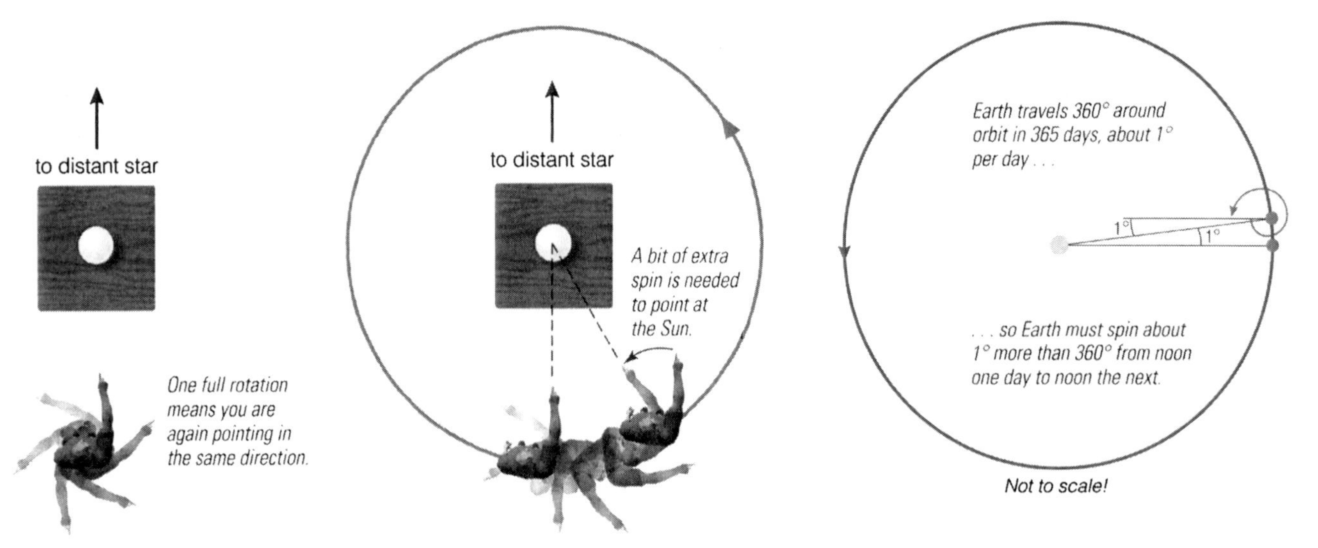

a One full rotation represents a sidereal day and means that you are again pointing to the same distant star.

b While you are "orbiting" the Sun, one rotation still returns you to pointing at a distant star; but you need slightly more than one full rotation to return to pointing at the Sun.

c Earth travels about 1° per day around its orbit, so a solar day requires about 361° of rotation.

the sidereal month gets its name because it describes how long it takes the Moon to complete an orbit relative to the positions of distant stars.

The Length of the Year

A year is related to Earth's orbital period, but again there are two slightly different definitions for the length of the year. The time it takes for Earth to complete one orbit relative to the stars is called a sidereal year. But our calendar is based on the cycle of the seasons, which we measure as the time from the spring equinox one year to the spring equinox the next year. This time period, called a tropical year, is about 20 minutes shorter than the sidereal year. A 20-minute difference might not seem like much, but it would make a calendar based on the sidereal year get out of sync with the seasons by 1 day every 72 years—a difference that would add up over centuries.

The difference between the sidereal year and the tropical year arises from Earth's 26,000-year cycle of axis precession [Section 2.3]. Precession not only changes the orientation of the axis in space but also changes the locations in Earth's orbit at which the seasons occur. Each year, the location of the equinoxes and solstices among the stars shifts about 1/26,000 of the way around the orbit. If you do the math, you'll find that 1/26,000 of a year is about 20 minutes, which explains the 20-minute difference between the tropical year and the sidereal year.

Planetary Periods

Although planetary periods are not used in our modern timekeeping, they were important to many ancient cultures. For example, the Mayan calendar was based in part on the apparent motions of Venus. In addition, Copernicus's ability to determine orbital periods of planets, with his Sun-centered model played an important role in keeping the model alive long enough for its ultimate acceptance [Section 3.3].

A planet's **sidereal period** is the time the planet takes to orbit the Sun. (As usual, it has the name *sidereal* because it

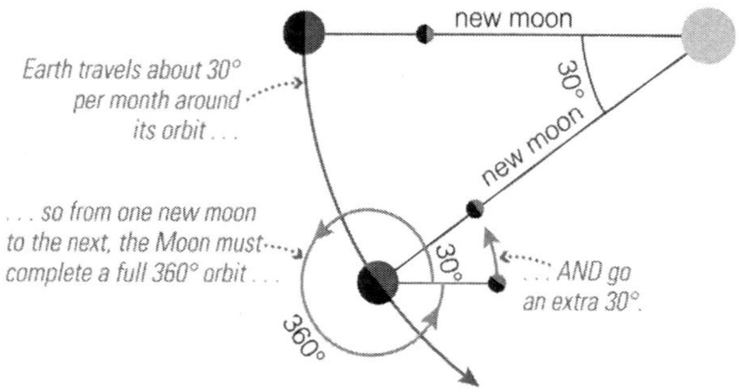

Figure 1.3 Interactive figure: The Moon completes one 360° orbit in about 27 1/3 days (a sidereal month), but the time from new moon to new moon is about 29 1/3 days (a synodic month).

is measured relative to distant stars.) For example, Jupiter's sidereal period is 11.86 years, so it takes about 12 years for Jupiter to make a complete circuit around the constellations of the zodiac. Jupiter therefore appears to move through roughly one zodiac constellation each year. If Jupiter is currently in Sagittarius (as it is for much of 2008), it will be in Capricorn at this time next year and Aquarius the following year, returning to Sagittarius in 12 years.

A planet's **synodic period** is the time between when it is lined up with the Sun in our sky at one time and the next similar alignment. (As with the Moon, the term *synodic* refers to the planets "meeting" the Sun in the sky.) Figure 1.4 shows that the situation is somewhat different for planets nearer the Sun than Earth (that is, Mercury and Venus) and planets farther away (all the rest of the planets).

Look first at the situation for the more distant planet in Figure 1.4. As seen from Earth, this planet will sometimes line up with the Sun in what we call a **conjunction**. At other special times, it will appear exactly opposite the Sun in our sky, or at **opposition**. We cannot see the planet during conjunction with the Sun because it is hidden by the Sun's glare and rises and sets with the Sun in our sky. At opposition, the planet moves through the sky like the full moon, rising at sunset, reaching the meridian at midnight, and setting at

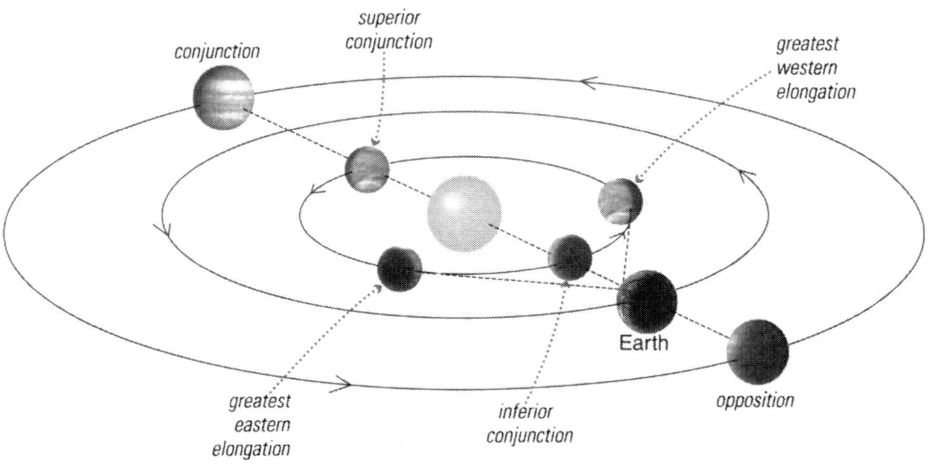

Figure 1.4 This diagram shows important positions of planets relative to Earth and the Sun. For a planet farther from the Sun than Earth (such as Jupiter), conjunction occurs when the planet appears aligned with the Sun in the sky, and opposition occurs when the planet appears on our meridian at midnight. Planets nearer to the Sun (such as Venus) have two conjunctions and never get farther from the Sun in our sky than at their greatest elongations. (Adapted from *Advanced Skywatching*, by Burnham et al.)

dawn. Note that the planet is closest to Earth at opposition and hence appears brightest in our sky at this time.

Now look at the planet that is *nearer* than Earth to the Sun in Figure 1.4. This planet never has an opposition but instead has two conjunctions—an "inferior conjunction" between Earth and the Sun and a "superior conjunction" when the planet appears behind the Sun as seen from Earth. Two other points are important for the inner planets: their points of **greatest elongation**, when they appear farthest from the Sun in our sky. At its greatest eastern elongation, Venus appears about 46° east of the Sun in our sky, which means it shines brightly in the evening. Similarly, at its greatest western elongation, Venus appears about 46° west of the Sun and shines brightly before dawn. In between the times when Venus appears in the morning sky and the times when it appears in the evening sky, Venus disappears from view for a few weeks with each conjunction. Mercury's pattern is similar, but because it is closer to the Sun, it never appears more than about 28° from the Sun in our sky. That makes Mercury difficult to see, because it is almost always obscured by the glare of the Sun.

THINK ABOUT IT

MATHEMATICAL INSIGHT 1.1: THE COPERNICAN LAYOUT OF THE SOLAR SYSTEM

Copernicus favored the Sun-centered model partly because it allowed him to calculate orbital periods and distances for the planets. Let's see how the Copernican system allows us to determine orbital (sidereal) periods of the planets.

We cannot directly measure orbital periods, because our own movement around the Sun means that we look at the planets from different points in our orbit at different times. However, we can measure synodic periods simply by seeing how much time passes between one particular alignment (such as opposition or inferior conjunction) and the next. Figure 1 shows the geometry for a planet *farther from the Sun than Earth* (such as Jupiter), under the assumption of circular orbits (which is what Copernicus assumed). Study the figure carefully to notice the following key facts:

- The dashed brown curve shows the planet's complete orbit. The time the planet requires for one complete orbit is its orbital (sidereal) period, P_{orb}.
- The solid brown arrow shows how far the planet travels along its orbit from one opposition to the next. The time between oppositions is defined' as its synodic period, P_{syn}.
- The dashed blue curve shows Earth's complete orbit; Earth takes P_{earth} = 1 yr to complete an orbit.
- The solid red curve (and red arrow) shows how far Earth goes during the planet's synodic period; it is *more* than one complete orbit because Earth must travel a little "extra" to catch back up with the other planet. The time it takes Earth to travel the "extra" distance (the thick part of the red curve) must be the planet's synodic period minus 1 year, or P_{syn} - 1 yr.
- The angle that the planet sweeps out during its synodic period is equal to the angle that Earth sweeps out as it travels the "extra" distance. Thus, the *ratio* of the planet's complete orbital period (P_{orb})to its synodic period (P_{syn}) must be equal to the ratio of Earth's orbital period (1 yr) to the time required for the "extra" distance (see Appendix C.S. for a review of ratios). Since we already found that the time required for this extra distance is P_{syn} - 1 yr, we write:

$$\frac{P_{orb}}{P_{syn}} = \frac{1\text{ yr}}{(P_{syn} - 1\text{ yr})}$$

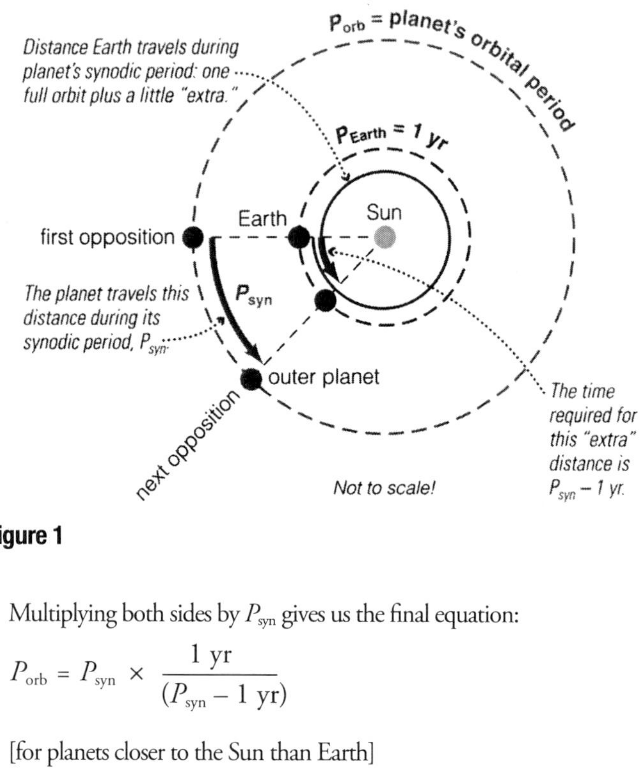

Figure 1

Multiplying both sides by P_{syn} gives us the final equation:

$$P_{orb} = P_{syn} \times \frac{1\text{ yr}}{(P_{syn} - 1\text{ yr})}$$

[for planets closer to the Sun than Earth]

The geometry is slightly different when Earth is the outer planet, as shown in Figure 2. In this case the two equal ratios are 1 yr/P_{syn} = P_{orb}/(P_{syn} - P_{orb}). With a little algebra, you can solve this equation for P_{orb}:

$$P_{orb} = P_{syn} \times \frac{1\text{ yr}}{(P_{syn} + 1\text{ yr})}$$

[for planets closer to the Sun than Earth]

Copernicus knew the synodic periods of the planets and was therefore able to use the above equations (in a slightly different form) to calculate their true orbital periods. He was then able to use the geometry of planetary alignments to compute the distances to the planets in terms of the Earth-Sun distance. (That is, he calculated distances in AU,

Do we ever see Mercury or Venus at midnight? Explain.

As you study Figure 1.4, you might wonder whether Mercury or Venus ever falls directly in front of the Sun at inferior conjunction, creating mini-eclipses as they block a little of the Sun's light. They do, but only rarely, because their orbital planes are slightly tilted compared to Earth's orbital plane (the ecliptic plane). As a result, Mercury and Venus usually appear slightly above or below the Sun at inferior conjunction. But on rare occasions, we do indeed see Mercury or Venus appear to pass directly across the face of the Sun during inferior conjunction (Figure 1.5). Such events are called **transits.** Mercury transits occur an average of a dozen times per century, with the first two of this century on November 8, 2006, and May 9, 2016. Venus transits come in pairs 8 years apart, with more than a century between the second of one pair and the first of the next. We are currently between the two transits of a pair: The first occurred on June 8, 2004, and the second will occur on June 6, 2002. After that, it will be 105 years until the first of the next pair of Venus transits, which will occur in 2117 and 2125.

HOW DO WE TELL THE TIME OF DAY?

[redacted]

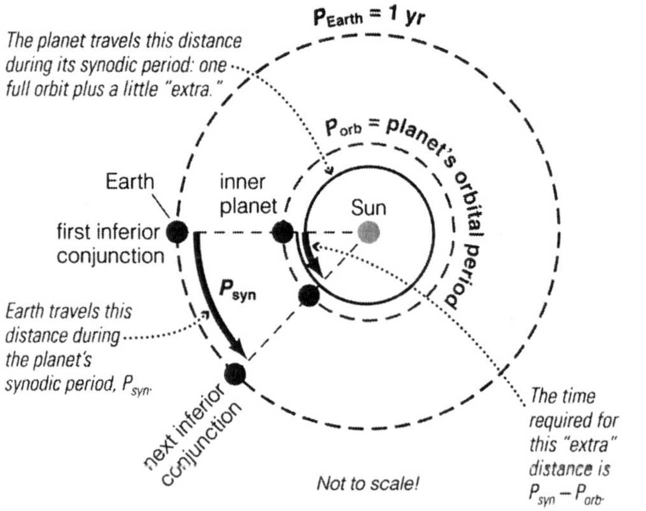

Figuro 2

but he did not have a method for determining how far the Earth is from the Sun.) His results, which were quite close to modern values, made so much sense to him that he felt he must have uncovered some deep truth about nature.

Example 1: Jupiter's synodic period is 398.9 days, or 1.092 years. What is its actual orbital period?

Solution:

Step 1 Understand: We are given Jupiter's synodic period (Psyn in the above equations) and are asked to find its orbital period (P_{orb})' Because Jupiter is farther from the Sun than Earth, we can use the first equation.

Step 2 Solve: We use the equation for a planet farther from the Sun than Earth with P_{syn} = 1.092 yr (Jupiter's synodic period):

$$P_{orb} = P_{syn} \times \frac{1\text{ yr}}{(P_{syn} - 1\text{ yr})}$$
$$= 1.092\text{ yr} \times \frac{1\text{ yr}}{(1.092 - 1\text{ yr})}$$
$$= 11.87\text{ yr}$$

Step 3 Explain: We have found that Jupiter's orbital period is 11.87 years. In other words, simply by measuring the time that passes from when Jupiter is opposite the Sun in one year to when it is again opposite the next year (which is Jupiter's synodic period), we have learned that Jupiter takes a little less than 12 years to orbit the Sun. Notice that the answer makes sense in that it is longer than Earth's orbital period of 1 year, just as we expect for a planet that is farther than Earth from the Sun.

Example 2: Venus's synodic period is 583.9 days. What is its actual orbital period?

Solution:

Step 1 Understand: We are given Venus's synodic period, P_{syn} = 583.9 days, and are asked to find its orbital period. Because Venus is closer to the Sun than Earth, we need the second equation. Also, because we are given Venus's synodic period in days, for unit consistency we need to convert it to years; you should confirm that 583.9 days = 1.599 yr.

Step 2 Solve: We simply plug the value of P_{syn} into the equation for a planet closer to the Sun than Earth:

$$P_{orb} = P_{syn} \times \frac{1\text{ yr}}{(P_{syn} + 1\text{ yr})}$$
$$= 1.599\text{ yr} \times \frac{1\text{ yr}}{(1.599 - 1\text{ yr})}$$
$$= 0.6152\text{ yr}$$

Step 3 Explain: We have found that Venus takes 0.6152 year to orbit the Sun. This number is easier to interpret if we convert it to days or months; you should confirm that it is equivalent to 224.7 days, or about 7 1/2 months. Notice that the answer makes sense in that it is shorter than Earth's orbital period of 1 year, just as we expect for a planet that is closer than Earth to the Sun.

Telling time seems simple, but in fact there are several different ways to define the time of day, even after we agree that time should be based on the 24-hour solar day. Let's explore some of the ways of telling time and see how they ultimately led to our modern system in which we can synchronize clocks anywhere in the world.

Apparent Solar Time

If we base time on the Sun's actual position in the local sky, as is the case when we use a sundial (Figure 1.6), we are measuring **apparent solar time**.

Noon is the precise moment when the Sun is highest in the sky (on the meridian) and the sundial casts its shortest shadow. Before noon, when the Sun is rising upward through the sky, the apparent solar time is *ante meridiem* ("before the middle of the day"), or A.M. For example, if the Sun will reach the meridian 2 hours from now, the apparent solar time is 10 A.M. After noon, the apparent solar time is *post meridiem* ("after the middle of the day"), or P.M. If the Sun crossed the meridian 3 hours ago, the apparent solar time is 3 P.M. Note that, technically, noon and midnight are neither A.M. nor P.M. However, by convention we usually say that noon is 12 P.M. and midnight is 12 A.M.

THINK ABOUT IT

Is it daytime or nighttime at 12:01 A.M.? 12:01 P.M.? Explain.

Mean Solar Time

Suppose you set a clock to precisely 12:00 when a sundial shows noon today. If every solar day were precisely 24 hours, your clock would always remain synchronized with the sundial. However, while 24 hours is the *average* length of the solar day, the actual length of the solar day varies throughout the year. As a result, your clock will not remain perfectly synchronized with the sundial. For example, your clock is likely to read a few seconds before or after 12:00 when the sundial reads noon tomorrow, and within a few weeks your clock time may differ from the apparent solar time by several minutes. Your clock (assuming it is accurate) will again be synchronized with the sundial on the same date next year, since it keeps track of the average length of the solar day.

If we average the differences between the time a clock would read and the time a sundial would read, we can define **mean solar time** (*mean* is another word for *average*). A clock set to mean solar time reads 12:00 each day at the time that the Sun crosses the meridian on *average*. The actual mean solar time at which the Sun crosses the meridian varies over the course of the year in a fairly complex way (see "Solar Days and the Analemma"). The result is that, on any given day, a clock set to mean solar time may read anywhere from about 17 minutes before noon to 15 minutes after noon (that is, from 11:43 A.M. to 12:15 P.M.) when a

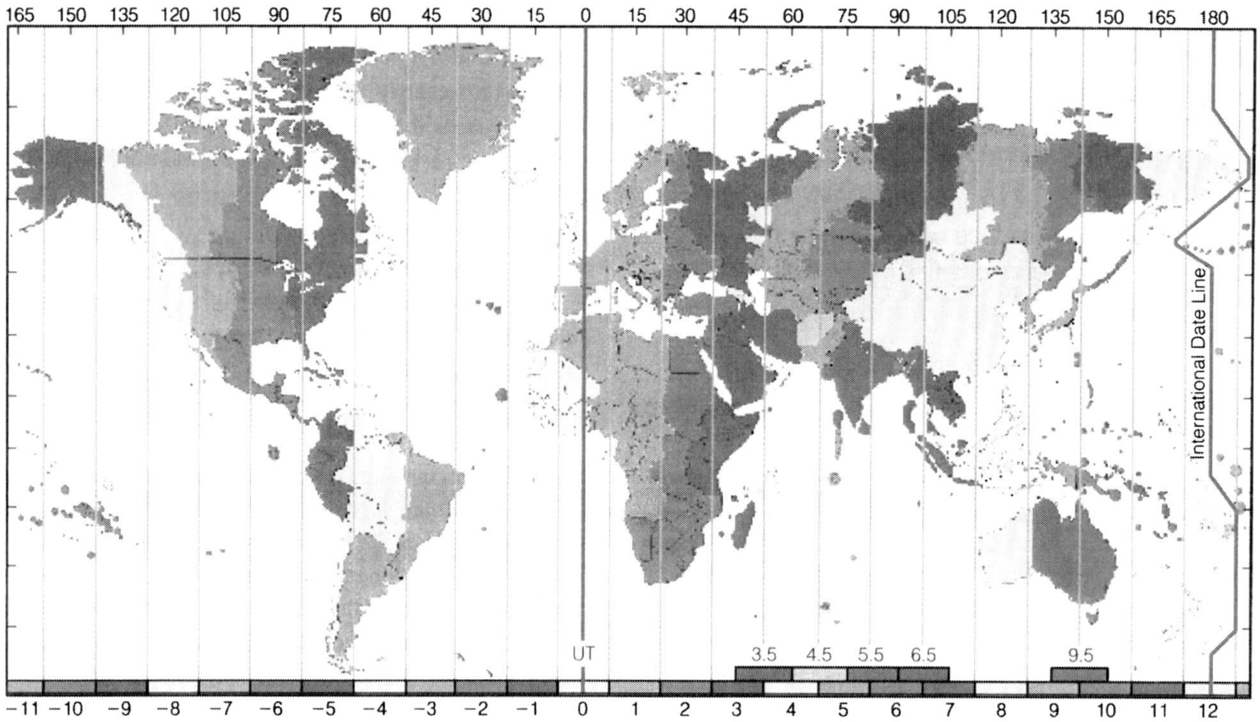

Figure 1.7 Time zones around the world. The numerical scale at the bottom shows hours ahead of (positive numbers) or behind (negative numbers) the time in Greenwich, England; the scale at the top is longitude. The vertical lines show standard time zones as they would be in the absence of political considerations. The color-coded regions show the actual time zones. Note, for example, that all of China uses the same standard time, even though the country is wide enough to span several time zones. Note also that a few countries use time zones centered on a half-hour (the upper set of four colored bars), rather than an hour, relative to Greenwich time.

sundial indicates noon.

Although the lack of perfect synchronization with the Sun might at first sound like a drawback, mean solar time is actually more convenient than apparent solar time (the sundial time)—as long as you have access to a mechanical or electronic clock. Once set, a reliable mechanical or electronic clock can always tell you the mean solar time. In contrast, precisely measuring apparent solar time requires a sundial, which is useless at night or when it is cloudy.

Like apparent solar time, mean solar time is a *local* measure of time. That is, it varies with longitude because of Earth's west-to-east rotation. For example, clocks in New York are set 3 hours ahead of clocks in Los Angeles. If clocks were set precisely to local mean solar time, they would vary even over relatively short east-west distances. For example, mean solar clocks in central Los Angeles would be about 2 minutes behind mean solar clocks in Pasadena, because Pasadena is slightly to the east.

Standard, Daylight, and Universal Time

Clocks displaying mean solar time were common during the early history of the United States. However, by the late 1800s, the growth of railroad travel made the use of mean solar time increasingly problematic. Some states had dozens of different "official" times, usually corresponding to mean solar time in dozens of different cities, and each railroad company made schedules according to its own "railroad time." The many time systems made it difficult for passengers to follow the scheduling of trains.

On November 18, 1883, the railroad companies agreed to a new system that divided the United States into four time zones, setting all clocks within each zone to the same time. That was the birth of **standard time**, which today divides the world into time zones (Figure 1.7). Depending on where you live within a time zone, your standard time may vary somewhat from your mean solar time. In principle, the standard time in a particular time zone is the mean solar time in the *center* of the time zone, so that local mean solar time within a 1-hour-wide time zone would never differ by more than a half-hour from standard time. However, time zones often have unusual shapes to conform to social, economic, and political realities, so larger variations between standard time and mean solar time sometimes occur.

In most parts of the United States, clocks are set to standard time for only part of the year. Between the second sunday in March and the first sunday in November,* most of the United States changes to **daylight saving time**, which is 1 hour ahead of standard time. Because of the 1-hour advance with daylight saving time, clocks read around 1 P.M. (rather than around noon) when the Sun is on the meridian.

*These dates for daylight saving time took effect in 2007 (thanks to legislation passed in 2005); before that, daylight saving time ran between the first Sunday in April and the last Sunday in October.

For purposes of navigation and astronomy, it is useful to have a single time for the entire Earth. For historical reasons, this "world" time was chosen to be the mean solar time in Greenwich, England—the place that also defines longitude 0° (see Figure 2.11). Today, this *Greenwich mean time (GMT)* is often called **universal time (UT)**. (Outside astronomy, it is more commonly called *universal coordinated time* [UTC]). Many airlines and weather services call it "Zulu time," because Greenwich's time zone is designated "z" and "zulu" is a common way of phonetically identifying the letter *Z*.)

SEASONS TUTORIAL, LESSON 2

WHEN AND WHY DO WE HAVE LEAP YEARS?

Our modern calendar is based on the length of the tropical year, which is the amount of time from one spring equinox to the next. The calendar is therefore designed to stay synchronized with the seasons. Getting this synchronization just right was a long process in human history.

The origins of our modern calendar go back to ancient Egypt. By 4200 B.C., the Egyptians were using a calendar that counted 365 days in a year. However, because the length of a year is about 365 $^1/_4$ days (rather than exactly 365 days), the Egyptian calendar drifted out of phase with the seasons by about 1 day every 4 years. For example, if the spring equinox occurred on March 21 one year, 4 years later it occurred on March 22, 4 years after that on March 23, and so on. Over many centuries, the spring equinox moved through many different months. To keep the seasons and the calendar synchronized, Julius Caesar decreed the adoption of a new calendar in 46 B.C. This *Julian calendar* introduced the concept of **leap year**: Every fourth year has 366 days, rather than 365, so that the average length of the calendar year is 365 $^1/_4$ days.

The Julian calendar originally had the spring equinox falling around March 24. If it had been perfectly synchronized with the tropical year, this calendar would have ensured that the spring equinox occurred on the same date every 4 years (that is, every leap-year cycle). It didn't work perfectly, however, because a tropical year is actually about 11 minutes short of 365 $^1/_4$ days. As a result, the moment of the spring equinox slowly advanced by an average of 11 minutes per year. By the late 1500s, the spring equinox was occurring on March 11.

Concerned by this drift in the date of the spring equinox, Pope Gregory XIII introduced a new calendar in 1582. This *Gregorian calendar* was much like the Julian calendar, with two important adjustments. First, Pope Gregory decreed that the day in 1582 following October 4 would be October 15. By eliminating the 10 dates from October 5 through October 14, 1582, he pushed the date of the spring equinox

SPECIAL TOPIC: SOLAR DAYS AND THE ANALEMMA

The average length of a solar day is 24 hours, but the precise length varies over the course of the year. Two effects contribute to this variation.

The first effect arises from Earth's varying orbital speed. Recall that, in accord with Kepler's second law, Earth moves slightly faster when it is closer to the Sun in its orbit and slightly slower when it is farther from the Sun. Thus, Earth moves slightly farther along its orbit each day when it is closer to the Sun. This means that the solar day requires more than the average amount of "extra" rotation (see Figure Sl.2) during these periods—making these solar days longer than average. Similarly, the solar day requires less than the average amount of "extra" rotation when it is in the portion of its orbit farther from the Sun—making these solar days shorter than average.

The second effect arises from the tilt of Earth's axis, which causes the ecliptic to be inclined by 23° to the celestial equator on the celestial sphere. Because the length of a solar day depends on the Sun's apparent eastward motion along the ecliptic, the inclination would cause solar days to vary in length even if Earth's orbit were perfectly circular. To see why, suppose the Sun appeared to move exactly 1° per day along the ecliptic. Around the times of the solstices, this motion would be entirely eastward, making the solar day slightly longer than average. Around the times of the equinoxes, when the motion along the ecliptic has a significant northward or southward component, the solar day would be slightly shorter than average.

Together, the two effects make the actual length of solar days vary by up to about 25 seconds (either way) from the 24-hour average. Because the effects accumulate at particular times of year, the apparent solar time can differ by as much as 17 minutes from the mean solar time. The net result is often depicted visually by an analemma (Figure 1), which looks much like a figure 8. You'll find an analemma printed on many globes (Figure 2.16 shows a photographic version).

By using the horizontal scale on the analemma (Figure 1), you can convert between mean and apparent solar time for any date. (The vertical scale shows the declination of the Sun, which is discussed in Section 1.2.) For example, the dashed line shows that on November 10, a mean solar clock is about 17 minutes "behind the Sun" or behind apparent solar time. Thus, if the apparent solar time is 6:00 P.M. on November 10, the mean solar time is only 5:43 P.M. The discrepancy between mean and apparent solar times is called the equation of time. It is often plotted as a graph (Figure 2), which gives the same results as reading from the analemma.

The discrepancy between mean and apparent solar time also explains why the times of sunrise and sunset don't follow seasonal patterns perfectly. For example, the winter solstice around December 21 has the shortest daylight hours (in the Northern Hemisphere), but the earliest sunset occurs around December 7, when the Sun is still well "behind" mean solar time.

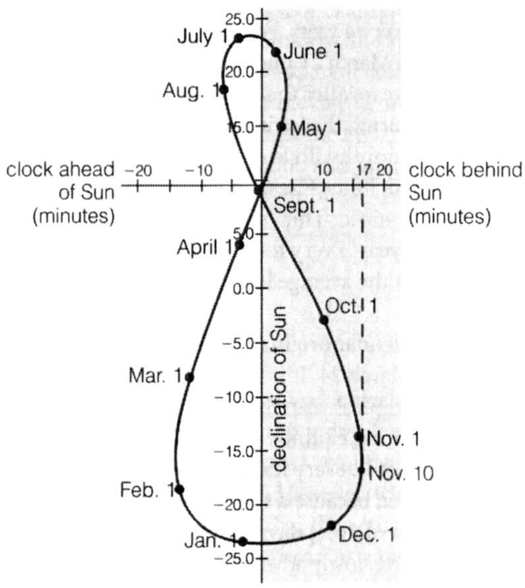

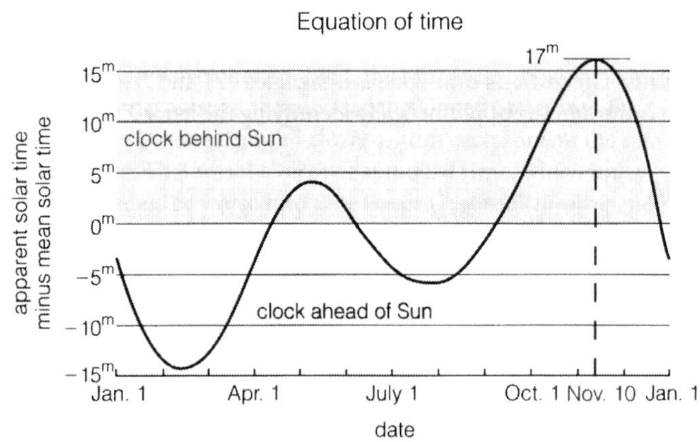

Figure 1 The analemma shows the annual pattern of discrepancies between apparent and mean solar time. For example, the dashed red line shows that on November 10, a mean solar clock reads 17 minutes behind (earlier than) apparent solar time.

Figure 2 The discrepancies can also be plotted on a graph as the equation of time.

in 1583 from March 11 to March 21. (He chose March 21 because it was the date of the spring equinox in A.D. 325, which was the time of the Council of Nicaea, the first ecumenical council of the Christian church.) Second, the Gregorian calendar added an exception to the rule of having leap year every 4 years: Leap year is skipped when a century changes (for example, in years 1700, 1800, 1900) unless the century year is divisible by 400. Thus, 2000 was a leap year because it is divisible by 400 (2,000 ÷ 400 = 5), but 2100 will *not* be a leap year. These adjustments make the average length of the Gregorian calendar year almost exactly the same as the actual length of a tropical year, which ensures that the spring equinox will occur on March 21 every fourth year for thousands of years to come.

Today, the Gregorian calendar is used worldwide for international communication and commerce. (Many countries still use traditional calendars, such as the Chinese, Islamic, and Jewish calendars, for cultural purposes.) However, as you might guess, the pope's decree was not immediately accepted in regions not bound to the Catholic Church. For example, the Gregorian calendar was not adopted in England or in the American colonies until 1752, and it was not adopted in China until 1912 or in Russia until 1919.

1.2: CELESTIAL COORDINATES AND MOTION IN THE SKY

We are now ready to turn our attention from timekeeping to navigation. The goal of celestial navigation is to use the Sun and the stars to find our position on Earth. Before we can do that, we need to understand the apparent motions of the sky in more detail. We'll begin in this section by discussing how we locate objects on the celestial sphere, which will then allow us to explore how positions on the celestial sphere determine motion in the local sky. With this background, we'll be ready to explore the principles of celestial navigation in the final section of this chapter.

How do we locate objects on the celestial sphere?

The celestial sphere is an illusion, but one that is quite useful when looking at the sky. We can make the celestial sphere even more useful by giving it a set of **celestial coordinates** that function much like the coordinates of latitude and longitude on Earth. Just as we can locate a city on Earth by its latitude and longitude, we will use an object's celestial coordinates to describe its precise location on the celestial sphere.

We have already discussed the basic features of the celestial sphere that will serve as starting points for our coordinate system: the north and south celestial poles, the celestial equator, and the ecliptic. Figure 1.8 shows these locations on a schematic diagram. The arrow along the ecliptic indicates the direction in which the Sun appears to move over

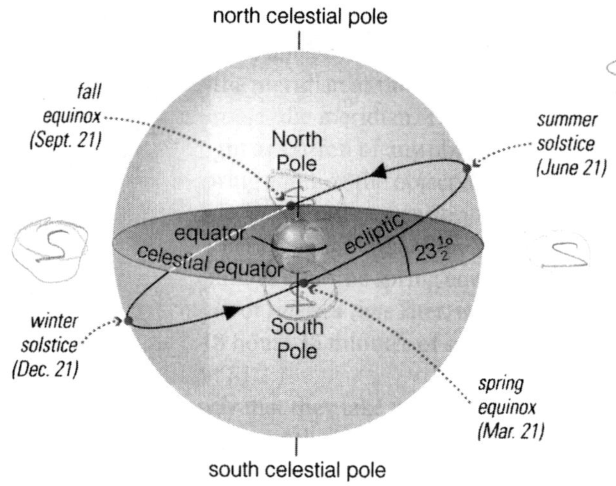

Figure 1.8 Schematic diagram of the celestial sphere, shown without stars. As a study aid, you should use a plastic ball as a model of the celestial sphere, marking it with the same special locations.

the course of each year. It is much easier to visualize the celestial sphere if you make a model with a simple plastic ball. Use a felt-tip pen to mark the north and south celestial poles on your ball, and then add the celestial equator and the ecliptic. Note that the ecliptic crosses the celestial equator on opposite sides of the celestial sphere at an angle of 23 ½° (because of the tilt of Earth's axis).

Equinoxes and Solstices

Remember that the equinoxes and solstices are special moments in the year that help define the seasons [Section 2.2]. For example, the *spring equinox*, which occurs around March 21 each year, is the moment when spring begins for the Northern Hemisphere and fall begins for the Southern Hemisphere. These moments correspond to positions in Earth's orbit (see Figure 2.15) and hence to apparent locations of the Sun along the ecliptic. As shown in Figure 1.8, the spring equinox occurs when the Sun is on the ecliptic at the point where it crosses from south of the celestial equator to north of the celestial equator. This point is also called the spring equinox. Thus, the term *spring equinox* has a dual meaning: It is the *moment* when spring begins and also the *point* on the ecliptic at which the Sun appears to be located at that moment.

Figure 1.8 also shows the points marking the summer solstice, fall equinox, and winter solstice, with the dates on which the Sun appears to be located at each point. Remember that the dates are approximate because of the leap-year cycle and because a tropical year is not exactly 365 ¼ days. (For example, the spring equinox may occur anytime between about March 20 and March 23.)

Although no bright stars mark the locations of the equinoxes or solstices among the constellations, you can find them with the aid of nearby bright stars (Figure 1.9). For example, the spring equinox is located in the constellation

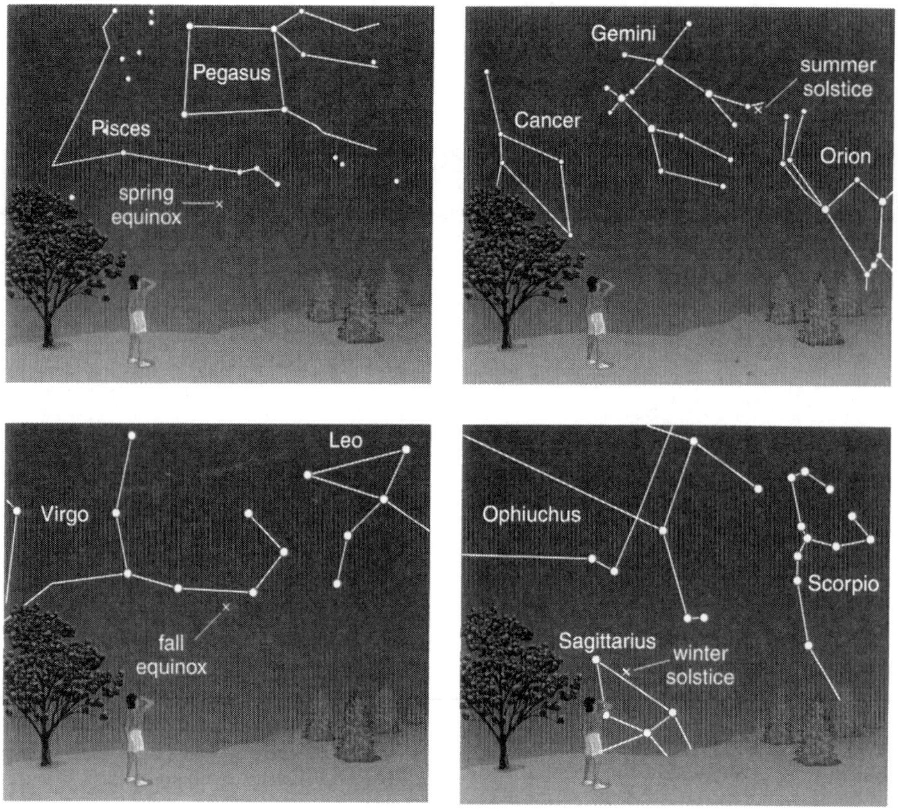

Figure 1.9 These diagrams show the locations among the constellations of the equinoxes and solstices. No bright stars mark any of these points, so you must find them by studying their positions relative to recognizable patterns. The time of day and night at which each point is above the horizon depends on the time of year.

Pisces and can be found with the aid of the four bright stars in the Great Square of Pegasus. Of course, when the Sun is located at this point around March 21, we cannot see Pisces or Pegasus because they are close to the Sun in our daytime sky.

Celestial Coordinates

We can now add our system of celestial coordinates to the celestial sphere. Because this will be the third coordinate system we've used in this book, it's easier to understand if we first review the other two. Figure 1.10a shows the coordinates of *altitude* and *direction* (or *azimuth**) we use in the local sky. Figure 1.10b shows the coordinates of *latitude*

*Azimuth is usually measured clockwise around the horizon from due north. By this definition, the azimuth of due north is 0°, of due east is 90°, of due south is 180°, and of due west is 270°.

SEE IT FOR YOURSELF

Using your plastic ball as a model of the celestial sphere (which you have already marked with the celestial poles, equator, and ecliptic), mark the locations and approximate dates of the equinoxes and solstices. Based on the dates for these points, approximately where along the ecliptic is the Sun on April 21? On November 21? How do you know?

Figure 1.10 Celestial coordinate systems.

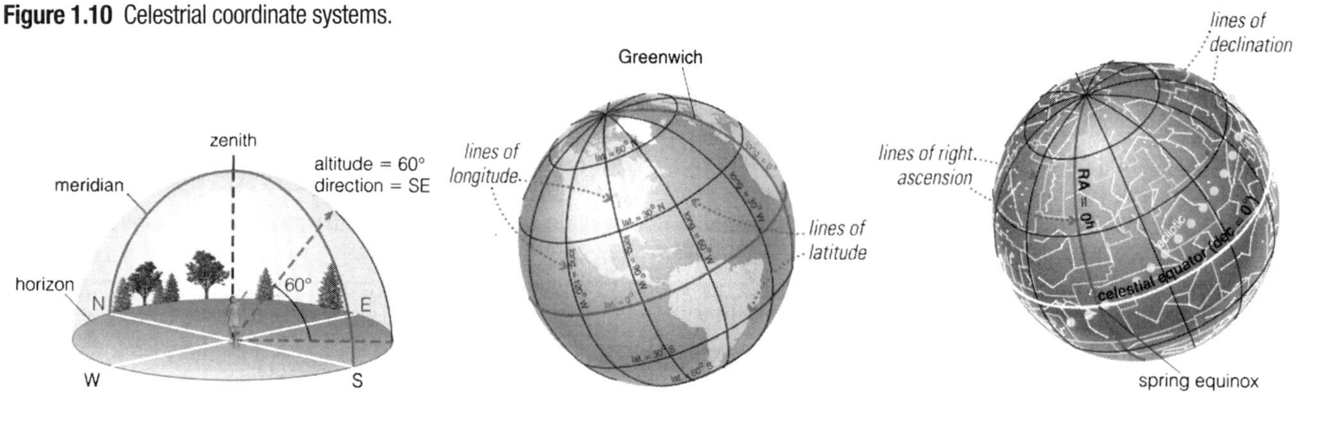

a We use altitude and direction to pinpoint locations in the local sky.

b We use altitude and direction to pinpoint locations on Earth.

c We use declination and right ascension to pinpoint locations on the celestial sphere.

and *longitude* we use on Earth's surface. Finally, Figure 1.10c shows our new system of celestial coordinates. As you can see in the figure, these coordinates are called **declination (dec)** and **right ascension (RA)**.

If you compare Figures 1.10b and c, you'll see that declination on the celestial sphere is similar to latitude on Earth and right ascension is similar to longitude. Let's start with declination; notice the following key points:

- Just as lines of latitude are parallel to Earth's equator, lines of declination are parallel to the celestial equator.
- Just as Earth's equator has lat = 0°, the celestial equator has dec = 0°.
- Latitude is labeled *north* or *south* relative to the equator, while declination is labeled *positive* or *negative*. For example, the North Pole has lat = 90°N, while the north celestial pole has dec = +90°; the South Pole has lat = 90°S, while the south celestial pole has dec = -90°.

Next, notice the close correspondence between right ascension and longitude:

- Just as lines of longitude extend from the North Pole to the South Pole, lines of right ascension extend from the north celestial pole to the south celestial pole.
- Just as there is no natural starting point for longitude, there is no natural starting point for right ascension. By international treaty, longitude zero (the prime meridian) is the line of longitude that runs through Greenwich, England. By convention, right ascension zero is the line of right ascension that runs through the spring equinox.
- Longitude is measured in *degrees* east or west of Greenwich, while right ascension is measured in *hours* (and minutes and seconds) east of the spring equinox. A full 360° circle around the celestial equator goes through 24 hours of right ascension, so each hour of right ascension represents an angle of 360° ÷ 24 = 15°.

SEE IT FOR YOURSELF

On your plastic ball model of the celestial sphere, add a scale for right ascension along the celestial equator and also add a few circles of declination, such as declination 00, ±300, ±600, and ±900. Where is Vega on your model?

As an example of how we use celestial coordinates to locate objects on the celestial sphere, consider the bright star Vega. Its coordinates are dec = +38°44' and RA = 18^h35^m (Figure 1.11). The positive declination tells us that Vega is 38°44' *north* of the celestial equator. The right ascension tells us that Vega is 18 hours 35 minutes east of the spring equinox. Translating the right ascension from hours to angular degrees, we find that Vega is about 279° east of the spring equinox (because IS hours represents 18 × 15° = 270° and 35 minutes represents 35/60 × 15° ≈ 9°).

The Vega example also shows why right ascension is measured in units of time: It is because time units make it easier to track the daily motions of objects through the local sky. All objects with a particular right ascension cross the meridian at the same time, which means that all stars with RA = O^h cross the meridian at the same time that the spring equinox crosses the meridian. Thus, when we use time units, the right ascension of any object tells us how long *after* the spring equinox the object crosses the meridian. For example, Vega's right ascension, 18^h35^m, tells us that on any particular day, Vega crosses the meridian about 18 hours 35 minutes after the spring equinox. (This is 18 hours 35 minutes of *sidereal time* later, which is not exactly the same as 18 hours 35 minutes of solar time; see Mathematical Insight 1.2.)

Stars are so far away that they take thousands of years or more to move noticeably on the celestial sphere. Nevertheless, the celestial coordinates of stars are not quite constant,

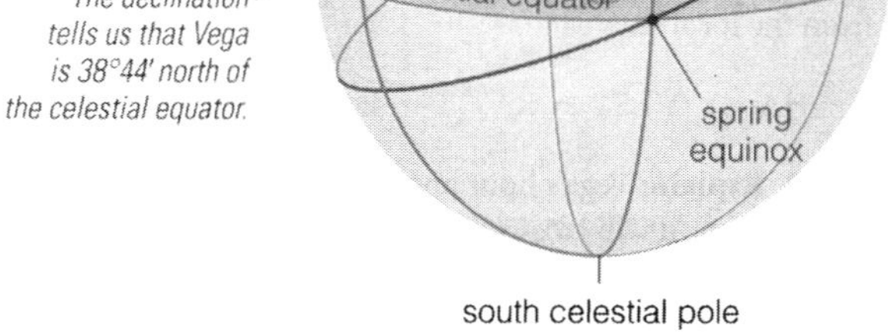

Figure 1.11 This diagram shows how we interpret the celestial coordinates of Vega.

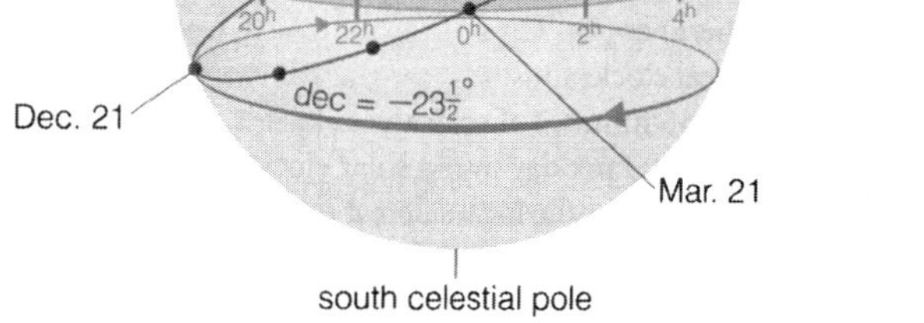

Figure 1.12 We can use this diagram of the celestial sphere to determine the Sun's right ascension and declination at monthly intervals.

MATHEMATICAL INSIGHT 1.2: TIME BY THE STARS

The clocks we use in daily life are set to solar time, ticking through 24 hours for each day of mean solar time. In astronomy, it is also useful to have clocks that tell time by the stars, or **sidereal time.** Just as we define *solar time* according to the Sun's position relative to the meridian, *sidereal time* is based on the positions of stars relative to the meridian. We define the **hour angle (HA)** of any object on the celestial sphere to be the time since it last crossed the meridian. (For a circumpolar star, hour angle is measured from the *higher* of the two points at which it crosses the meridian each day.) For example:

- If a star is crossing the meridian now, its hour angle is 0^h.
- If a star crossed the meridian 3 hours ago, its hour angle is 3^h.
- If a star will cross the meridian 1 hour from now, its hour angle is -1^h or, equivalently, 23^h.

By convention, time by the stars is based on the hour angle of the spring equinox. That is, the local sidereal time (LST) is

$$LST = HA_{spring\ equinox}$$

For example, the local sidereal time is 00:00 when the spring equinox is *on* the meridian. Three hours later, when the spring equinox is 3 hours west of the meridian, the local sidereal time is 03:00.

Note that, because right ascension tells us how long after the spring equinox an object reaches the meridian, the local sidereal time is also equal to the right ascension (RA) of objects currently crossing your meridian. For example, if your local sidereal time is 04:30, stars with RA = 4^h30^m are currently crossing your meridian. This idea leads to an important relationship among any object's current hour angle, the current local sidereal time, and the object's right ascension:

$$HA_{object} = LST - RA_{object}$$

This formula will make sense to you if you recognize that an object's right ascension tells us the time by which it trails the spring equinox on its daily trek through the sky. Because the local sidereal time tells us how long it has been since the spring equinox was on the meridian, the difference LST − RA_{obj} must tell us the position of the object relative to the meridian.

Sidereal time has one important subtlety: Because the stars (and the celestial sphere) appear to rotate around us in one sidereal day (23^h56^m), sidereal clocks must tick through 24 hours of sidereal time in 23 hours 56 minutes of solar time. That is, a sidereal clock gains about 4 minutes per day over a solar clock. As a result, you cannot immediately infer the local sidereal time from the local solar time, or vice versa, without either doing some calculations or consulting an astronomical table. Of course, the easiest way to determine the local sidereal time is with a clock that ticks at the sidereal rate. Astronomical observatories always have sidereal clocks, and you can buy moderately priced telescopes that come with sidereal clocks.

Example 1: Suppose the local apparent solar time is 9:00 P.M. on the spring equinox (March 21). What is the local sidereal time?

Solution:

Step 1 Understand: The trick to this problem is understanding exactly what we are looking for. We are asked to find the local sidereal time, which is defined as the hour angle of the spring equinox in the local sky. Thus, we need to know where the spring equinox is located in the local sky. We are given the key clue: The date is the day of the spring equinox, the one day of the year on which the Sun is located in the same position as the spring equinox in the sky.

Step 2 Solve: We can now find the hour angle of the spring equinox from the hour angle of the Sun. We are told that the local apparent solar time is 9:00 P.M., which means that the Sun is 9 hours past the meridian and thus has an hour angle of 9 hours. Because the spring equinox and the Sun are located in the same place on this one date of the year, the hour angle of the spring equinox is also 9 hours.

Step 3 Explain: The hour angle of the spring equinox is 9 hours, which means the local sidereal time is LST = 09:00.

Example 2: Suppose the local sidereal time is LST = 04:00. When will Vega cross the meridian?

Solution:

Step 1 Understand: We are given the local sidereal time, which tells us the hour angle of the spring equinox in the local sky. To determine when Vega will cross the meridian, we need to know its hour angle, which we can calculate from its right ascension and the formula given above. Figure 1.11 shows us that Vega has RA = 18^h35^m, so we have all the information we need. (Appendix F also gives RA and dec for Vega and other stars.)

Step 2 Solve: We now use the formula to find Vega's hour angle from the local sidereal time and Vega's right ascension:

$$HA_{Vega} = LST - RA_{Vega} = 4{:}00 - 18{:}35 = -14{:}35$$

Step 3 Explain: Vega's hour angle is −14 hours 35 minutes, which means it will cross your meridian 14 hours and 35 minutes from now. This also means that Vega crossed your meridian 9 hours and 25 minutes ago (because $14^h35^m + 9^h25^m = 24^h$). (Note that these are intervals of sidereal time.)

because they are tied to the celestial equator and the celestial equator gradually moves relative to the constellations with Earth's 26,000–year cycle of axis precession [Section 2.2]. (Axis precession does not affect Earth's orbit, so it does not affect the location of the ecliptic among the constellations.) Even over just a few decades, the coordinate changes are significant enough to make a difference in precise astronomical work—such as aiming a telescope at a particular object. As a result, careful observations require almost constant updating of celestial coordinates. Star catalogs therefore always state the year for which coordinates are given (for example, "epoch 2000"). Astronomical software can automatically calculate day-to-day celestial coordinates for the Sun, Moon, and planets as they wander among the constellations.

Celestial Coordinates of the Sun

Unlike stars, which remain fixed in the patterns of the constellations on the celestial sphere, the Sun moves gradually along the ecliptic. It takes a year for the Sun to make a full circuit of the ecliptic, which means it moves through all 24 hours of right ascension over the course of the year. Each month, the Sun moves approximately one-twelfth of the way around the ecliptic, meaning that its right ascension changes by about 24 ÷ 12 = 2 hours per month. Figure 1.12 shows the ecliptic marked with the Sun's monthly position and a scale of celestial coordinates. From this figure, we can create a table of the Sun's month-by-month celestial coordinates.

Table 1.1 starts from the spring equinox, when the Sun

Table 1.1 The Sun's Approximate Celestial Coordinates at 1-Month Intervals

Approximate Date	RA	Dec
Mar. 21 (spring equinox)	0^h	0°
Apr. 21	2^h	+12°
May 21	4^h	+20°
June 21 (summer solstice)	6^h	+23 ½°
July 21	8^h	+20°
August 21	10^h	+12°
Sept. 21 (fall ~~solstice~~)	12^h	0°
Oct. 21	14^h	-12°
Nov. 21	16^h	-20°
Dec. 21 (winter solstice)	18^h	-23 ½°
Jan. 21	20^h	-20°
Feb. 21	22^h	-12°

SEE IT FOR YOURSELF

On your plastic ball model of the celestial sphere, add dots along the ecliptic to show the Sun's monthly positions. Based on your model, what are the Sun's approximate celestial coordinates on your birthday?

has declination 0° and right ascension O^h. You can see in the shaded areas of the table that while RA advances steadily through the year, the Sun's declination changes much more rapidly around the equinoxes than around the solstices. For example, the Sun's declination changes from –12° on February 21 to 12° on April 21, a difference of 24° in just 2 months. In contrast, during the 2 months around the summer solstice (that is, between May 21 and July 21), the declination varies only between +20° and its maximum of +23°. This behavior explains why the number of daylight hours increases rapidly in spring and decreases rapidly in fall, while the number of daylight hours remains long and nearly constant for a couple of months around the summer solstice and short and nearly constant for a couple of months around the winter solstice.

MA SEASONS TUTORIAL, LESSON 3

HOW DO STARS MOVE THROUGH THE LOCAL SKY?

We can use our system of celestial coordinates to gain a better understanding of the way stars move through the local sky. Earth's rotation makes all celestial objects appear to circle around Earth each day (see Figure 2.9), but what we actually see in the local sky is more complex because we see only half the celestial sphere at one time (the ground blocks our view of the other half). Let's explore the appearance of the local sky. As we'll see, the path of any star through your local sky depends on only two things: (1) your latitude and (2) the declination of the star whose path you want to know.

The Sky at the North Pole

The daily paths of stars are easiest to understand for the local sky at the North Pole, so let's begin there before moving on to other latitudes. Figure 1.13a shows the rotating celestial sphere and your orientation relative to it when you are standing at the North Pole. Your "up" points toward the north celestial pole, which therefore marks your zenith. Earth blocks your view of anything south of the celestial equator, which therefore runs along your horizon. To make it easier for you to visualize the local sky, Figure 1.13b shows your horizon extending to the celestial sphere. The horizon is marked with directions, but remember that all directions are south from the North Pole. We therefore cannot define a

Figure 1.13 Interactive Figure: The sky at the North Pole

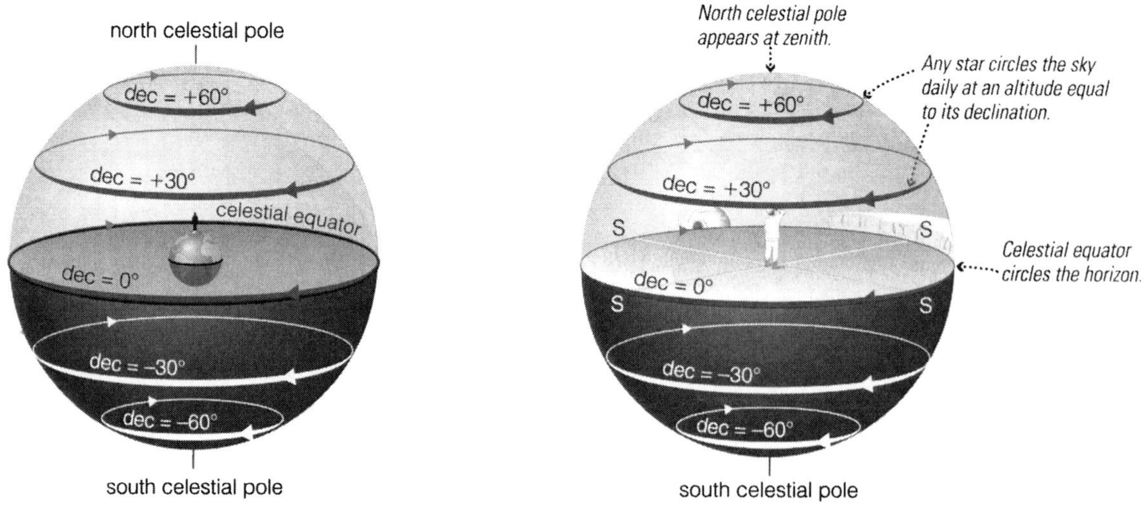

a The orientation of the local sky, relative to the celestial sphere, for an observer at the North Pole.

b Extending the horizon to the celestial sphere makes it easier to visualize the local sky at the North Pole. (Note: To understand the extension, think of Earth in part (a) as being extremely small compared to the celestial sphere.)

meridian for the North Pole, since a meridian would have to run from the north to the south points on the horizon and there are no such unique points at the North Pole.

Notice that the daily circles of the stars keep them at constant altitudes above or below the North Polar horizon.

Moreover, the altitude of any star is equal to its declination. For example, a star with declination +60° circles the sky at an altitude of 60°, and a star with declination -30° remains 30° below your horizon at all times. As a result, all stars north of the celestial equator are circumpolar at the North Pole, meaning that they never fall below the horizon. Stars south of the celestial equator can never be seen at the North Pole. If you are having difficulty visualizing the star paths, it may help you to watch star paths as you rotate your plastic ball model of the celestial sphere.

You should also notice that right ascension does not affect a star's path at all: The path depends only on declination. As we'll see shortly, this rule holds for all latitudes. Right ascension affects only the *time* of day and year at which a star is found in a particular position in your sky.

The Sky at the Equator

After the Poles, the equatorial sky is the next easiest case to understand. Imagine that you are standing somewhere on Earth's equator (lat = 0°), such as in Ecuador, in Kenya, or on the island of Borneo. Figure 1.14a shows that "up" points directly away from (perpendicular to) Earth's rotation axis. Figure 1.14b shows the local sky more clearly by extending the horizon to the celestial sphere and rotating the diagram so the zenith is up. As everywhere except at the poles, the meridian extends from the horizon due south, through the zenith, to the horizon due north.

Look carefully at how the celestial sphere appears to rotate in the local sky. The north celestial pole remains stationary on your horizon due north. As we should expect, its altitude of 0° is equal to the equator's latitude [Section 2.1]. Similarly, the south celestial pole remains stationary on your horizon due south. At any particular time, half the celestial equator is visible, extending from the horizon due east, through the zenith; to the horizon due west. The other half lies below the horizon. As the equatorial sky appears to turn, all star paths rise straight out of the eastern horizon and set straight into the western horizon, with the following features:

- **Stars with dec = 0°** lie on the celestial equator and therefore rise due east, cross the meridian at the zenith, and set due west.
- **Stars with dec > 0°** rise north of due east, reach their highest point on the meridian in the north, and set north of due west. Their rise, set, and highest point depend on their declination. For example, a star with dec = +30° rises 30° north of due east, crosses the meridian 30° to the north of the zenith—that is, at an altitude of 90° - 30° = 60° in the north—and sets 30° north of due west.
- **Stars with dec < 0°** rise south of due east, reach their highest point on the meridian in the south, and set south of due west. For example, a star with dec = -50° rises 50° south of due east, crosses the meridian 50° to the south of the zenith—that is, at an altitude of 90° - 50° = 40° in the south—and sets 50° south of due west.

Notice that exactly half of any star's daily circle lies above the horizon, which means that every star at the equator is above the horizon for exactly half of each sidereal day, or just under 12 hours, and below the horizon for the other

Figure 1.14 Interactive Figure: The sky at the equator

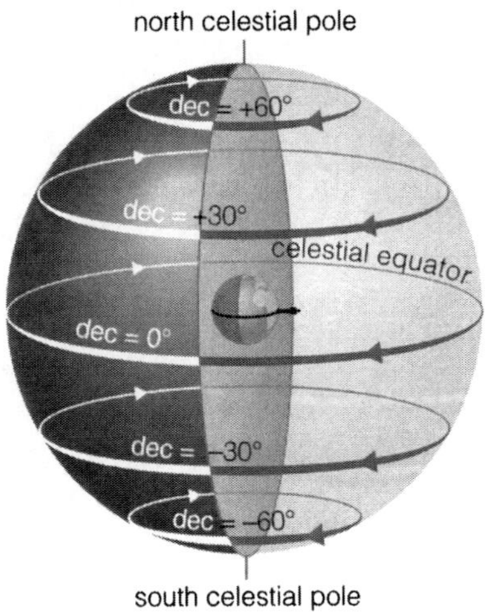

a The orientation of the local sky, relative to the celestial sphere, for an observer at Earth's equator.

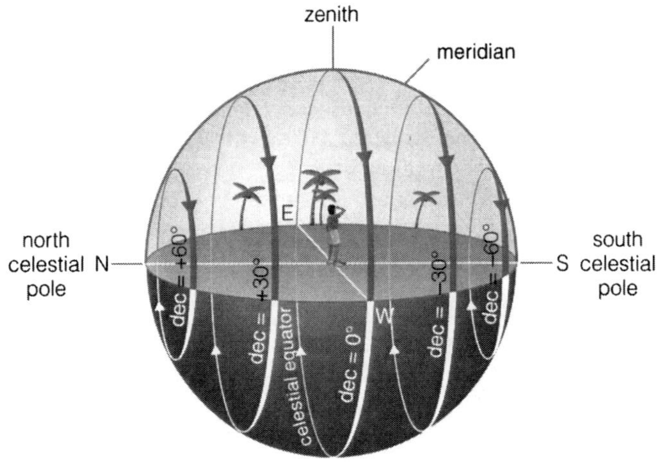

b Extending the horizon and rotating the diagram make it easier to visualize the local sky at the equator.

half of the sidereal day. Also, notice again that right ascension does not affect star paths, although it does affect the time of day and year at which a star will be in a particular place along its path.

THINK ABOUT IT

Are any stars circumpolar at the equator? Are there stars that never rise above the horizon at the equator? Explain.

Skies at Other Latitudes

Star tracks may at first seem more complex at other latitudes, with their mixtures of circumpolar stars and stars that rise and set. However, they are easy to understand if we apply the same basic strategy we've used for the North Pole and equator. Let's consider latitude 40°N, such as in Denver, Indianapolis, Philadelphia, or Beijing. First, as shown in Figure 1.15a, imagine standing at this latitude on a basic diagram of the rotating celestial sphere. Note that "up" points to a location on the celestial sphere with declination +40°. To make it easier to visualize the local sky, we next extend the horizon and rotate the diagram so the zenith is up (Figure 1.15b).

As we would expect, the north celestial pole appears 40° above the horizon due north, since its altitude in the local sky is always equal to the latitude. Half the celestial equator is visible. It extends from the horizon due east, to the meridian at an altitude of 50° in the south, to the horizon due west. By comparing this diagram to that of the local sky for the equator, you can probably notice a general rule for the celestial equator at any latitude except the poles:

The celestial equator always extends from due east on your horizon to due west on your horizon, crossing the meridian at an altitude of 90° *minus* your latitude.

The celestial equator crosses the meridian south of the zenith for locations in the Northern Hemisphere and north of the zenith for locations in the Southern Hemisphere.

If you study Figure 1.15b carefully (or, better yet, rotate your plastic ball model of the celestial sphere with the same orientation), you'll notice the following features of the sky for latitude 40°N:

- **Stars with dec = 0°** lie *on* the celestial equator and therefore follow the path of the celestial equator through the local sky. That is, for latitude 40°N, these stars rise due east, cross the meridian at altitude 90° - 40° = 50° in the south, and set due west.
- **Stars with dec > (90° - lat)** are circumpolar. Thus, for latitude 40°N, stars with declination greater than 90° - 40° = 50° are circumpolar, because they lie *within* 40° of the north celestial pole.
- **Stars with dec > 0°** but that are not circumpolar follow paths parallel to but north of the celestial equator: They rise north of due east and set north of due west, and cross the meridian to the north of the place where the celestial equator crosses it by an amount equal to their declination. For example, because the celestial equator at latitude 40° crosses the meridian at altitude 50° in the south, a star with dec = +30° crosses the meridian at altitude 50° + 30° = 80° in the south. Similarly, a star with dec = +60° crosses the meridian 60° farther north than the celestial equator, which means at altitude 70° in the north. (To calculate this result, note that the sum

Figure 1.15 Interactive Figure: The sky at 40°N latitude

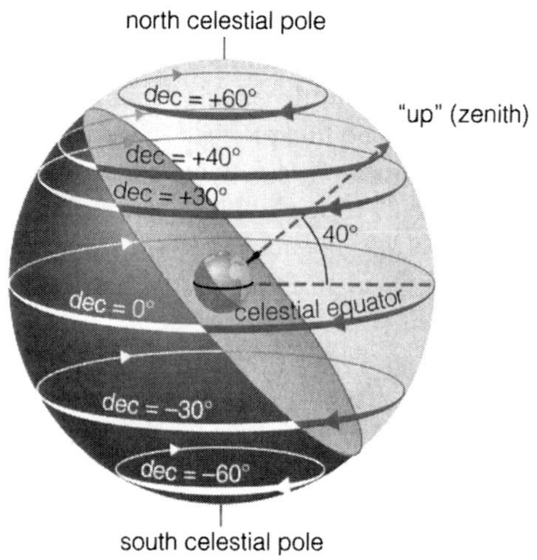

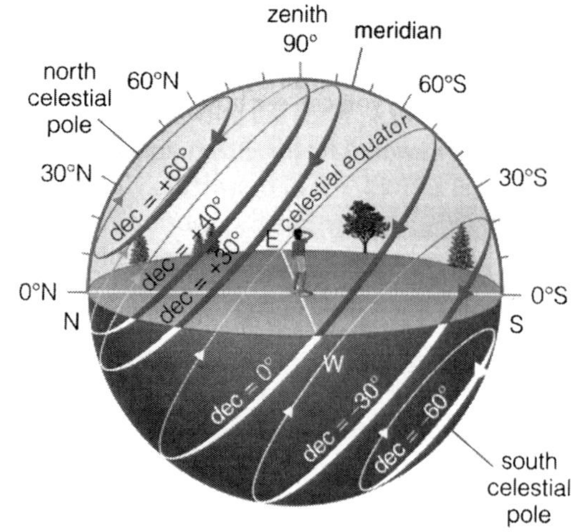

a The orientation of the local sky, relative to the celestial sphere, for an observer at latitude 40°N. Because latitude is the angle to Earth's equator, "up" points to the circle on the celestial sphere with declination +40°.

b Extending the horizon and rotating the diagram so the zenith is up make it easier to visualize the local sky. The blue scale along the meridian shows altitudes and directions in the local sky.

50° + 60° = 110° goes 20° past the zenith altitude of 90°, making it equivalent to 90° - 20° = 70°.)

- **Stars with dec < (-90° + lat)** never rise above the horizon. Thus, for latitude 40°N, stars with declination less than -90° + 40° = -50° never rise above the horizon, because they lie within 40° of the south celestial pole.
- **Stars with dec < 0° but that are sometimes visible** follow paths parallel to but south of the celestial equator: They rise south of due east and set south of due west, and cross the meridian south of the place where the celestial equator crosses it by an amount equal to their declination. For example, a star with dec = -30° crosses the meridian at altitude 50° - 30° = 20° in the south.

You should also notice that the fraction of any star's daily circle that is above the horizon—and hence the amount of time it is above the horizon each day—depends on its declination. Because exactly half the celestial equator is above the horizon, stars on the celestial equator (dec = 0°) are above the horizon for about 12 hours per day. For northern latitudes like 40°N, stars with positive declinations have more than half their daily circles above the horizon and hence are

Figure 1.16 Interactive Figure: The sky at 30°S latitude.

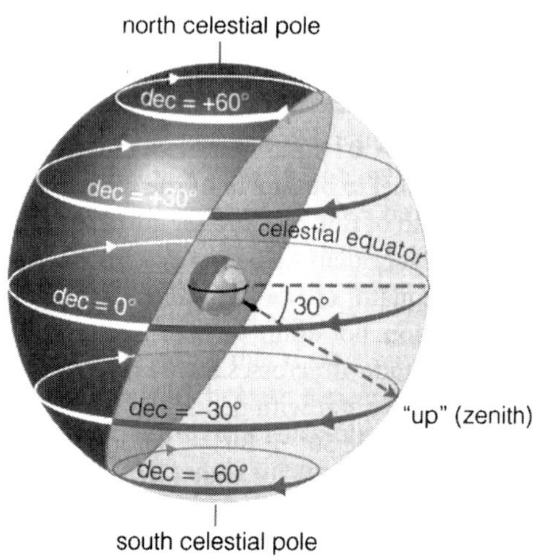

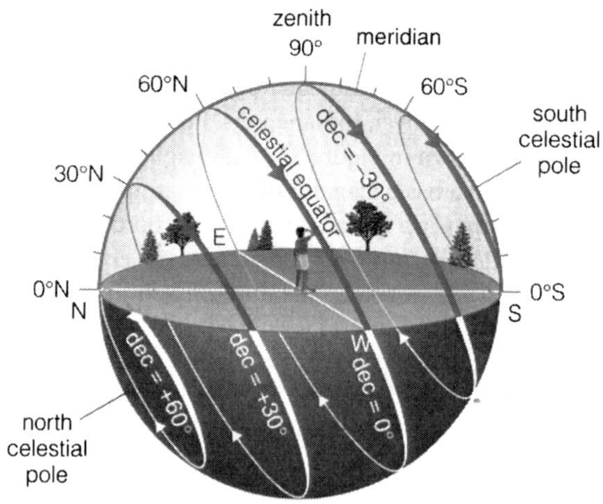

a The orientation of the local sky for an observer at latitude 30°5, relative to the celestial sphere. "Up" points to the circle on the celestial sphere with dec = -30°.

b Extending the horizon and rotating the diagram so the zenith is up make it easier to visualize the local sky. Note that the south celestial pole is visible at altitude 30° in the south, while the celestial equator stretches across the northern half of the sky.

above the horizon for more than 12 hours each day (with the range extending to 24 hours a day for the circumpolar stars). Stars with negative declinations have less than half their daily circles above the horizon and hence are above the horizon for less than 12 hours each day (with the range going to zero for stars that are never above the horizon).

We can apply the same strategy we used in Figure 1.15 to find star paths for other latitudes. Figure 1.16 shows the process for latitude 30°S. Note that the south celestial pole is visible to the south and that the celestial equator passes through the northern half of the sky. If you study the diagram carefully, you can see how star tracks depend on declination.

THINK ABOUT IT

Study Figure 1.6 for latitude 30°S. Describe the path of the celestial equator. Does it obey the 90° - lat rule given earlier? Describe how star tracks differ for stars with positive and negative declinations. What declination must a star have to be circumpolar at this latitude?

HOW DOES THE SUN MOVE THROUGH THE LOCAL SKY?

Like the stars and other objects on the celestial sphere, the Sun's path depends only on its declination and your latitude. However, because the Sun's declination changes over the course of the year, the Sun's path also changes.

Figure 1.17 shows the Sun's path on the equinoxes and solstices for latitude 40°N. On the equinoxes, the Sun is on the celestial equator (dec = 0°) and therefore follows the celestial equator's path: It rises due east, crosses the meridian at altitude 50° in the south, and sets due west. Like any object on the celestial equator, it is above the horizon for 12 hours. On the summer solstice, when the Sun has dec = +23 ½° (see Table 1.1), the Sun rises well north of due east,* reaches an altitude of 50° + 23 ½° = 73 ½° when it crosses the meridian in the south, and sets well north of due west. The daylight hours are long because much more than half the Sun's path is above the horizon. On the winter solstice, when the Sun has dec = -23 ½°, the Sun rises well south of due east, reaches an altitude of only 50° - 23 ½° = 26 ½° when it crosses the meridian in the south, and sets well south of due west. The daylight hours are short because much less than half the Sun's path is above the horizon.

We could make a similar diagram to show the Sun's path on various dates for any latitude. However, the 23 ½° tilt of Earth's axis makes the Sun's path particularly interesting at the special latitudes shown in Figure 1.18. Let's investigate.

*Calculating exactly how far north of due east the Sun rises is beyond the scope of this book, but *Sky Gazer*, *Starry Night*, and other astronomical software packages can tell you exactly where (and at what time) the Sun rises and sets each day.

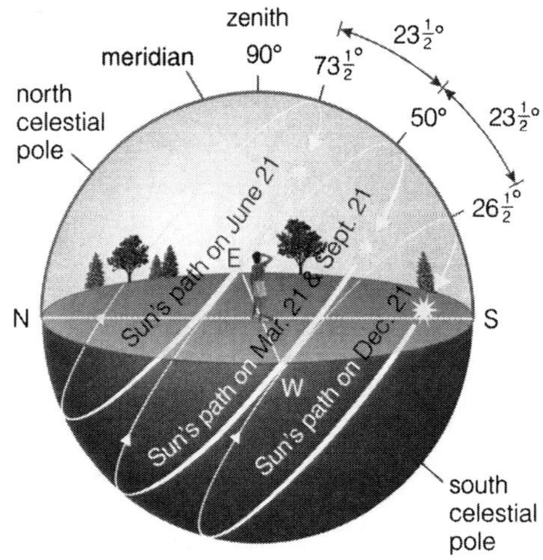

Figure 1.17 Interactive Figure: The Sun's daily paths for the equinoxes and solstices at latitude 40°N.

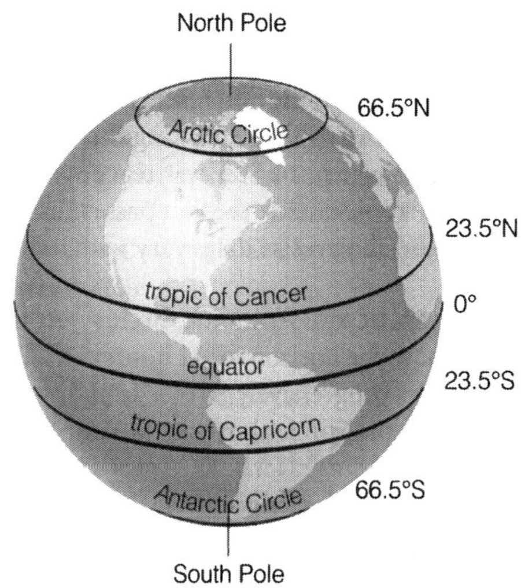

Figure 1.18 Interactive Figure: Special latitudes defined by the Sun's path through the sky.

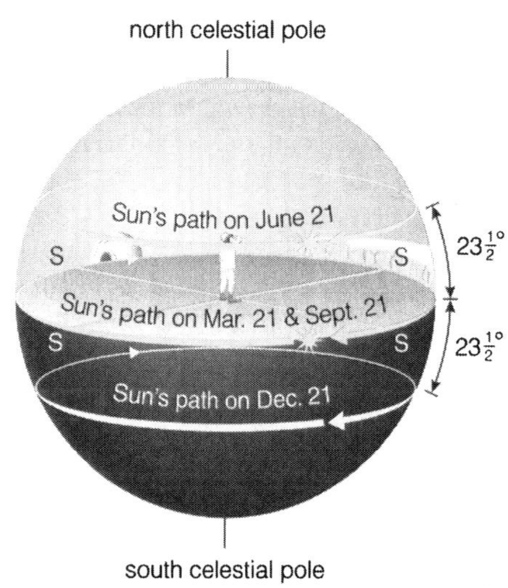

Figure 1.19 Interactive Figure: Daily paths of the Sun for the equinoxes and solstices at the North Pole.

The Sun at the North and South Poles

Remember that the celestial equator circles the horizon at the North Pole. Figure 1.19 shows how we use this fact to find the Sun's path in the North Polar sky. Because the Sun appears *on* the celestial equator on the day of the spring equinox, the Sun circles the North Polar sky *on the horizon* on March 21, completing a full circle in 24 hours (1 solar day). Over the next 3 months, the Sun continues to circle the horizon each day, circling at gradually higher altitudes as its declination increases. It reaches its highest point on the summer solstice, when its declination of +23 1/2° means that it circles the North Polar sky at an altitude of 23 1/2° After the summer solstice, the daily circles gradually fall lower over the next 3 months, reaching the horizon on the fall equinox. Then, because the Sun's declination is negative for the next 6 months (until the following spring equinox), it remains below the North Polar horizon. Thus, the North Pole essentially has 6 months of daylight and 6 months of darkness, with an extended twilight that lasts a few weeks beyond the fall equinox and an extended dawn that begins a few weeks before the spring equinox.

The situation is the opposite at the South Pole. Here the Sun's daily circle first reaches the horizon on the fall equinox. The daily circles then rise gradually higher, reaching a maximum altitude of 23 ½° on the *winter* solstice, and then slowly fall back to the horizon on the spring equinox. Thus, the South Pole has the Sun above the horizon during the 6 months it is below the north polar horizon.

Although we've correctly described the Sun's true position in the polar skies over the course of the year, two effects complicate what we actually see at the poles around the times of the equinoxes. First, the atmosphere bends light enough so that the Sun *appears* to be slightly above the horizon even when it is actually slightly below it. Near the horizon, this bending makes the Sun appear about 1° higher than it would in the absence of an atmosphere. Second, the Sun's angular size of about ½° means that it does not fall below the horizon at a single moment but instead sets gradually. Together, these effects mean that the Sun appears above each polar horizon for slightly longer (by several days) than 6 months each year.

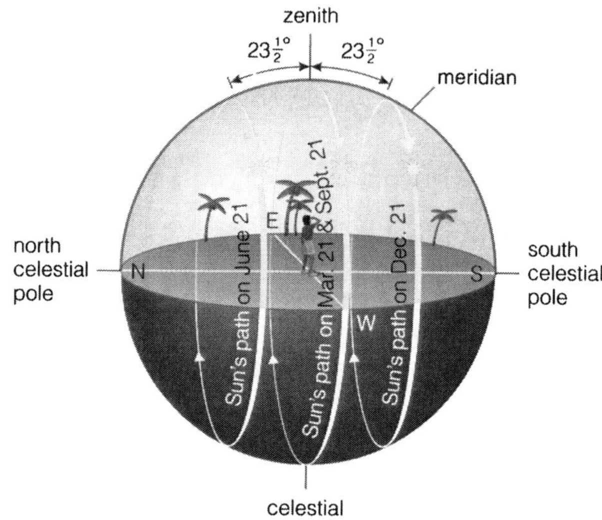

Figure 1.20 Interactive figure: Daily paths of the Sun for the equinoxes and solstices at the equator.

The Sun at the Equator

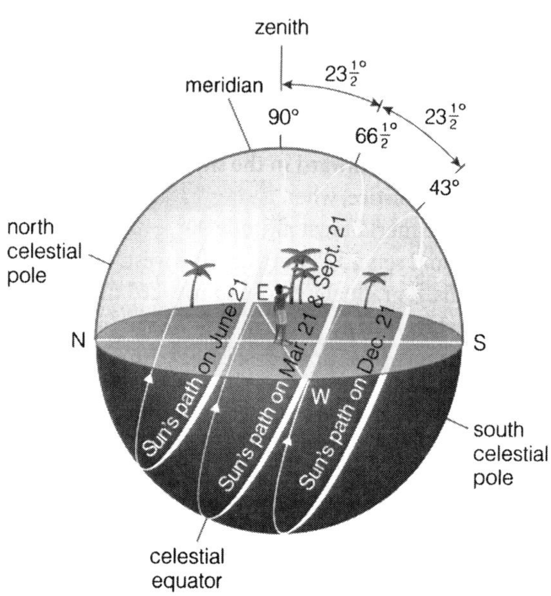

Figure 1.21 Interactive figure: Daily paths of the Sun for the equinoxes and solstices at the tropic of Cancer.

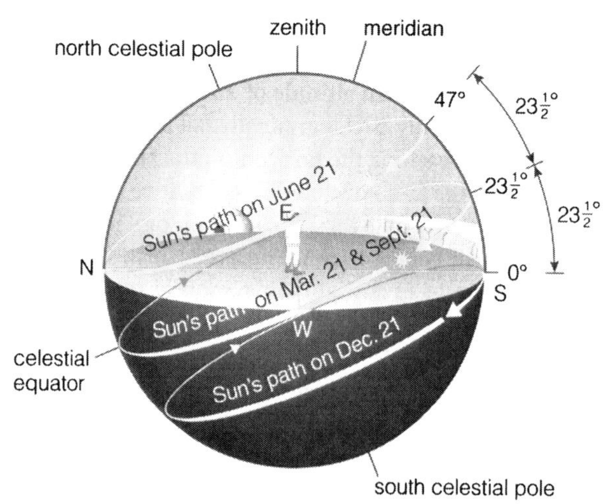

Figure 1.22 Interactive figure: Daily paths of the Sun for the equinoxes and solstices at the Arctic Circle.

At the equator, the celestial equator extends from the horizon due east, through the zenith, to the horizon due west. The Sun therefore follows this path on each equinox, reaching the zenith at local noon (Figure 1.20). Following the spring equinox, the Sun's increasing declination means that it follows a daily track that takes it gradually northward in the sky. It is farthest north on the summer solstice, when it rises 23 ½° north of due east, crosses the meridian at altitude 90° - 23 $^{1}/_{2}$° = 66 ½° in the north, and sets 23 ½° north of due west. Over the next 6 months, it gradually tracks southward until the winter solstice, when its path is the mirror image (across the celestial equator) of its summer solstice path.

Like all objects in the equatorial sky, the Sun is always above the horizon for half a day and below it for half a day. Moreover, the Sun's track is highest in the sky on the equinoxes and lowest on the summer and winter solstices. That is why equatorial regions do not have four seasons like temperate regions [Section 2.2]. The Sun's path in the equatorial sky also makes it rise and set perpendicular to the horizon every day of the year, making for a more rapid dawn and a briefer twilight than at other latitudes.

The Sun at the Tropics

The circles of latitude 23.5°N and 23.50S are called the **tropic of Cancer** and the **tropic of Capricorn**, respectively (see Figure 1.18). The region between these two circles, generally called the *tropics*, represents the parts of Earth where the Sun can sometimes reach the zenith at noon.

Figure 1.21 shows why the tropic of Cancer is special. The celestial equator extends from due east on the horizon to due west on the horizon, crossing the meridian in the south at an altitude of 90° - 23 ½° (the latitude) = 66 ½°. The Sun follows this path on the equinoxes (March 21 and September 21). As a result, the Sun's path on the summer solstice, when it crosses the meridian 23 ½° northward of the celestial equator, takes it to the zenith at local noon. Because the Sun has its maximum declination on the summer solstice, the tropic of Cancer marks the northernmost latitude at which the Sun ever reaches the zenith. Similarly, at the tropic of Capricorn the Sun reaches the zenith at local noon on the winter solstice, making this the southernmost latitude at which the Sun ever reaches the zenith. Between the two tropic circles, the Sun passes through the zenith twice a year; the precise dates vary with latitude.

The Sun at the Arctic and Antarctic Circles

At the equator, the Sun is above the horizon for 12 hours each day year-round. At latitudes progressively farther from the equator, the daily time that the Sun is above the horizon varies progressively more with the seasons. The special latitudes at which the Sun remains continuously above the horizon for a full day each year are the polar circles: the **Arctic Circle** at latitude 66.5°N and the **Antarctic Circle** at latitude 665°S (see Figure 1.18). Poleward of these circles, the length of continuous daylight (or darkness) increases beyond 24 hours, reaching the extreme of 6 months at the North and South Poles.

Figure 1.22 shows why the Arctic Circle is special. The celestial equator extends from due east on the horizon to due west on the horizon, crossing the meridian in the south at an altitude of 90° - 66 ½° (the latitude) = 23 ½°. As a result, the Sun's path is circumpolar on the summer solstice: It skims the northern horizon at midnight, rises through the eastern sky to a noon maximum altitude of 47° in the south (which is the celestial equator's maximum altitude of 23 ½° plus the Sun's summer solstice declination of 23 ½°), and then gradually falls through the western sky until it is back on the horizon at midnight (see the photograph of this path in Figure S2.18). At the Antarctic Circle, the Sun follows the same basic pattern on the winter solstice, except that it skims the horizon in the south and rises to a noon maximum altitude of 47° in the north.

However, as at the North and South Poles, what we actually see at the polar circles is slightly different from this idealization. Again, the bending of light by Earth's atmosphere and the Sun's angular size of about ½° make the Sun *appear* to be slightly above the horizon even when it is slightly below it. Thus, the Sun seems not to set for several days, rather than for a single day, around the summer solstice at the Arctic Circle (the winter solstice at the Antarctic Circle). Similarly, the Sun appears to peek above the horizon momentarily, rather than not at all, around the winter solstice at the Arctic Circle (the summer solstice at the Antarctic Circle).

SUMMARY OF KEY CONCEPTS

1.1 Astronomical Time Periods

- **How do we define the day, month, year and planetary periods?** Each of these is defined in two ways. A **sidereal day** is Earth's rotation period, which is about 4 minutes shorter than the 24-hour **solar day** from noon one day to noon the next day. A **sidereal month** is the Moon's orbital period of about 27 1/3 days, a **synodic month** is the 29 $^1/_2$ days required for the Moon's cycle of phases. A **sidereal year** is Earth's orbital period, which is about 20 minutes longer than the **tropical year** from one spring equinox to the next. A planet's **sidereal period** is its orbital period, and its **synodic period** is the time from one opposition or conjunction to the next.

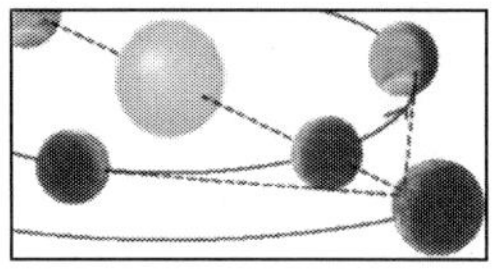

- **How do we tell the time of day?** There are several time measurement systems. **Apparent solar time** is based on the Sun's position in the local sky. **Mean solar time** is also local, but it averages the changes in the Sun's rate of motion over the year. **Standard time** and **daylight saving time** divide the world into time zones. **Universal time** is the mean solar time in Greenwich, England.

- **When and why do we have leap years?** We usually have a leap year every 4 years because the length of the year is 365 ¼ days. However, it is not exactly 365 ¼ days, so our calendar skips a leap year in century years not divisible by 400.

1.2 Celestial Coordinates and Motion in the Sky

- **How do we locate objects on the celestial sphere? Declination** is given as an angle describing on object's position north or south of the celestial equator. **Right ascension**, usually measure in hours (and minutes and seconds.), tells us how far east an object is located relative to the spring equinox.

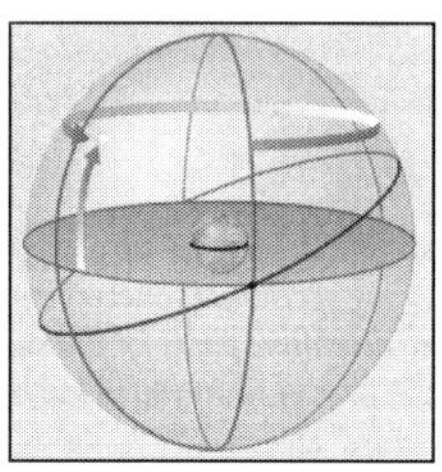

- **How do stars move through the local sky?** A star's path through the local sky depends on its declination and your latitude. Latitude tells you the orientation of your sky relative to the celestial sphere, while declination tells you how a particular star's path compares to the path of the celestial equator through your sky.

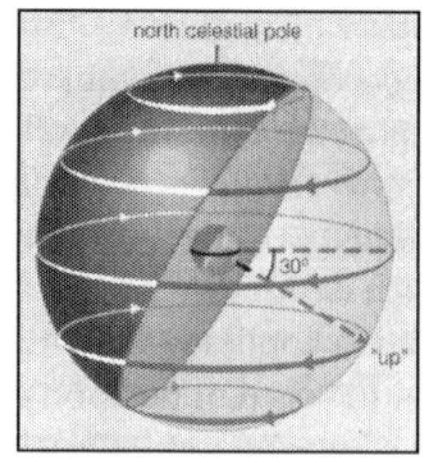

- **How does the Sun move through the local sky?** The Sun's path also depends on its declination and your latitude, but it varies throughout the year because its declination varies. The Sun's varying path helps define special latitudes, including the **tropics of Cancer** and **Capricorn** and the **Arctic** and **Antarctic Circles**.

CHAPTER 2
WHAT DO WE CALL THE STARS?

THE MAKING OF MYTHS

Human beings have always been fascinated with the night sky, which is hardly surprising when one considers the beauty of the celestial pageant and the many practical uses to which the stars have been put. To explain the existence of these celestial bodies, humans have created myths of all descriptions—tales of love, hate, bravery, and cowardice—involving gods and heroes, animals and monsters, and various treasures.

The stars have always influenced human beings: early peoples planted their crops and navigated by the stars, and the earliest sciences and religions generally correlated with stellar observations. Early civilizations created mythologies to give meaning to the constellations and other astonishing phenomena of the universe. There is even a myth about the myths: Aeschylu's Prometheus tells us that he taught man the lore about the stars. Today, these tales tell us much about the inner and outer worlds both of human beings and of civilization in general. This book explores these worlds by recounting and comparing the tales that for countless generations people of diverse cultures have projected onto the constellations.

Myths, legends, parables or fables, folktales, and fairy tales all contribute to the lore of the stars. Myths serve many functions, from the sacred to the secular. The word *myth* derives from the Greek *mythos* and means "spoken word". Stories are handed down from generation to generation, often told by parents to their children. They have many variations. Legends are grounded in history. Folktales are often humorous and entertaining, highlighting a clever human hero or heroine. The parable, or fable, teaches. In folktales, characters are often general types, rather than the individualized personalities of myth. Fairy tales are like folktales but involve the supernatural (if not the actual presence of fairies). The folktale is usually grounded in a specific time, whereas the myth tends to be timeless. Modern fantasies about the stars can be found in works of science fiction, from Asimov's *Foundation* trilogy to *Star Wars* and *Alien* to *No Escape*. One can see how all these categories overlap.

There are many types of myths, including the religious tale, the dream (wish fulfillment) tale, the explanatory tale, and the entertaining tale. Obviously, these categories are flexible. Some myths endorse a claim or an institution; others are religious incantations meant to secure the status quo or to bring about change; some myths sublimate anxieties and assuage traumas; and still others are meant simply to amuse. Myths can order nature in human terms, allowing us to understand it better by identifying with it. Sometimes a myth is recited to exert control over nature—for instance, to ensure that the Sun rises (in his nightly voyage, the Egyptian Sun god, Ra, has to defeat the serpent Apophis in order to rise, and prayers were offered daily to help the god on his way). Myths are often associated with the supernatural, with gods, and with ritual. They represent a sacred endorsement and are often used to legitimize people's claims to land or inheritance ("charter myths"). Many myths deal with death and the individual's place in the underworld. Fertility— of humans, animals, and land—is a frequent theme. Men and women also need to be organized and ranked in their activities and institutions, and myths can help impose social order. Myths can pose problems (and at times, solve them) or articulate (and in some cases, allay) human fears. They can appeal to the intellect and arouse feelings. Myths can be dreams that explore the meaning of life. As Prospero tells us, "We are such stuff/As dreams are made on, and our little life/Is rounded with a sleep" (*The Tempest*).

Some would claim, as Jung did, that certain archetypes are common to all people. Joseph Campbell carried on a comparative study of mythologies and found many ties between them: he viewed heroic tales, for instance, as symbolic rites of passage a person must undergo to achieve a productive adult life. Sometimes Campbell errs by seeing too many similarities and forcing tales to fit his procrustean ideology, which is firmly rooted in Eastern philosophy. Sir James G. Frazer and Robert Graves also find unifying themes, such as fertility rituals and "the white goddess." Friedrich Max Mueller sees the Sun behind every myth.

To explain similarities in myths, one could say that there is a universality in human perception and imagination—perhaps because of human brain structure or because of common experiences—that led diverse cultures to develop the same myths independently. But there are many possible explanations for the similarities among myths. One could speculate that travelers at the dawn of civilization established links that are now forgotten. Perhaps it is obvious that humans should draw parallels between themselves and the universe in which they live, such as noting the reflection of the human cycle of life in the passing of the seasons. There is also the theory that some divine or supreme force in the universe has communicated these ideas (myths) to mankind. Perhaps it is simply a function of neural biology: some people claim that humans share certain myths because of the bicameral structure of the human brain. Individual differences and needs, however, should not be minimized in these studies.

Human beings must define their status—not only their relation to one another, but to gods and animals and also to the heavens and the earth. Mythologizing helps define these relations. Not surprisingly, myths change with each telling, particularly if they are not written down (as is the case with most myths). Furthermore, these stories are often modified for specific reasons: to suit history, for instance, or to accommodate sacred influences involving cult and ritual. Besides the oral tradition itself, the literary tradition has altered myths as well. All these factors have a hand in altering myths over time.

Myths differ from land to land, and we learn from the comparison of cultures the ways in which myths can operate. Bronislaw Malinowski studied the myths of the Trobriand Islanders; Lévi-Strauss, the South American Indians; Clyde Kluckhohn, the Navajos; and the Cambridge school, including Jane Harrison, the Greeks. Each of these researchers is inclined toward an overarching theory that may work for the specific culture he studied, but often breaks down when applied to other cultures. Malinowski sees myths as inexplicable narratives having no explanatory function, instead serving as sacred stories that accompany rituals. Lévi-Strauss sees myths as a strategy for mediating between opposites, such as raw and cooked, fresh and rotten, high and low, old and young, or primitive and civilized (nature/culture). Lévi-Strauss emphasizes structure over content, and shows how myths can be understood better through juxtaposition with other myths than they can be understood in themselves.

Lévi-Strauss disagrees with E. Durkheim, who sees myths expressing the collective ideas of a social group. Paul Ricoeur objected to Lévi-Strauss' generalizing from the myths of North and South American Indian societies rather than from those of Semitic, Indo-European, or Mediterranean areas. The myths Lévi-Strauss investigated often confused the human and the animal, exhibiting totemism in contradistinction to myths that centered upon anthropomorphic or abstract gods. Ricoeur also pointed out that Lévi-Strauss used myths from countries that emphasized structure and arrangement over content (they were based on *bricolage*, or assembling the things they found at hand rather than choosing objects or subjects that would convey a predetermined meaning). Lévi-Strauss also claims that a myth consists of all its versions, not taking note of the historical changes that can radically alter the meaning and significance of a myth. Thorkild Jacobsen saw the myths of Sumeria as concerned with creation, organization, and evaluation: one learned about gods, man, and the structure of the universe.

The mythical stage can also precede or contribute toward the scientific. As F. M. Cornford has said, "After the primitive stage of genuine myth-making, there is a transitional period, in which the old images and symbols are retained, but with a nascent consciousness that they do go beyond the meaning proper ... Finally there may come a time when rational thinking consciously asserts itself, and the foremost intellects of the race awaken out of the dream of mythology ... This happened in sixth-century Ionia, and what the Western world calls philosophy or science was born."

Kluckhohn and the Cambridge school saw an essential relationship between myths and rituals: the *legomenon* was the myth (what was said) and the *dromenon* was what was done (the ritual). Rituals could be rites of passage that enabled a person to leave one stage in his or her existence and go on to another, usually by means of isolation from the community and a series of tests. Or there could be rituals to ensure fertility, sometimes involving a sacrifice or payment to a god. The Cambridge school also spoke of the Year-Spirit, which clearly implied the change of seasons that coincided with the spirit's life, death, and rebirth. Myths can also be interpreted as star maps. For instance, the Mesopotamian Gilgamesh epic can be seen as representing the yearly transit of the Sun through the constellations of the zodiac (obviously, it is the earth that moves, but for centuries it was thought that the Sun itself traveled along a yearly path). There are also psychological interpretations of myths, as we find handed on by Karl Kerenyi, who is influenced by Jung, or Philip Slater, who shows a Freudian bias. Freud claimed that all myths were symbolic. The "unconscious" can shape myths, and Freud saw this as the "unconscious" of the individual, whereas Jung saw it as that of the race.

All these theories are useful, but one should be careful not to take one system of interpretation as absolute. We should paraphrase Lévi-Strauss and say that the proper interpretation of a myth consists of all interpretations. It is obvious, however, that in some cases one interpretation fits a particular myth better than another. The Freudian interpretation of the myth of Oedipus is more satisfying than the structural analysis of the same myth proposed by Lévi-Strauss, but both are interesting. Freud's association of the myth of Oedipus with the Primal Horde that first assassinated "the father" as the first act in an elaborate rite of passage does ask the average person to make a leap of faith. Also, the last thing that the ancient mythical Oedipus wanted to do was marry his mother, but Freud took the story to signify a universal drive of man to mate with his mother. To Freud, the myth of Electra describes the complex relationship between a daughter and a father. Once again, his interpretation is an extrapolation of the myth, which shows a daughter interested in seeing that her murdered father is avenged. Freud took myths and used them like a grammar, a framework for the meaning that he wanted to convey.

Ernst Cassirer focuses on the language of myth that is symbolically used to translate mythic thought. He also takes over L. Levy-Bruhl's idea of a primitive mentality (*la mentalité primitive*). According to this view, one becomes possessed, and myths write themselves. As he says, "The primitive mentality does not invent myths, it *experiences* them."

Many American Indian cultures view myths as dreams. Author Carlos Castenada writes about "dreaming" as something essential for a warrior, the ideal of the living human being. These dreams are like waking myths, sacred keys to a second reality. Similarly, Australian aborigines also speak of "Dreamtime" as an intensification of the present (consider Calderon's play *La Vida es Sueno* [Life Is a Dream]). It is also worth noting the Hindu creation/reincarnation myth of the god Vishnu dreaming the universe dream as he rides on the cosmic serpent who floats on the Milky Ocean.

Through dreams, one can sometimes approach the truth more clearly than through rational thought. Accordingly, the myths of a given culture should be taken as seriously as that culture's philosophic and scientific canons. Even if these tales do not necessarily contain cosmic truths, at least they can yield important information about a people.

Astronomy is an ancient science. The Egyptians practiced astronomy as early as the third millennium B.C. Soon, the movements of the stars were associated with human activity, and astrologers started to predict what would happen to humans based on the places the stars occupied in the sky. In many cultures, various predictions, both scientific and fanciful, were made based on the movements of the heavenly bodies. Thales of Miletus is said to have shown his astronomical prowess in predicting an eclipse in 585 B.C. Already in the works of Homer (c. 900 B.C.) some constellations are mentioned: Pleiades, Hyades, Bear/Wagon, Orion with his dog. According to Plato, the Greeks gave the name of their gods to five planets in the fifth century B.C., which in their Latin form are Jupiter, Saturn, Mars, Mercury, and Venus (their Babylonian equivalents are, respectively, Marduk, Ninib, Nergal, Nebo, and Ishtar). As other planets were discovered, they were given names of more gods in the Greek and Roman pantheon (e.g., Uranus, Neptune, and Pluto). The etymology of *planet* yields "wanderer." Perhaps the mobility of the planets led to this use of the gods' names for the planets; divine beings, more than any others, after all, have the ability to come and go as they wish. Western culture has given mainly Latin or Latinate Greek names to the planets, and Greek names to the constellations; Arabic names are mainly given to individual stars.

A myth can be a profound prayer, or it can be a simple wish. A star led the shepherds to Bethlehem to find Jesus Christ. For years humans have cast horoscopes based on the stars; if one is born under a certain constellation, one is said to have certain characteristics and to experience life events, influenced by planetary movements. The Babylonians worshiped the Sun, Moon, planets, constellations, and stars. It was easy to infer from the regularity of the movement of the heavenly spheres that humankind was controlled by some outside force. This enhanced the idea of fate (and even necessity) governing a person's life. Obviously, if one could foretell life's events, then one had great power.

As we stand on earth looking at the night sky, from most places we can see about five thousand stars (probably closer to two thousand, actually). There are about seven thousand that we could potentially see if we could travel around the earth. In the Milky Way, there are about four hundred billion stars, and in the entire universe as we know it about one hundred billion billion stars. So we see only a small fraction of our galactic surroundings. As we have discussed, people of most cultures have been devising stellar ordering systems for millenia, right up until the present day. E. C. Krupp notes that in 1928 there were eighty-eight constellations accepted by the International Astronomical Union. The Chinese have different sets of stars, numbering twenty-eight, called *Sieu*, through which the Moon passes each month.

Hesiod (c. 700 B.C.) certainly spoke of the birth of the universe in his *Theogony*, and in *Works and Days* he gives an almanac of favorable and unfavorable days. His lost *Astronomia*, which gave the risings and settings—and possibly myths—of the stars only survives in a few fragments. Eudoxus in the fourth century B.C. gives us an account of some of the signs of the zodiac, and some of the names of the planets. Aratus (270 B.C.) supplies more such information in his *Phainomena*. A generation later, Eratosthenes fills in the associated myths in the *Katasterismoi*, a work that was condensed by Pseudo-Eratosthenes, who wrote two to three centuries later. The Greek Stoics (from about 300 B.C.) prized star lore. They believed in fate and that the whole universe was organized in a cosmic unity. The Persians also believed in astral divinity.

Astrologers abounded in ancient Rome. In the first century A.D. it was a capital offense in Rome to cast the emperor's horoscope without the emperor's explicit authorization. Beginning in the second century A.D. Roman emperors were identified with the Sun. Hyginus' *Poetica Astronomica* deals with some of the lore. Astrology continued to be popular during the Byzantine period, and Averroes (twelfth century) also considered it of scientific value, as did the thirteenth-century English philosopher Roger Bacon. Pope Leo X (sixteenth century) established a chair of astrology in Rome. By 1650 the belief in astrology diminished, at least in part because of scientific observations by Galileo, the development of a scientific method (based on observation rather than faith) by Francis Bacon, and the rationalist philosophy of Rene Descartes. Not surprisingly, the Enlightenment was a difficult time for the followers of astrology, but the practice nonetheless survived. As soon as new planets were discovered (Uranus in 1781 and Neptune in 1845), their influences were incorporated into astrological lore. There is a London Faculty of Astrological Studies that is world-renowned. From the many astrological publications alone, it is obvious that this science is alive and well. (Indeed, Nancy Reagan conducted her life according to astrology's dictates, and no doubt influenced her husband while he was president of the United States.

Some people also wanted their religious beliefs confirmed by the stars, and thus rewrote the ancient text of the heavens. This process is a reflection of the ways in which ascendent cultures have established ownership and dominion through the renaming of conquered territories, religions, and even ideas. In Utah, the rock formations called The Four Apostles by the agents of Christendom had been given entirely different names by the regional Native American populations. The Irish had their own Gaelic place names, but these were brushed aside by the conquering Britons and replaced with English names to ensure domination (well commemorated in Brian Friel's play *Translations*). New stories and renamings often accompany a shift in power, and are a way for the new regime to dominate. One way to rule is to rewrite history, and the heavens were not immune to this phenomenon.

Myths deal with mankind's most profound concerns, desires, and fears, including (but not limited to) life and death and the individual's place in the universe. Indeed, some myths constitute an attempt to transcend the mortal coil. Although death is a sentence with no appeal, at least stories that tell of eternal life—or stories that are themselves eternal—allow human beings to take their place in the heavens, next to the gods in the form of constellations.

Heroes and heroines were frequently elevated to the stars, usually for having accomplished some exceptional deed—or many such deeds, as Heracles did. Great actions in themselves can be a source of eternal fame, but *katasterism*, or elevation to the stars, was the Nobel prize of heroism. But such immortalization can also be a punishment: the wrongs of Cassiopeia were punished by the inversion of her image in the heavens, thus making her appear ridiculous. Beside heroes, we also find among the constellations animals and monsters, and such things as a crown, a lyre, or a humble casserole in the heavens. All the objects associated with human beings can benefit from divine translation.

This book is an examination of the similarities and differences among different cosmologies. The text cites twenty-seven of the primary Greek constellations, using them as the starting point for a discussion of the related constellations of other cultures from around the world. Interestingly, many cultures picked out the same constellations, and some even interpreted those star groupings in the same way. For instance, the seven (or six) stars of the Pleiades represent seven (or six) young girls in several cultures, from Greece to Australia. Other constellations are interpreted differently by different cultures: the three bright stars that make up the belt of Orion from Greek mythology are Freya's spinning wheel of Scandinavian legend.

What do the stars tell us about the similarities and differences among the world's peoples? This will be the major thrust of this investigation. We shall concentrate on Greek myths because in the Western world, the Greek influence is at the foundation of many areas of cultural expression, from art and literature to philosophy and medicine. Additionally, the Greek mythological canon is the largest and most complete such compendium of thought about the human condition that any people has ever produced; and it has been inscribed on the stars.

Similarities among star myths probably occurred when diverse peoples came into contact with one another or when the shape of a certain star formation was so suggestive that it inspired a particular interpretation. In general, however, different cultures developed different constellation myths—for any number of reasons, including geographic, economic, political, or religious. For instance, seafaring nations of antiquity had a full complement of star myths, just as did civilizations that survived by farming. These canons differed according to the functions of the myths: the former depended on the stars for plotting voyages, and the latter used the stars to determine when to plant and when to harvest. Merchants had their own myths as well. The shapes the various constellations assumed also differed: sailors saw the Big Dipper as a ship, farmers saw it as a plow, and merchants saw it as a wagon. In the Arabic tradition, which emphasizes strictly observed rituals, the Big Dipper is seen as a funeral procession (the dipper part is the coffin). In China, where the emphasis is on communal harmony, the Big Dipper is *Ten Li*, or "Heavenly Reason". Obviously, there will be differences in the myths of a country that experienced yearly flooding and was structured around a hierarchy of kings and gods (Egypt, for example) and one that had a more regular supply of rain and was more democratic (such as Greece). The reasons for the differences in these traditions are many.

Greek myths have their own defining characteristics, with an emphasis on heroes and a tendency to underline the rational element. The Greeks put themselves at the center of the universe, and thus did not place abstractions in the heavens; instead, they placed there gods, heroes, and the animals associated with both. In addition to mythological stories, the Greeks gave scientific explanations for movement in the universe. Many Native American Indians had matriarchal societies (for instance, the Ajumawi band of Pit River Indians of northeastern California), and in their myths the Sun is female and the Moon is male The ancient Japanese worshiped the goddess of the Sun, Amaterasu, reflecting how, in the early history of the culture (before the militaristic Tokugawan regime came to power), women had a large degree of power.

One of the most famous Japanese epics, the *Tale of Genji* (twelfth century), is by Lady Murasaki. The Karraru aborigines of South Australia also claim that the Sun is a woman, the Sun Mother who created life as she walked. Patriarchal societies favor a male Sun god, and in some societies male rulers trace their lineage from the Sun: the Chinese emperor is the son of heaven; Egypt's pharaoh is the son of the Sun; and the Inca dynasty in Peru is descended from the Sun. The Sumerians depict Shamash, the Sun god, overcoming Siduri, the goddess of knowledge; Shamash takes over Siduri's powers and forces her to become a slave and work in a tavern.

The human race has traditionally written its story in the heavens, sometimes even using an existing body of myths as the components of the narrative. As Seamus Deane says in a poem, "Homer Nods":

And could it be
We have here an autobiography
Pretending to epic for authority?
Protean creature, who has come to be
The proof that greatness lives vicariously
In the lies swapped round from men to God.
And when we hear the truth,
Not subject naturally, to proof,
We allow for Homer and we let him nod.

HOW DIFFERENT CULTURES CREATED CONSTELLATIONS

Clearly, there are major differences in the interpretations of star myths throughout the world, and at the same time many similarities. The similarities can be attributed to the more-or-less fixed patterns of the stars and also to the wanderings of various peoples. There are nevertheless particular characteristics in the stories told by shepherds, who see sheep and pastures, and by fishermen, who see fishes and rivers. In the north, there are stories about snow (in Sweden, the Milky Way is Winter Street, and in Siberia, a ski track left by a god); in the tropics, there are stories about water, rain, and floods (in Sanskrit, the Milky Way is "The Bed of the Ganges"); and stories about winds and dust are told in the mountainous areas (the Peruvians see the Milky Way as star dust).

The Chinese differ the most from Westerners in what they see; they routinely placed their bureaucracy in the sky, including all the various classes, from king to peasant. They added gods, animals, objects, and abstract principles, such as *Ten Li*, or "heavenly wisdom". Confucianism led to an ordered society that has found its way into the heavens. The constellations of China do not tell the stories of violence that the Western ones do, but reflect the order and control of their society. They have stories of love, such as that of the Weaver (the star Vega) and the Shepherd (the star Altair), rather than Western tales of rape (for instance, those surrounding the constellations Orion and Cygnus). There is also an acceptance of the natural that is not as idealizing as in the West: thus "Piled-up Corpses" (Algol in Perseus) can be found in the heavens, as can the stars called "Heavenly Sewer" (Cetus). Jesuits brought stories from the West, and so some constellations have similar names, but if one digs deeply, one usually finds another story that explained a particular constellation originally.

The Arabs have different representations, which are usually nonhuman in accordance with certain precepts of Islam after the seventh century A.D. But these constellations, too, have taken over some these common Western images. One also finds mundane representations, such as of camels and the paraphernalia associated with living in desert climes.

Native Americans often found tales of the hunt in the heavens. They also put some of their gods, who were so important in governing the weather, in the heavens. The Polynesians often had tales of fishing and voyages (the Milky Way is a big shark). African myths speak of hunting and the animals that are indigenous to their various countries. Indians see the universe as resting first upon a snake, then a turtle, and then four elephants (all indigenous animals). One tribe of Native Americans, the Chumash, have myths about the Sky Coyote in a yearly gambling match with the Sun; if he wins, the earth flourishes, whereas if the Sun wins, there is drought and death. Again, these tales are populated by local animals. The perception among the Australian aborigines of the Corona Borealis as a boomerang is another illustration of the elevation of day-to-day paraphernalia to heavenly status.

The Jews see their tribes and their symbols in the sky. Christians see the twelve apostles in the constellations of the zodiac, and recount Christian stories where they can, taking whatever in the Greek canon that resembles some Christian account.

The Greeks have put their heroes in the sky (for example, Perseus) and sometimes their criminals (Cassiopeia) and sometimes the method of punishment (Ixion's wheel in Ursa Major). Their passions are embodied there, with tale after tale of anger and rape—and also some tales about noble achievements. The violence of the tales contrasts with the more restrained stories that China has put in its heavens. Some of the Greek myths are tales of human life and death, and of the alternation of the vital crop in its cycle of life and death.

Celtic interpretations usually relate to heroes—legendary, epic, or historical. The Indians placed their gods in the sky, especially those associated with myths that deal with humanity's survival—for instance, Indra's victory over Vritra, representing the triumph of light over dark. On the more practical side, India is a country of many jewels, which have made their way into their heavens.

So we conclude our brief tour of the heavens. We have witnessed the struggles of heroes and the loves of men and gods commemorated for all time in these shining memorials. Saint-Exupery located the Little Prince in the heavens after the little intergalactic traveler left earth. In the heavens, humans share immortality with the gods. The threatening monsters they conquered and the beasts that helped them are still to be seen at their side. Human life continues in a stellar form. Some chases—and some loves—will continue for eternity.

We can apply Keats's words from "Ode on a Grecian Urn":

Bold lover, never, never canst thou kiss,
Though winning near the goal—yet, do not grieve;
She cannot fade, though thou hast not thy bliss,
For ever wilt thou love, and she be fair!

The constellations shall always be fair, and they will no doubt continue to inspire great passion in the peoples of the world. In the heavens, human beings achieve the ultimate catharsis—through the constellations and the myths attributed to them. These groups of stars can be the loci of prayers and the source of great happiness—indeed, the presence of the stars makes it a joy to be alive. The stars are also our protectors. For every threat (scorpion or bear), there is a hunter or a hero to save us. Our life unfolds with the stars, which themselves announce the seasons that punctuate our existence. Among the stars, there is always the promise of life and a new dawn, even if preceded by a desperate struggle. In the theater of the sky, these dramas are replayed every night, with new installments that unfold throughout the year.

THE STORY BEHIND ONE CONSTELLATION: ORION

Orion is one of the most famous Greek constellations. The three bright stars in the belt are often the only part of a constellation that people can identify. Orion was a great hunter; he is usually depicted holding a club or a sword that he swings at the bull, Taurus. He is also near a unicorn and a rabbit, so he does not lack for game. Finally he has Canis Major, his trusty dog, nearby.

Orion was the son of Euryale and Poseidon or Hyrieus (actually he may have had several fathers). His mother may have been Gaia; so if his father were Poseidon, he was a combination of the traits of the earth and the sea; some versions of his tale depict him walking on water, a gift from his father. Other versions depict him as being so tall that when he walked along the ocean floor, his head rose above the waves.

It is said that Zeus, Poseidon, and Hermes were visiting Hyrieus, who gave them all the hospitality he could offer. The gods asked him what he wanted in return, and he said a son. The gods then urinated into the skin of the ox that they had been fed, and Hyrieus was told to bury it. (Some versions state that Hyrieus himself was told to urinate in it.) From that ox-skin, Orion was born. His name is related to the Greek word for "urine." He was also very handsome.

Orion went to Chios and fell in love with Merope, the daughter of the king. In return for Merope's hand, the king requested that Orion slay all the wild beasts on the island. When the slaughter was over, however, the king reneged. Orion, not unlike Heracles in being impatient and hot-tempered, drank too much one night and raped Merope. Oenopion put out Orion's eyes as punishment. Hephaestus' apprentice Cedalion guided the blind hero to Delos, where he could be cured by Helius, the Sun. In ancient Greece, the Sun was thought to create sight. Eos, the dawn, saw the handsome Orion and seduced him. It is said that she blushes when she rises because she remembers his embraces.

Ultimately, Orion was killed by Artemis, either because he challenged her hunting abilities or tried to rape some maidens who followed her (see Pleiades). Some say that he tried to rape Artemis herself, who sent a giant scorpion after him in revenge. This scorpion stung him mightily, eventually killing him, and was rewarded by being transformed into a constellation.

Other versions tell the story differently: Apollo, protecting his sister's honor lest she succumb to Orion's attractions, sent the scorpion. Apollo even went so far as to trick his sister into aiming an arrow at a black spot swimming in the ocean that happened to be Orion. She was duped and was the inadvertent source of Orion's death. His constellation was mentioned in Homer's writings. Other versions of the myth state that on one occasion Orion was hunting with Artemis and boasted that he would kill every animal on earth. Gaia overheard him and sent a giant scorpion up to kill him. It stung him and he died, and both were placed in the heavens—not only as a warning to others but as a tribute to Mother Earth. The chase continues: whenever Scorpion rises, Orion sets, so the scorpion continues to be victorious.

The Egyptian gods Osiris and the star Isis were also seen in the constellation Orion and the star Sirius. The Egyptians also saw Orion as Horus, in a boat, followed by a cow (Sirius), also in a boat (Isis was often represented with cows' horns, possibly referring to her relation to the crescent moon). Indians saw this constellation as the god Praja-pati in the form of a deer. The deer pursues a doe, his daughter Rohini (represented by the star Aldebaran), but the hunter Lubdhka (Sirius) prevents him from catching the doe by shooting an arrow (the three stars of Orion's belt) into the deer. The Jews had various names for this constellation, from "Impious" to "Foolish" to "Violence" to *Gibbor*, "Giant" (recalling the foolish Nimrod, who rebelled against Jehovah). Christians saw it as Joshua, Jacob; or Saint Joseph. Some Christians saw the three stars of the belt as "Our Lady's Wand," while others saw the grouping as the Three Magi.

The Irish called this constellation *Caomai*, "The Armed King." A modern appellation dubs it "California of the

The Constellation of Orion

Sky." In the nineteenth century, some French called the stars of the belt "Napoleon"; the English promptly retaliated, calling the stars "Nelson." French farmers called the three stars "The Rake," and in Germany the same three stars were "Three Mowers." (This is proof, perhaps, that constellations often take the shape of what one wants to see, or what one is familiar with; in a sense, they are the Rorschach blots of the sky.)

The Chinese pragmatically called the belt stars "Three Stars." They had many other names for the other stars in Orion, including "Three Flags," "Golden Well," "Head of the Tiger," "Sharp Edge," "Middleman," and even "Punishment." Some Arabic cultures saw a belt, a string of pearls, or nuggets of gold. The Hindus saw the three stars as a three-jointed arrow. Some Native Americans saw the sword of Orion as an arrow aimed at three mountain sheep (the belt stars). Other Native Americans saw other kinds of game in these stars: antelope, deer, and sheep. Some aborigines in Australia saw these stars as three men in a heavenly canoe; others saw them as three young men trying to attract some girls (the Pleiades). In Greenland, the constellation was *Siktut*, "Seal Hunters," who were lost while hunting and transposed to the heavens. The Norse saw the stars in the belt as *Fiskikallar*, meaning "Staff"; Scandinavians called it *Frigge Rok*, "Frigg's (or Freya's) Distaff," from which she spun the thread to create the clouds that float over Midgard, where the mortals dwell. Other Scandinavians called the constellation "Kaleva's Sword," and still others "Tavern."

Betelgeuse, the constellation's main star, which is located in Orion's shoulder, gets its name from the Arabic name *Ibt al-Jauzah*, "The Armpit of He Who is in the Center." It gleams bright orange. The star Bellatrix, "Warrior Woman," is on the left shoulder of Orion. People from the Amazon River called Bellatrix "Young Boy in a Canoe." Some Arabic cultures called it "One Who Roars," referring perhaps to the lion skin that is sometimes depicted in Orion's left hand. Rigel, another famous star from this constellation, is the brightest in this grouping; the name comes from the Arabic word *Rijl*, from *Rijijauzah al Yusra*, "Left Leg of the Jauzah." Some stars were the Herdsman of the Jauzah, complete with camels. The Norse considered it the thumb of the giant Orwandil. The other thumb, which had been frozen, was cut off by Thor; it became the star Alcor in Ursa Major.

Orion has also been associated with the Sumerian epic of Gilgamesh, the legendary king of Uruk. He abused his power and lived in a dissolute way, seducing the maidens of the land. The people complained. Elsewhere, a double, Enkidu—who was covered in hair—was created to provide a rival for Gilgamesh. The Sun-god Shamash used a priest-

The illustration of Orion the Hunter that was published in Johannes Bayer's *Uranometria* in 1603. Such illustrations helped to personify the constellation figures, which did not always resemble in shape the people, animals, and objects they were meant to represent. (Courtesy of University of Arizona Library Special Collections)

ess of Ishtar to seduce Enkidu and bring him away from the wild country in which he thrived. She took him to Uruk. He learned the ways of civilization, eating cooked food for the first time. He rubbed off his hair and was very handsome. He was shocked to hear of Gilgamesh's wanton ways, and he stopped the profligate as Gilgamesh was on his way to an erotic rendezvous and wrestled with him. They became fast friends and undertook various adventures. They first tried to slay the giant Humbaba. Enkidu tried to talk Gilgamesh out of this, but Gilgamesh said that if he could not obtain immortality (and what mortal could?), the least he could do would be to achieve great deeds (similar sentiments are found in Homer's *Iliad*, which may have been influenced by this epic; Achilles and Patrodus resemble Gilgamesh and Enkidu). After a great struggle, the friends were able to kill Humbaba. Then they purified themselves.

Gilgamesh was so handsome that Ishtar pursued him. Gilgamesh refused her indignantly, pointing out the disastrous ends that her lovers suffered. Because she was rejected, she begged Anu, her father, to send the celestial bull against them (the zodiacal constellation Taurus is near this constellation).

Gilgamesh and Enkidu slew the bull and threw the pieces at Ishtar in defiance. Ishtar was rightly furious and demanded that the gods avenge her: one of the duo would have to die to compensate for her loss. Enkidu died and Gilgamesh suffered leprosy. Gilgamesh mourned at his friend's side for many days, until a worm crawled out of his nose.

Gilgamesh went to the desert and continued his mourning, growing his hair long and wearing a lion skin. He was terrified by death. He decided to consult Utnapishtim, his great-uncle (and analogous to Noah), to find out how he could become immortal, since the gods had made Utnapishtim himself immortal. On the way, he had to overcome scorpion men, go through great darkness, and was rewarded by a vision of paradise—a forest where the trees gleamed with brilliant multicolored gems.

Siduri, a divine maiden, tried to dissuade him, saying that man should be satisfied with enjoying music, dance, celebrations, and a loving wife and children—but these were not for our hero. He continued his treacherous voyage over the sea of death, until he finally met Utnapishtim. Utnapishtim also tried to discourage him, saying that all men must die, that only very few were exempted: in his case, it was for his good and pious deeds. Utnapishtim cured Gilgamesh's leprosy and decided to test Gilgamesh by getting him to stay awake for seven nights. He failed miserably, not even being able to stay awake for one night. But Utnapishtim told him about a tree that grew branches that would make him immortal if he could keep a branch with him. Gilgamesh succeeded in finding the tree and acquiring the branch, but he set it down when he took a swim, and a serpent swallowed it.

Gilgamesh was forced to accept his mortality. He buried his beloved friend. He even persuaded Nergal, the god of the dead, to let him talk to his friend once more. Enkidu told Gilgamesh that if a man dies in battle and is buried with honor, then that person has some comfort, but the one whose body is abandoned in the fields has no rest.

This tale contrasts civilization with the life of the primitive and concludes, as does the *Iliad*, with an affirmation of mankind's mortality. Man's life is symbolically traced in

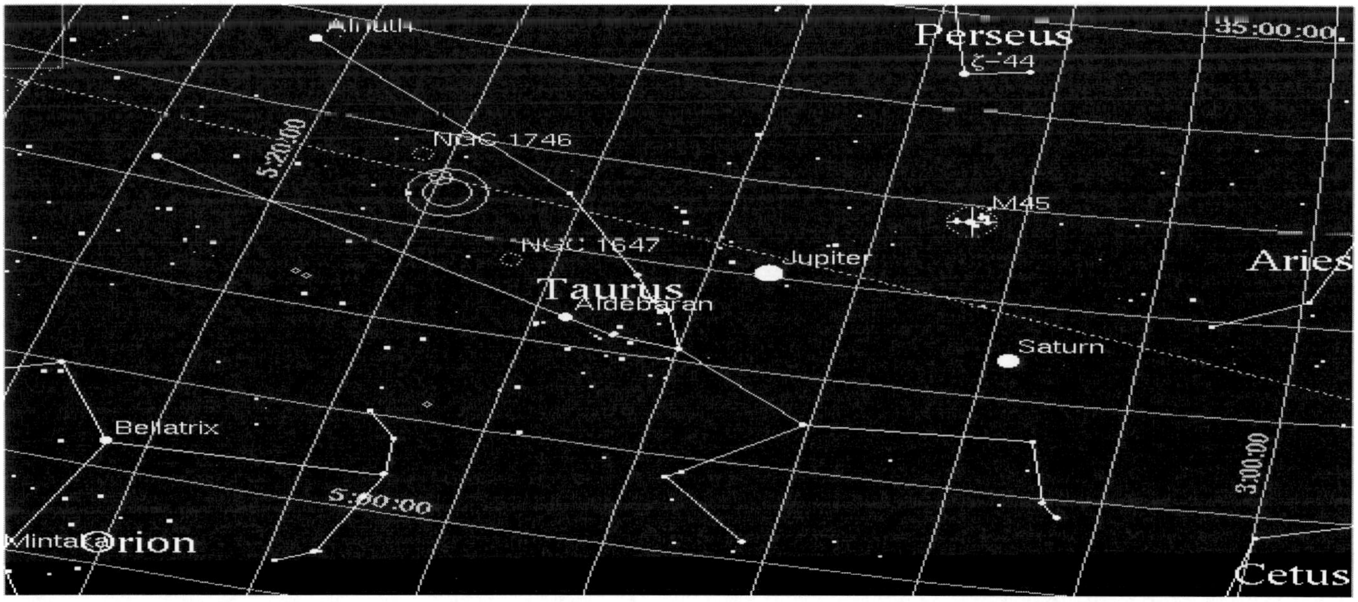

The Constellation of Taurus

this epic: we are all savages as children, and when overcome by disease and old age, we return to a primitive state, as Shakespeare states us in the speech about the "Seven Ages of Man," from *As You Like It.*

The Iliad and the tale of Gilgamesh celebrate not only life, but also death achieved in the midst of youth and glory. These young glorious heroes have found their way to the stars in Perseus, Heracles, and Orion. It is rare to find the aged immortalized in the Greek and Western myths about the stars. For such representations it is necessary to refer to the heavens as seen in China, where old age is venerated.

As Thomas Kinsella puts it, describing an old man looking at himself in a mirror:

In slow distaste,
I fold my towel with what grace I can,
Not young and not renewable, but man.

THE ANCIENT CONSTELLATIONS: STORIES IN THE SKY

With some very interesting exceptions to be examined later, the stars are randomly distributed across the heavens. Randomly, however, does not mean uniformly. They are spread quite irregularly, groups of bright ones separated by areas with few stars at all. The eye will naturally tend to pick out patterns. A newcomer to astronomy will automatically select such groupings, or constellations, in the same way our ancestors did thousands of years ago. Already our neophyte is, as the Greeks would have it, an *astronomos*, an arranger of stars. In many cases, the configurations are so prominent that the beginner's patterns will in fact quite well match the old. As many as 8000 stars can be seen with the unaided eye over the entire celestial sphere, and the novice will rapidly discover that the simplest way to organize them for easy identification and for later study is through these groupings.

All societies, everywhere, have invented constellations. The ones with which we are now familiar in our western civilization were developed so long ago that we have lost their true origins; we know they go back to the peoples of Mesopotamia of perhaps 2000 B.C. or even earlier. These figures were adopted first by the ancient Greeks, who renamed many of them, and then by the Romans, who gave them the Latin names we use. The Arabs of the Middle Ages, preservers of classical Greek culture, then applied their own distinctive patina. The sky was becoming populated with a wonderful mixture of people, animals, and artifacts, a reflection of human history.

Forty-eight ancient constellations have been handed down to us. They were first formally recorded by the Greek mathematician Eudoxus (403-350 B.C.), and the list given its final form some 600 years later by the great Alexandrian astronomer Ptolemy in his *Syntaxis*. The constellation names commemorated myths and stories and honored gods and great heroes. People often remark on how little most (but definitely not all) of the figures look like what they are supposed to be. We should not expect them to: They are not meant to *portray*, but to *represent*.

By far the most important group is the Zodiac, the set of 12 constellations that lie along the ecliptic. These were the residences not only of Apollo, the Sun god, but also of the planets that personified other Olympian gods such as the warrior Mars, the fertile Venus, and even the great king Jupiter. There are 12 because the Moon, which also traverses the Zodiac, orbits the Earth a bit over a dozen times a year, so that with every pass through its phases, the Sun occupies a different figure. (A Chinese equivalent of the Zodiac has 28 "moon stations," roughly one for each day of the lunar voyage around the Earth.)

Some of these figures are among the most familiar of the heavens. From our vantage point in Greece, Taurus glares down in late autumn and winter with a bright red eye set within his V-shaped head, really reminding us of a great bull. In Gemini you can still see the warrior twins Castor and Pollux. Leo, the lion, with his giant sickle-shaped foreparts that terminate in the bright star Regulus, regally treads the otherwise drab spring skies. And in summer, striking Scorpius, the superb red jewel Antares at its heart, awaits his prey, looking for all the world like a real scorpion basking above the southern landscape. In Homeric times Aries, the ram, an appropriate fertility symbol for the start of the growing season, was home to the vernal equinox. As the days warmed, eyes at night turned oppositely to Libra, the scales, which balanced the autumnal equinox in its pans (though at that time it was part of Scorpius). Dim Capricornus, a mysterious water-goat, held the winter solstice, and Cancer the crab, so obscure it would probably never have been an ancient constellation at all had not the path of the Sun demanded it, curled his claws around the summer. As a result, the northern terrestrial tropic, where the Sun stands overhead on June 21, is called the Tropic of Cancer, and its counterpart the Tropic of Capricorn. (Although their names have been retained, the locations of the four points have changed. The explanation awaits you below.)

Of the constellations outside the band of the Zodiac, undoubtedly the most beloved of groupings are the large and small bears, Ursa Major and Ursa Minor, which stalk perpetually around the north pole. They respectively contain the Big and Little Dippers, each a seven-starred figure that almost everyone in the northern hemisphere outside big cities recognizes from childhood. The Dippers are not formal constellations but are asterisms, small prominent segments of larger figures. The Big Dipper, known in England as the Plough, is singled out in many cultures: In Arabia part of it symbolized a funeral bier, and for some Native Americans it represented six brothers and their little sister. At the end of the handle of the smaller Dipper, which is really quite hard to see as most of its stars are so faint, is Polaris, the North Star; the Big Dipper's front bowl stars always point to it.

In our century Polaris is less than a degree from the north celestial pole, and for those of us north of the equator, it is a signpost that allows us to find our nightly way. The small bear is circumpolar for the majority of people in the northern hemisphere, and the larger one mostly so. Behind Ursa Major is Boötes, the herdsman, with brilliant orange Arcturus—the bear-driver—ever on his tracks, always chasing, never reaching.

Then as the sky turns, and the great bear walks low beneath the pole, enters majestic Orion, placed by the gods in the winter sky opposite deadly Scorpius so that the celestial hunter need never look upon his killer. This magnificent figure so dominates northern winter heavens that the Arabs called it "the Central One." Down and to the left of Orion is his larger dog, Canis Major, which features the brightest star of the sky, Sirius, and to the upper left is the small dog, Canis Minor, lit by fair Procyon.

Sometimes several constellations considered together illustrate a familiar tale. Six figures of northern autumn concern the story of Perseus. Proud *Cassiopeia*, the queen of King *Cepheus*, boasted that she was more beautiful than even the sea-nymphs. In anger, Neptune sent the sea-monster Cetus the whale to ravage Cepheus' kingdom. The king was forced to offer his daughter *Andromeda* as a sacrifice to propitiate Neptune, chaining her to a rock by the sea as fodder for Cetus. To the rescue comes Perseus, riding on his flying horse, *Pegasus*. Fresh from beheading the frightful gorgon Medusa, Perseus brandishes the hideous snake-haired head before Cetus, and the sea-monster is turned to stone. Celestial Cassiopeia, humiliated, spends half the year with her head downward. A lively adventure story, and an aid to remembering the placements of the stars.

The constellations and their stars revealed the dispositions of the gods and told their stories; they also held a key to worldly success. When Sirius first appeared in the morning's bright glare, the Egyptians knew the Nile would soon rise to bring them life. And listen to Hesiod, who nearly 3000 years ago advised Greek farmers:

At the time when the Pleiades, the daughters of Atlas, are rising,
begin our harvest, and plow when they are setting.
The Pleiades are hidden for forty nights and forty days,
and then, as the year reaches that point
they show again, at the time you first sharpen your iron.

THE MODERNS: FILLING THE GAPS

The classical constellations hardly fill the sky. In between many of the old figures are broad areas with few bright stars, but many faint ones, which the Greeks called the *amorphotoi*, the unformed. Furthermore, the Near Eastern and Mediterranean peoples could not see a huge portion of the far southern skies, that below a declination of about 50°S, and perforce left its stars and figures unnamed.

The new scientists of the Renaissance, however, needed better organization. They looked again at the *amorphotoi* and, aided by travelers and explorers, they went on to complete the sky. For a period of some 200 years in the seventeenth and eighteenth centuries they lavishly competed for space, inventing dozens of new constellations that reflected their own interests and discoveries.

The German astronomer Johann Hevel, or Hevelius (1611–1687), artfully filled the *amorphotoi* of the northern heavens with animals and artifacts, and to him we owe such creations as Canes Venatici, the hunting dogs south of the Big Dipper, a sextant, a lizard, a lynx, and a little lion. Around the year 1600, fancy and fanciful animals were set in the skies of the southern hemisphere by the Dutch travelers Pieter Keyser and Frederick de Houtman; these new figures are immortalized in Johannes Bayer's famous atlas, the definitive *Uranometria* of 1603. The southern population was finally completed in the mid-1700s when the Abbé Nicolas de Lacaille saluted the inventions of a blossoming technology with Microscopium, Telescopium, and Fornax (the furnace). The most famous of all the modem constellations is Crux, the Southern Cross, made from the feet of Centaurus by the seventeenth-century French astronomer Augustine Royer. Nothing dominates and represents the southern skies quite so much as this exquisite four-star figure that lies to the west of the two luminaries of the centaur. It is honored on the flags of Australia and New Zealand, and even has a "theme song"—Richard Rodgers's haunting tango "Beneath the Southern Cross," written for the television series *Victory at Sea*.

But these represent only a fraction of the inventions. A large number (a great many very fortunately) fell by the wayside. Chief among the "defunct constellations" are those with a nationalistic bent, designed to curry favor with (and extract money from) kings and princes: Robur Carolinium, "Charles's Oak," was made from stars of Argo by no other than Edmund Halley to remind the world of Charles II, who was so impressed that he "allowed" Cambridge University to grant the great mathematician and astronomer a master's degree. The only political one that made it into the modem array is Scutum, beautifully set into the northern summer Milky Way, which represents the shield of the Polish hero and king John Sobieski, who saved Europe from the invading Turks. Others were unneeded or just plain silly. One that stands out, however, is Antinoüs, which the emperor Hadrian set below Aquila to honor his young companion, who committed suicide in the belief that his remaining years would go to his master.

The confusion of competing claims to the sky was finally

straightened out at the 1922 convention of the International Astronomical Union, when the organization of professional research astronomers adopted the classical 48—actually 50, since the ship Argo was redesignated by its parts as Carina (keel), Puppis (stern), and Vela (sails)—plus 38 of the moderns, to total the currently recognized 88, all listed in Appendix 1 and Appendix 2. Subsequently, formal rectilinear constellation boundaries were also adopted in order at last to organize the heavens.

THE MILKY WAY

One ancient image names our Galaxy itself. A sky full of stars is awesome indeed, but it is dwarfed by a great white band of light that completely encircles the heavens, poetically and aptly called the Milky Way. Like the constellations that it graces, the lore of this milky circle is ancient and deep. Its English name simply translates the Greek galaxias, from gala, "milk"—celestial milk from the breast of Hera as she nursed the vigorous, squirming infant Heracles. It was the pathway of souls to heaven, the road to paradise, a serpent, a celestial river, a sparkling cloud of dust. Imagine Galileo's astonishment when he turned his telescope onto it in 1609 to find that it was actually faint stars, countless numbers of them, as we know now literally billions, the combined light of the Galaxy itself. All stars are members of galaxies of many different types. In the first analysis ours is a flat, disk-shaped system of some 200 billion stars roughly 100,000 light-years across (a light-year is the distance that light travels in a year at a speed of 300,000 km/s). If we could look at it face on we would see a beautiful system of spiral arms, best appreciated by viewing other, similar, systems.

The Milky Way is hardly uniform, but instead is broken up by enormous thick clouds of opaque interstellar dust that irregularly block starlight. The result is a band of great character, so much so that the Incas of Peru made constellations from the dark patches. Furthermore, the Sun is not at the center but is offset about three-fifths of the way to the edge, so that the Milky Way's visibility changes continuously: It is vastly brighter toward the galactic center in Sagittarius than it is in the opposite direction toward Taurus and Auriga. On summer evenings in the northern hemisphere the Milky Way streams out of the northeast from Cassiopeia and overhead through Cygnus, where a dust cloud, the Great Rift, splits it in two. The western branch descends through Ophiuchus and into Scorpius, where it produces an extraordinary cascade of stars; the eastern branch plunges even more brightly through Aquila and Scutum (where Sobieski's shield is made by an intense star cloud) on into the galactic core in Sagittarius, where it slips below the southern horizon. In northern winter, by contrast, we can hardly see the Milky Way in Auriga, and it brightens only some as it passes south through Canis Major, giving but a hint below the dog of the glories hidden by the Earth. The true majesty of the Milky Way is reserved for the lucky dwellers of the southern hemisphere, where Sagittarius arches his bow overhead, and spectacular cataracts of stars tumble down the heavens to either side, a singular sight that no one who sees it could ever forget.

We who live in the cities and towns forget this magnificent display of nature, one that no photograph, however skillfully or artistically made, can possibly match. Go out into the country, dozens of miles away from the lights, into the mountains if need be, to see for yourself.

THE NAMING OF STARS

The constellations parcel the celestial vault into convenient segments. Now let us narrow our view to individual stars. For any kind of study, from simple sky-watching to astrophysical research, they must be identified. A principal qualifier is their brightness, ranging from brilliant Sirius, which can be seen through a window of a modestly lit room, to as faint as you can see by eye and by telescope. Before anything else we must have a method of measurement. Over 2000 years ago, as part of the best star catalogue of the time, the Greek astronomer Hipparchus divided stellar brightnesses into six divisions called magnitudes, the first representing the most luminous stars and the sixth the faintest that could be seen. Polaris and most of the stars of the Big Dipper are second magnitude, the stars that flank Altair and represent the eagle are third, the handle stars of the Little Dipper are fourth, and the faintest star of its bowl fifth. A truly dark, moonless night away from city lights is needed to see fifth and sixth. The fainter the category, the more stars there are in it, and when the weakest come into view, the sky becomes wonderfully filled.

Nineteenth-century astronomers found that Hipparchus' scale was logarithmic in nature, and that the first-magnitude stars were as a class about 100 times brighter than those of the sixth; the obvious step was to make the ratio exact by quantifying the system. If five successive changes in magnitude correspond to a factor in brightness of 100, then each magnitude division increases or decreases the brightness by the fifth root of 100, or approximately 2.512; second magnitude, for example, is 2.5 times brighter than third. To establish the scale, the astronomers selected a set of faint, stars near the sky's north pole, around Polaris, and fixed their average magnitude at 6.00 (the general term "sixth magnitude" stretches from 5.51 to 6.50 and so on). Twenty-two stars are brighter than magnitude 1.5. Although they are loosely called "first magnitude," their range is so large that eight are of zeroth magnitude (between -0.49 and +0.50), and two, Canopus in Carina (Argo) and Sirius, are even of the minus first. The nearby planets are even brighter: Jupiter and Mars reach -3, and Venus is sufficiently radiant at -5 to be seen in broad daylight with the naked eye. Continuing the scale, the full Moon, under which you cannot quite read, is -12, and the Sun, which will destroy the eye, is -27. Telescopically, of course, we can see fainter than sixth, and the largest instruments are capable of recording stars down to the 29th magnitude, 23 magnitudes less luminous than the eye can see alone, an astonishing factor of 6×10^8. To the physicist the system may seem bizarre and needlessly complicated, but it works well in a celestial context and one rapidly gets used to it.

Now we assign names. About 2000 naked-eye stars still carry their ancient proper names, adopted from a variety of cultures and languages (the flavor of them is given in the list of the brightest 40 in Appendix 3, all those down to and including Polaris). A few derive from Greek, like Sirius, which means "scorching," an obvious reference to its brilliance, and Procyon, from *pro kuon*, "before the dog" (it rises ahead of the dog-star Sirius). A few more, like Capella, "she-goat," and Spica, "ear of wheat," are Latin. The overwhelming majority, however, are of Arabic extraction. Much of Greek astronomy was translated into Arabic before it returned to Europe in the Middle Ages. The Arabs embraced the Greek images at the expense of their own, but named the stars in their language largely according to their positions within the Greek figures. (We have inherited Greek constellations with Latin names made up of Arabic stars—a wonderful amalgam!) Thus Deneb, which means "tail," pops up in Cygnus at the end of the swan and also appears as Denebola, in the hind quarters of Leo, as Deneb Kaitos, the tail of Cetus the whale, and Deneb Algedi in Capricornus. Occasionally, the original Arabic figure shines through. Alkaid in Ursa Major refers to the "chief of mourners" because the Arabic image of this constellation was not a bear but a bier. Most of these names would be quite unintelligible to a modern Arab, as they were highly distorted and corrupted upon retranslation into the European languages. Betelgeuse, originally Bet al Jauza, "the right hand of the Central One," was mangled and contracted and finally retranslated as the "armpit" of the giant we know as Orion.

The proper names, however, are neither unique nor easy to remember. Bayer solved the problem in his *Uranometria* by assigning Greek-letter names to stars within a constellation more or less in order of stellar brightness and combining the letter with the genitive form of the figure's name: Betelgeuse is thus "*a* of Orion," or "*a* Orionis." Another popular method originated with the English astronomer John Flamsteed (1646–1719), who created the best celestial maps of the time. He ordered all the naked-eye stars by position from west to east within the constellations, and a short time later the Frenchman Joseph Lalande applied numbers to them. By this system Betelgeuse is also 58 Orionis. These Flamsteed numbers are generally used for the stars that do not carry Greek-letter names; proper names are usually employed only within the first magnitude. Commonly, just the three-letter abbreviation of the constellation name is used, so Betelgeuse is simply *a* Ori, and 61 Cygni, the first star whose distance from Earth was found, is 61 Cyg.

For the telescopic stars, which require designations as well, we resort to various catalogue numbering systems. In all, an extraordinary 15 million stars carry some kind of formal label, yet these are but a tiny fraction of the billions of stars that can actually be observed. Even today, the vast majority of stars are anonymous points of light. However, with various maps and atlases the named ones are sufficient to find our way easily and accurately around the sky—you can begin with the star maps in Appendix 4.

CHAPTER 3
HOW DO WE GET INFORMATION FROM THE STARS?

Early in the seventeenth century, several opticians were said to have noticed that combinations of lenses could make distant things look larger. The telescope was not Galileo's invention, but in his hands it took life. His real contribution was not so much that he looked and commented, but that he recorded and thought about what he saw. What followed was an explosion of discovery. Telescopes were made ever larger and more powerful. By the nineteenth century, astronomers were able to analyze starlight and thus could begin to dissect these distant suns. In our own century we have expanded the celestial science beyond light into radio, even x-ray, and have left the confining bounds of the planet to float our telescopes freely in space. This is the story.

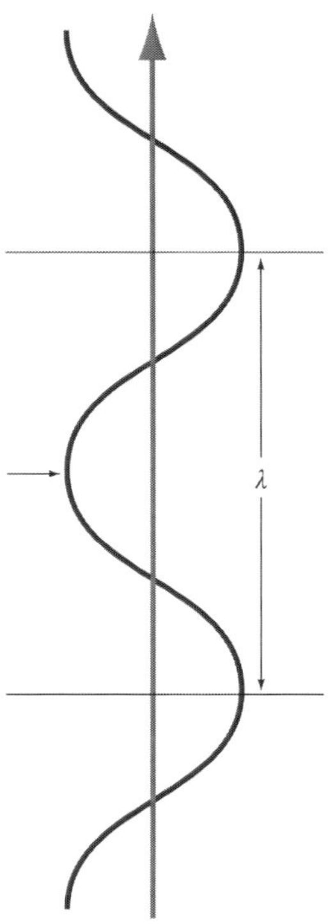

The wave is moving at the speed of light, c (300,000 km/sec). The wavelength is the distance between crests; the frequency is the number of crests that pass a particular point (the arrow) per second.

LIGHT

We cannot go to the stars; we can only observe and analyze, with the aid of telescopes, the radiation they send us. However, before we can understand how they operate we must examine the nature of this radiation and how it can be manipulated.

Light is generally thought of as an oscillating electromagnetic (EM) wave, similar in a loose sense to a wave moving in water. The difference is that light waves do not travel in a substance. In a vacuum, the waves move at a velocity, *c*, equal to 300,000 km/sec. Have respect for this number: a light beam could travel eight times around Earth in 1 second and fly from here to the Moon in 1.3. It is the maximum speed allowable anywhere, a fact fully backed up by countless experiments and solid theory.

Any wave is characterized by two dependent numbers, wavelength (λ), the physical distance between crests, and frequency (ν), the number of crests that pass a point per second (measured in Hertz, abbreviated Hz). These two numbers multiplied give the speed, so in the case of light $\lambda\nu = c$. Light waves to which the eye is sensitive come in a variety—a spectrum—of wavelengths, which we identify as color. Red light has a wavelength of about 7×10^{-5}cm, or 7000 Ångstroms (1 Å=10^{-8}cm). As the waves become shorter, we progress from red through orange, yellow, green, blue, and then violet at about 4000 Å.

Nature does not stop at these limits. The entire electromagnetic spectrum stretches to over a factor of 10^{15} in wavelength, and different domains carry different names. The visual or optical region contains the observed colors. Longer than red lie the infrared and radio, in which wavelengths extend to kilometers. Shorter than violet, we encounter the ultraviolet, x-rays, and then gamma (γ) rays, the last with wavelengths less than an Ångstrom. In a vacuum, all of them move at *c*. The stars and their associated phenomena send radiation across the entire spectrum, and to understand processes in space, we have to be able to observe it all.

The wave description of light (commonly used as a synonym for all EM radiation) is strangely inadequate. In some circumstances, light behaves not as a wave but as

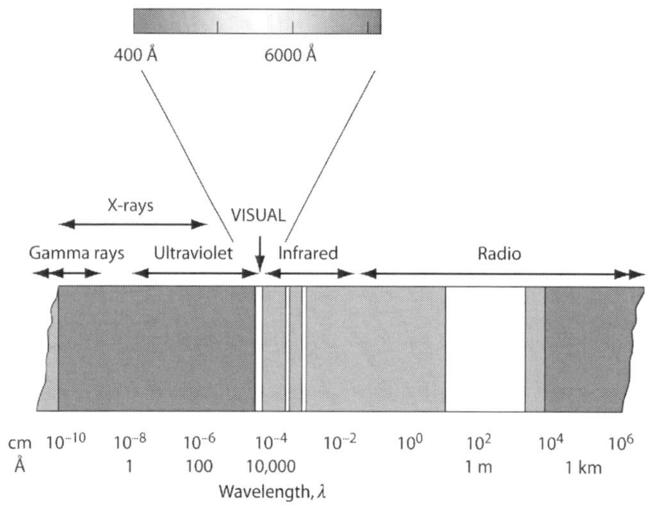

The EM spectrum, which runs from wavelengths shorter than 10^{-8} cm to kilometers, is loosely divided into broad, overlapping bands. Since many physical processes do not radiate energy in the optical, the only way to know of them is to study the relevant spectral regions. The lighter the shading, the greater the transparency of the Earth's atmosphere.

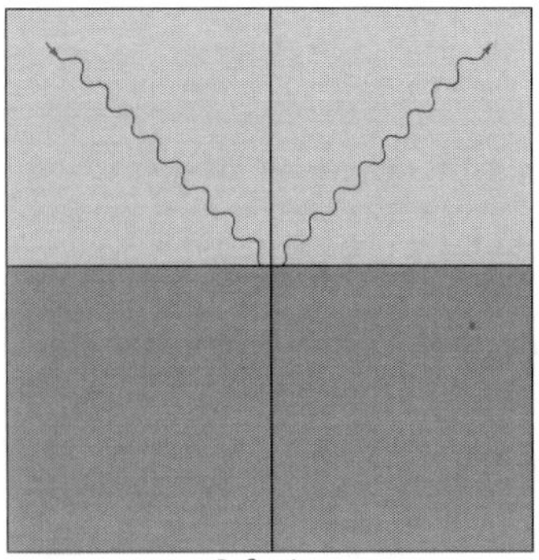

Reflection.

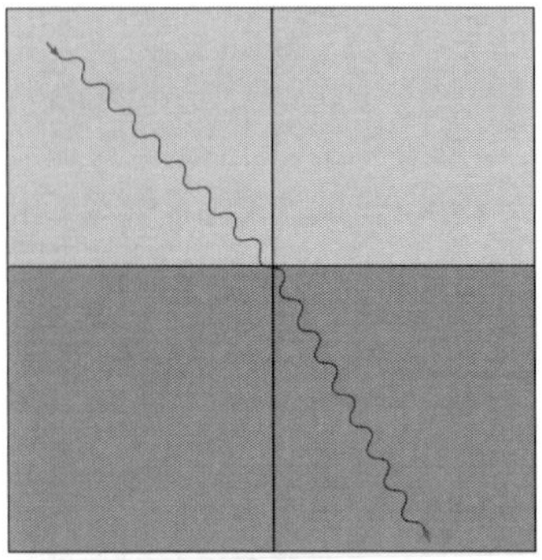

Reflection.

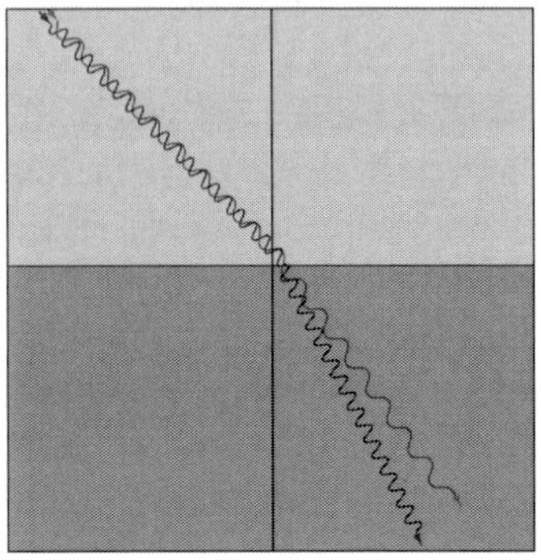

Dispersion.

a particle. Such a particle, which also has the wave properties described above, is called a photon. Light is the principal way by which energy (the ability of a body, or radiation, to do work on another body, to accelerate it over a distance, or to heat it) is carried through the universe. The energy transported by a particular photon depends on its frequency according to the equation $E = h\nu$, where h is a universal constant of nature called Planck's constant. Gamma-ray photons then carry some 10^{15} times as much energy as radio photons, which is why a radio broadcast will not hurt you but an atomic bomb's *y* rays will. The amount of energy transported by a single photon is quite small. For yellow light, $\lambda = 5500$ Å, so $\nu = 5 \times 10^{14}$. Since h (in common meter-kilogram-second units) is a mere 6×10^{-34}, E is only 3×10^{-19} joules (a joule is the unit of energy). A hundred-watt light bulb radiates 100 joules per second, so (very crudely) it sprays an amazing 10^{20} photons at you every second.

OPTICAL PRINCIPLES

Light interacts with matter in all sorts of ways. For example, it can be reflected. A ray of light aimed at a smooth reflecting surface, a common mirror, will bounce off at the same angle (measured against the perpendicular) at which it arrived. If not for that simple principle, your face would appear weirdly distorted in a bathroom mirror.

A light ray can also be bent by passing it through a substance. Send a beam hurtling at an angle from the air to a transparent material such as glass or water. As it enters, it slows, the change in velocity causing it to bend, or refract, toward the perpendicular. The greater the angle of incidence, the more the refraction. If the ray goes in the other direction and *leaves* the water, it will bend in the opposite direction—*away* from the perpendicular—as it speeds up. The result is

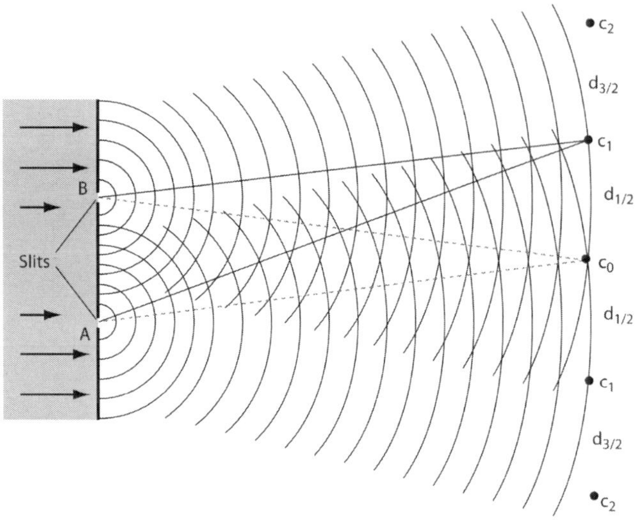

Light waves passing through a double slit produce interference fringes on a distant screen. At each bright fringe C, the difference in path length to each of the slits is an integer multiple of the wavelength; for each dark fringe d, it is an integer multiple of one-half the wavelength.

that anything under water will appear misplaced and distorted (just look at the fish in a tank). The degree to which the light is slowed and the beam refracted depends on the substance. The speed of light, c, divided by the internal velocity is called the index of refraction, n. Glass, with n equal to about 1.5, has a higher index than water, and diamond has the highest of all, nearly 2.5.

The index of refraction depends on wavelength and almost always increases as λ decreases. The result is that violet light is refracted more than green, and green more than red. When a ray of white light, which is a mixture of all the colors of the spectrum, enters or leaves a substance, rays with different wavelengths will bend in slightly different directions, and the white light becomes dispersed into its components. Dispersion is everywhere around you—in a rainbow, in colored sparkles from crisp snow, in the shower of hues from a well-cut diamond. It is the phenomenon that allows us to take starlight apart and learn how stars are put together. Pass a sunbeam through a prism, a triangular piece of glass, and it is dispersed both upon entry and exit to show the colors of the Sun.

Light can also be bent and dispersed by a more obscure phenomenon called diffraction. Send a beam of monochromatic (single-color) light at a plate with a pair of slits cut into it. Each slit acts as a new source, and the waves leave them in step—that is, in phase. When the light from the apertures falls upon a distant screen, the different sets of waves interfere with one another, producing a set of light and dark bands called fringes. The brightest fringe is centered exactly on a line between the two slits at a point where the distance from the fringe to each of the slits is the same. The two sets of waves then overlap one another and add together, a condition called constructive interference. A small distance away on either side of the first fringe the distance from one slit is a half-wavelength greater than the distance to the other. The crest of one wave then fills in the trough of the other, producing destructive interference and a zone of darkness. Farther away yet, the difference in distance is a full wavelength, and the waves again overlap and constructively interfere. The pattern continues through 3/2 wavelengths, 2 wavelengths, and so on. Remarkably, interference fringes are also seen with a single slit. In a sense, the single slit is made of an infinite number of pairs of double slits, all so close together that they produce a single interference pattern. You can see these fringes if you place your fingers very close together next to your eye and look at a bright light.

Diffraction also produces spectra. The greater the wavelength, the farther from the center a bright fringe will be. If white light falls upon the double slit, each of the fringes will have red light falling on one edge and blue or violet on the other. If a third slit is added, the conditions for constructive interference are more stringent, and the color separation will be clearer. The more slits, the better the spectra. If thousands of slits are used, the spectra are stunning. A device that produces such spectra is called a diffraction grating. Usually these are reflective plates with grooves cut into them. The most common example is a compact audio disk, which has a fine spiral groove. Hold one up to the light to see several spectra, one after the other. Spectra made this way are superior to those produced by prisms because diffraction spreads the light linearly relative to wavelength, while prisms compact redder light. Moreover, glass absorbs in the ultraviolet.

Finally, the wavelength and frequency of EM radiation depend on the relative velocity along the line of sight between the source and the observer, a phenomenon called the Doppler effect. If you move into a set of waves, such as those on a lake, the frequency with which you encounter them will be increased and their separations will appear to be diminished; if you move along with them you experience the opposite. The effect is common with sound. An airplane approaches at high velocity, and the pitch of the engine is higher than it would be at rest. As it passes overhead, the line-of-sight speed drops, and so does the pitch; as the craft recedes, the wavelengths become longer and the pitch drops even further. If a light source is coming toward you its radiation will be shifted to shorter wavelengths; that is, it will appear to be slightly bluer than it would be if the source were at rest: Recession will produce a shift toward the red end of the spectrum. The relative amount of the shift is directly proportional to speed, or $\Delta\lambda/\lambda = v/c$ (the Doppler formula), where λ is wavelength, Δ is the change in wavelength, and v is the line-of-sight speed between the source and observer.

The formula works only if v is much less than c; if v approaches c, a different equation based on the theory of relativity must be used. In most instances the speeds and the shifts are very small. The color of a source can change

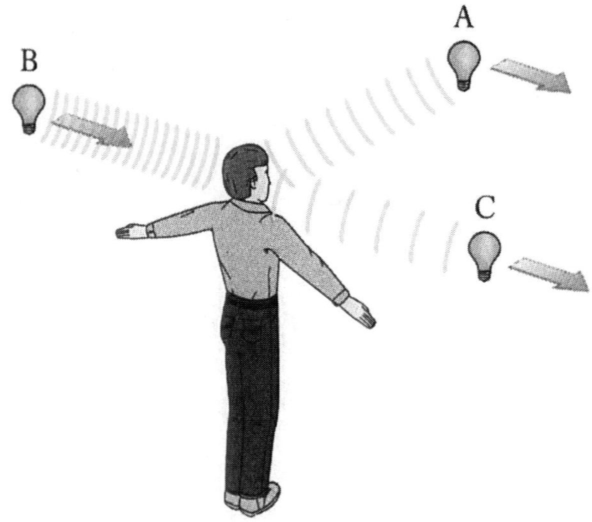

Light bulb A is moving past the observer, but because there is no motion along the line of sight, the observer sees the original wavelength. Bulb B is coming at the observer, who sees the waves Doppler-shifted to shorter wavelengths. Bulb C is moving away, and the waves are shifted longward. The effect is the same whether the source or the observer is moving.

only if the source is moving at a significant fraction of c. A common example of the EM Doppler effect is police radar. A radar gun transmits a radio signal at a precisely defined wavelength. When the radio waves bounce off an approaching car, their wavelengths are made shorter. The radar gun picks up the returned signal, compares its wavelength with that sent out, applies the Doppler formula, and (perhaps dismayingly) displays the vehicle's speed.

With these principles in hand we can build the tools necessary to study the stars.

REFRACTORS

Even in its most sophisticated form, a telescope is extremely simple in its basic operation. It is a collector of light with two distinctly different purposes: to make images of faint sources bright so that there is enough light for analysis, and to provide resolution, which is the ability to see the separation between things that are close together. Galileo's telescopes used lenses and employed the principle of refraction. Stars are effectively at infinity, and the light waves that fall upon the Earth are parallel to one another. They are bent by refraction at the spherically curved surfaces of the lens toward a focal point, where they meet. If we place a detector such as a photographic plate in the plane parallel to the lens that holds the focal point, we can record an image of the star. That is all there is to it: a lens and a detector.

The most critical parameter of a telescope is the diameter, or aperture, of the "objective," the light-gathering lens. The larger the aperture, the more light it can accumulate, and the brighter will be the image and the fainter the stars that can be seen or photographed. The amount of light that can be collected is proportional to the area of the objective lens, and consequently to the square of the aperture. Aperture is everything. Telescopes are named by aperture: We speak of the "Palomar 200-inch," (or "5-meter"), the "Kitt Peak 4-meter," or the "Lick 36-inch." The dark-adapted human eye has an aperture of about 7 mm. With a 70-mm telescope you could view stars of the 11th magnitude—100 times fainter than those you can see without aid.

Of course, with the telescope we see and record not one star but a whole field of them over a specific angular diameter. The scale of the field of view depends not on the aperture, but directly on the focal length (the distance from the lens to the focal point). The longer the focal length, the more widely separated two given stars will be. The single lens will also invert the image on the focal plane, east for west, north for south.

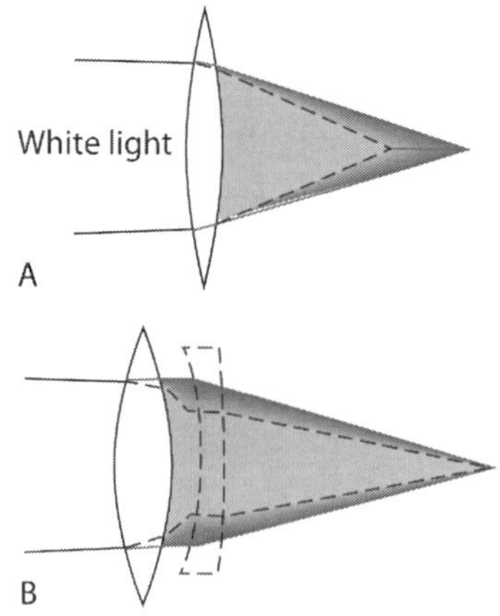

The image of a star made by a single lens will be a colored line along the optical axis, because blue rays bend more than red. Only a single color can be brought into focus at a time. A combination of convex and concave lenses, with different indices of refraction, can bring any two colors to a focus.

Galileo did not have photographic film. He looked through the instrument, just as the average backyard astronomer does today. If you were to place your eye in the focal plane, however, you would not see stellar images but instead the illuminated objective. In order to see the stars, you must use an eyepiece, which renders the rays parallel once again so they can be focused on the retina. There are two choices. Galileo placed a concave lens in front of the focal plane. This method results in a normal, uninverted image and is still used in inexpensive binoculars and opera glasses, but the resulting magnification and field of view are quite restricted. It is far better to use a convex lens behind the focal plane. In a loose sense, the eyepiece is used as a magnifier to examine the focal plane. The resulting angular magnification, or power, of the combination depends on the ratio of the focal lengths of the objective and the eyepiece, or $M = F_{\text{objective}}/F_{\text{eyepiece}}$. The power is easily changed by switching eyepieces. The one disadvantage of this system is that it still inverts the image. Astronomers get used to looking at things rotated upside down.

The professional astronomer rarely looks through the telescope. Direct viewing is done only for target acquisition and for guiding during long exposures, and even that can be accomplished automatically. The instrument is used instead to feed light to some kind of detector. Therefore, magnifying power is of little importance. What is significant is scale, and more important, resolution (also called resolving power), which involves diffraction. A telescope aperture acts as a single slit, and although it is very wide, it still produces an interference pattern. The stellar image is the bright central fringe. Since the lens is circular, this image is surrounded by a series of light and dark circular fringes. Although the star is effectively a mathematical point so far as the telescope is concerned, the central fringe, or diffraction disk, is not. The ability of a telescope to resolve fine detail depends only upon the diameter of this disk, which shrinks as aperture is increased. The only way to achieve a point image from

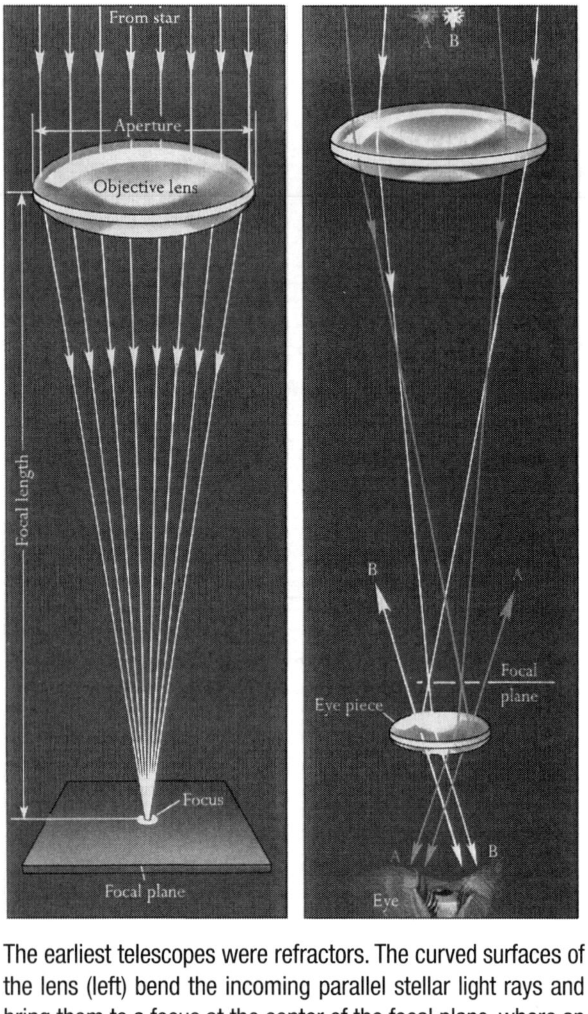

The earliest telescopes were refractors. The curved surfaces of the lens (left) bend the incoming parallel stellar light rays and bring them to a focus at the center of the focal plane, where an image of the star is created. A photographic plate placed there will record a picture of the star. The brightness of the image depends on the square of the diameter, or aperture, of the lens. The light from two stars enters the lens (right) and is brought to a focus at the focal plane. The scene is inverted on the focal plane: up for down, left for right. An eyepiece placed behind the focal plane renders the light rays parallel once again and magnifies the angular separation of the stars. The field of view is still inverted.

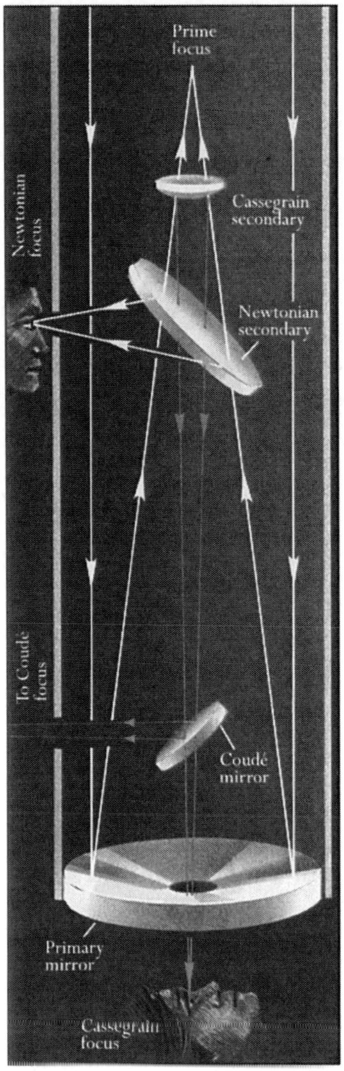

The reflecting telescope uses a large, paraboloidal primary mirror to collect and focus light. The prime focus is often used for direct imaging because nothing intervenes in the light path. Amateur instruments commonly employ the Newtonian focus. Large equipment—a spectrograph, for example—requires a more mechanically stable platform, so the Cassegrain focus is often used. For massive equipment, the light is further reflected to the coudé focus.

a point source, however, is to make the aperture infinitely large, which is impossible. As a standard rule of thumb, the resolving power of a telescope in seconds of arc is 5"/*D*, where *D* is in inches, or 13"/*D* if *D* is in centimeters.

Practically, however, resolution is not established by the telescope but by the Earth's atmosphere. The charming phrase "twinkle, twinkle, little star" is enough to spoil an astronomer's whole day, let alone night. The twinkling, or "seeing" as it is properly called, is caused by variable refraction in the air, and it smears the apparent point that is a star into a disk that can be several seconds of arc across. Twinkling is one of the principal reasons why professional telescopes are placed on high mountains, where the seeing is the best. But even at the choicest sites, the image size, or "seeing disk," never gets below about a half-second of arc. Once the aperture is above about 12 inches (30 cm), gains in resolution become increasingly limited without using highly advanced equipment designed to overcome the effects of atmospheric seeing.

Refractors have three debilitating problems. First, refractive dispersion causes the blue and violet rays from a star to converge faster to a focus than the red rays. Consequently, the location of the focal plane depends on color. If you try to focus on the red or yellow image of a star, there will be a big blue haze around it. The effect of this chromatic aberration is minimized by long focal length, and even better controlled by constructing the objective out of two or more elements. An achromatic objective (something of a misnomer) usually consists of two lenses with different curvatures and different indices of refraction placed together. In that way any two colors of choice can be brought to a point. Three colors can be brought to a focus by three lenses, but each added lens also absorbs and reflects some light, so except for some specialized cameras, two is generally the limit. Residual chromatic aberration is reduced by using a large focal length and filters that block the unfocused light.

The second problem is aperture. A refractor objective uses two lenses of high optical quality and has four surfaces to be ground and polished, and the cost goes up dramatically with size. The largest refractor ever built is only about 1 meter in diameter. Moreover, the tubes have to be so long that the supporting structure and the observatory dome must be huge and massive, adding further to the already great cost. The third problem is that glass absorbs the ultraviolet radiation that gets through our atmosphere, seriously reducing the instrument's research capability. The problem can be ameliorated with special—and costly—ultraviolet-transmitting glasses, or by using quartz, which is more expensive yet. Yet for all the difficulties, it is a testimony to the beauty and simplicity of these magnificent instruments that although the last great refractor was built nearly a century ago, they can still be used effectively for specific types of research.

THE REFLECTOR REVOLUTION

The solution to these difficulties was initiated in the seventeenth century by none other than Isaac Newton (1642–1727), who invented the reflecting telescope, which uses a mirror instead of a lens to focus light. It was not until the mid-1700s, however, that the technology was sufficiently developed that the reflector could compete with the refrac-

The 2 x 8.4-m Large Binocular Telescope (LBT) atop Mt. Graham, AZ.

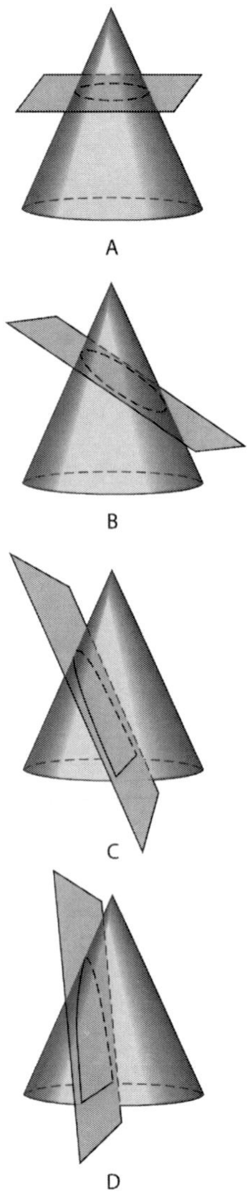

The conic sections are a set of curves produced by cutting a cone with a plane: (A) circle; (B) ellipse; (C) parabola; (D) hyperbola. Rotation of the parabola around its axis generates a solid paraboloid, the shape of a primary telescope mirror.

tor. Then, in the hands of William Herschel (1738–1822), who effectively founded modern astronomy, the reflector came alive. In this century, it finally won out.

Since the angle of reflection does not depend on color, there is no chromatic aberration, and since there is no transmission of light through anything, there is no ultraviolet absorption. Because there is only one optical surface to be ground, the interior of the mirror can even be flawed. The instrument can be made very large and still quite short, reducing the size of the containing building. As a result, reflectors are much cheaper to build than refractors, freeing money that can be used to increase the size.

A telescope objective mirror begins with a concave circular disk of glass or ceramic, the latter now used to reduce thermal expansion effects. A spherically shaped surface will not properly focus light, however, because radiation from the outer zones of the mirror comes to a focus before that from the inner zones. The resulting spherical aberration produces a fuzzy image. Instead, the mirror is worked into the shape of a paraboloid, which will bring all parallel light rays of a star on the optical axis to a point focus. The surface facing the incoming light is overcoated in a vacuum chamber with a very thin, bright layer of aluminum. The focal plane, or the prime focus, is in front of the mirror. How does the observer avoid blocking the light? Newton's solution was to place a small, flat secondary mirror at a 45° angle just in front of the focus, sending the light to the side. This Newtonian focus, common in older reflectors, is popular for amateur telescopes; it is inexpensive and places the viewer in a comfortable position.

There are a number of other possibilities. Most professional telescopes use the Cassegrain focus, in which light is reflected from the secondary through a hole drilled in the objective, or primary, mirror. The secondary is ground into the shape of a convex hyperboloid, making the light converge at a smaller angle and sending it to a focus that, like the Newtonian, is outside the tube. By adjusting the curvature and placement of the secondary, very long *effective* focal lengths—far greater than that of the primary—can be achieved in a very short tube. Moreover, the secondary is easy to change, so different focal lengths (and consequently different focal scales) are available for different purposes. The objective may have a focal ratio (focal length divided by aperture) as short as 3 (called f/3) and effective Cassegrain focal ratios of f/7, f/13, and f/30 or greater. Moreover, the focal position is on the mechanical axis of the instrument, allowing easy placement of heavy analyzing equipment.

In another variation on the theme, the light from the Cassegrain secondary can be intercepted by a flat mirror placed in front of the primary at one of the rotation points of the instrument. It is then shot to the side where additional mirrors send it to a fixed room in the basement of the observatory to the coudé focus (from the French for "bent"), where massive equipment can be mounted. Finally, if the telescope is large enough, we can actually use the prime focus. Several giant reflectors have a cage built into the instrument at that position, allowing the observer to ride inside them during the night. All these options make the reflecting telescope enormously flexible while remaining relatively inexpensive.

The first of the really great reflectors, completed in 1918, was the 100-inch telescope built by the Carnegie Institution of Washington on Mount Wilson above Los Angeles. It was followed 30 years later by the Hale 200-inch (5-meter), constructed by Carnegie and the California Institute of Technology on Palomar Mountain about halfway between Los Angeles and San Diego. Just think of the light-gathering power of this telescope.

It will collect over half a million times as much light as the unaided eye, and a view through an eyepiece would allow you to see 20th magnitude! A marvel of engineering, it is so delicately balanced on high-pressure oil pads that all its 50G tons can be moved by hand. Since its construction, several telescopes in the 3- to 4-meter class have been built, but except for a largely ineffective 6-meter in the Soviet Union, only recently has the Hale 200-inch been surpassed in size. A number of 8-meter instruments, with 2.5 times the light-gathering power of Palomar—a full magnitude of brightness—are planned. It is unlikely that the world will see monolithic mirrors much larger because of the difficulty of casting, polishing, and mounting them. The huge Keck 10-meter telescopes now operating in Hawaii consist of segmented mirrors, each individual reflecting surface only a meter across. With such innovative designs, there is no immediate limit to how large telescopes

The 10-m Heinrich Hertz Telescope atop Mt. Graham, AZ, receives signals at very high radio frequencies. (Used with permission from University of Arizona, T. W. Folkers, photographer.)

might become except that imposed by available money (a *very* significant constraint).

Reflectors have their own problems. A simple paraboloid will produce a beautiful image, but only on the optical axis. Stars that are not right in the center of the field of view suffer from an aberration called coma (from the Latin for "hair"), which draws the stellar images into diffuse lines that point radially outward. The effect greatly restricts the field of view: It would take lifetimes to survey the entire sky with the Palomar 5-meter. One solution was devised by Bernard Schmidt in the 1930s. The instrument he developed uses a nonfocusing objective lens called a corrector plate that is placed in front of a larger spherical mirror. The complex shape of the corrector plate changes the paths of the light rays to compensate for both spherical aberration and coma. The result is a camera with a huge field of view. The most famous of these instruments is the 48-inch Schmidt at Palomar. Another solution, the Ritchey-Chrétien design (used in the Keck), employs hyperboloidal primary and secondary mirrors.

DESIGN AND OPERATION

All telescopes must rotate on two axes in order to point anywhere on the two-dimensional surface of the celestial sphere. In the standard equatorial configuration, one axis points to the celestial pole, permitting the telescope to rotate along a daily path parallel to the equator. The other axis allows it to be pointed north and south, perpendicular to the equator. Each axis has a setting circle, one reading hour angle, the other declination. In classical or historic instruments these are iron rings with engraved graduations. The astronomer looks up the coordinates of a star, planet, or galaxy. The difference between α and the LST from the observatory clock gives the hour angle. When the positions are set, the object appears in the eyepiece of a low-power finding telescope attached to the telescope tube. When the observer centers the object in the finder, it appears in the main instrument. The polar axis is directly attached to a sidereal clock, which slowly moves the telescope along a parallel of declination, keeping the star in the field of view as the Earth rotates.

The telescope is generally housed in a rotating dome with a shutter that can be opened to the sky. The dome provides protection against inclement weather and, when the shutter is open, against wind and stray light. Telescopes are outdoor instruments that must be closely maintained at the ambient temperature in order to minimize the size of the seeing disk. No heat is allowed, and in the past the astronomer often endured rigorous working conditions. Temperature control is so critical that the outside of the dome is painted with a special titanium dioxide compound that reflects sunlight and radiates well in the infrared. Some telescopes even have chillers built into the floor so that daytime temperatures can be held low.

Professional astronomers have largely removed themselves from the telescope. The eye was long ago replaced by dispassionate instruments, except perhaps for an irresistible peek at a favorite object during twilight (the view through a 3- or 4-meter telescope is awesome). Setting circles have also been succeeded by electronic encoders that give the precise rotation of the telescope shafts, just type α, δ, and the epoch onto a keyboard. A computer will perform the precessional rotation to determine tonight's coordinates, read the LST from its internal sidereal clock, do the subtraction, and command motors to turn the rotational shafts to the desired setting. The computer contains information that can compensate for the flexure of the heavy telescope and for the lofting effect of refraction produced by the Earth's atmosphere. The control is so precise that the finder is not needed: A modern instrument can be set, and will track or follow a star, to the amazing precision of 1 second of arc. The sought-for object then appears almost magically at the center of a television monitor, allowing the astronomer to direct the light to the analyzing instruments. Since all this is done automatically, the astronomer no longer need be in the dome at all and can sit comfortably in a "warm room" sipping hot coffee while the data are acquired.

Computer control has also freed us from equatorial design.

Because the engineering is far simpler, it is less expensive to align a telescope mounting to the horizon so that one axis points to the zenith along the direction of the Earth's gravity. The computer continually converts the right ascension and declination to altitude and azimuth (the angle along the horizon measured from the north point through east), moving the telescope smoothly through the sky. All new large telescopes, such as the Keck, work this way.

Large telescopes are so costly and their design and control so complex that they must be run by highly trained operators. The astronomer provides the coordinates and is given command only of slow motions for guiding.

ANALYSIS

By itself, the telescope does little. Its purpose is to feed light into analytical detectors. Until nearly the end of the nineteenth century, all we had was the human eye, which cannot make permanent records. The first great breakthrough was photography. The astronomical camera is no more than a plate-holder. It is affixed to the focus, the shutter held open sometimes for hours to integrate and record a faint signal as the telescope tracks through the sky. Astronomical photographic emulsions are almost always spread on glass plates, which provide great durability and dimensional stability. Sophisticated chemistry—mostly by Eastman Kodak—-has gone into producing emulsions that record well at very low light levels. The plates range in size from small chips a centimeter wide to monsters half a meter across used in Schmidt systems. Photography has been the astronomical workhorse of the twentieth century. In the 1950s, the Palomar Schmidt was used to survey the entire sky visible from southern California in two colors, blue and red, down to about the 18th magnitude. (Although some magnificent color photography is shown in this book for illustration, research photography is always done in black and white, the different "colors" referring to filters used to make the images.) A companion instrument run by the United Kingdom and the European Southern Observatory in Australia has now done the same in the southern hemisphere, and the Palomar Schmidt is redoing the northern survey in *three* colors. These are all mammoth undertakings, search missions hunting for objects of interest to be examined by the big reflectors.

The Hubble Space Telescope, still tied to the shuttle arm, hovers above the Earth. It Now flies in near-Earth orbit about 600 kilometers up.

The photographic plate has some severe disadvantages. It is very inefficient, recording at best only about 2 percent of the light that falls upon it, and enormously long exposure times are required. It is also nonlinear. The blackening rate of the photographic grains does not respond directly to the intensity of radiation, so careful calibration is required. The result is considerable inaccuracy in the measurement of brightnesses. Very gradually, photography has been replaced by electronics. The revolution, which is now nearly complete, began about 1915 at the University of Illinois with the work of Joel Stebbins and the invention of the photoelectric photometer. This device makes use of the particle nature of light. A variety of substances will release electrons when struck by photons (the particles of light that carry energy), with an efficiency near 90 percent. These electrons can be accelerated by an electric field to produce a measurable current that is directly proportional to luminosity. In Stebbins's first successful experiment he was able to record the Moon; 50 years later, photoelectric photometers could measure the brightness of a 20th magnitude star.

The photoelectric photometer can observe only one star at a time and, although accurate, is again inefficient. Nor does it speed the process of imaging the sky. Video cameras used in the 1970s were an advance, and the ultimate is the charge-coupled device, or CCD, which also has an efficiency near 90 percent. The surface of the CCD is divided into tiny picture elements. When exposed to light, each of these pixels will build up an electric charge that is directly proportional to brightness: that is, each photon of light will produce the same slight increase in charge. The charge pattern in the chip will thus match the brightness pattern of the field of view. When the exposure is complete, the charges are drained out as an electric current, a row of pixels at a time. The computer reads the fluctuating current at a precisely timed rate so that each instant can be related to a specific pixel. The pattern is stored on disk or tape and can be reassembled, row by row, column by column, to re-create the image on a video monitor. A modern CCD system can record in one minute what used to take one to two hours with photography, enormously increasing the efficiency of the telescope and allowing us to reach an astounding 29th magnitude with the largest instruments. Moreover, the electronic linear recording permits us to measure the magnitudes of all the stars in the field of view at once.

The electronic revolution has allowed small 1-meter telescopes to do what was impossible a mere 20 years ago even for the giant 5-meter. That is the principal reason why the Palomar instrument was not exceeded in size for so many years. We were getting effectively larger and larger telescopes without having to build them. Now that we have reached nearly full efficiency in detection, we have to put our efforts back into glass and iron.

Photography, although down, is not yet dead. CCD chips are small, a 2048-by-2048 array the largest yet constructed. A mere 30 mm on a side, they can still image only restricted fields of view. Wide-angle cameras like the Schmidts must still resort to plates. But CCDs have so taken over the astronomical industry that Palomar has now banned photography from the 5-meter, a momentous historical watershed.

All observatories are equipped with far more than imaging devices. The spectrograph, for example, disperses light rays (rendered parallel by a collimator) into their component colors or wavelengths to record spectral signatures of individual atoms and to measure velocities. Prisms will do the job, but all modern spectrographs use the much more versatile diffraction gratings. In older days, the spectrum was recorded on a photographic plate. Spectrographs now come with their own CCD systems that allow the recording of spectral information on magnetic disk or tape.

Observing is actually the simple part; more difficult is the reduction of the images, magnitudes, or spectra to meaningful data. CCD pixels are not of uniform sensitivity, so the brightness of one part of an image may be distorted relative to that of another. The Earth's atmosphere dims the incoming radiation, making stars look fainter than they are, the effect dependent on altitude. The air also absorbs better in the blue than in the red and consequently garbles the relative brightnesses of the different colors in the spectrum, as do wavelength variations in the total transmission efficiency of the telescope optics. All these effects must be corrected for by various calibration techniques. One night's worth of good data can require several days at the computer terminal. Then, of course, the real work begins—the interpretation of the data, which might take the next year or even decade.

Observatories work on a tightly scheduled and organized basis. There are many astronomers and few large telescopes, so it is necessary to apply for time as much as six months to a year in advance for a few precious nights. The analyzing equipment is usually so complex that it is mounted on the telescope for an extended period, days or even weeks at a time, with various astronomers rotating in and out to use it. The scheduling is done in close synchronization with the

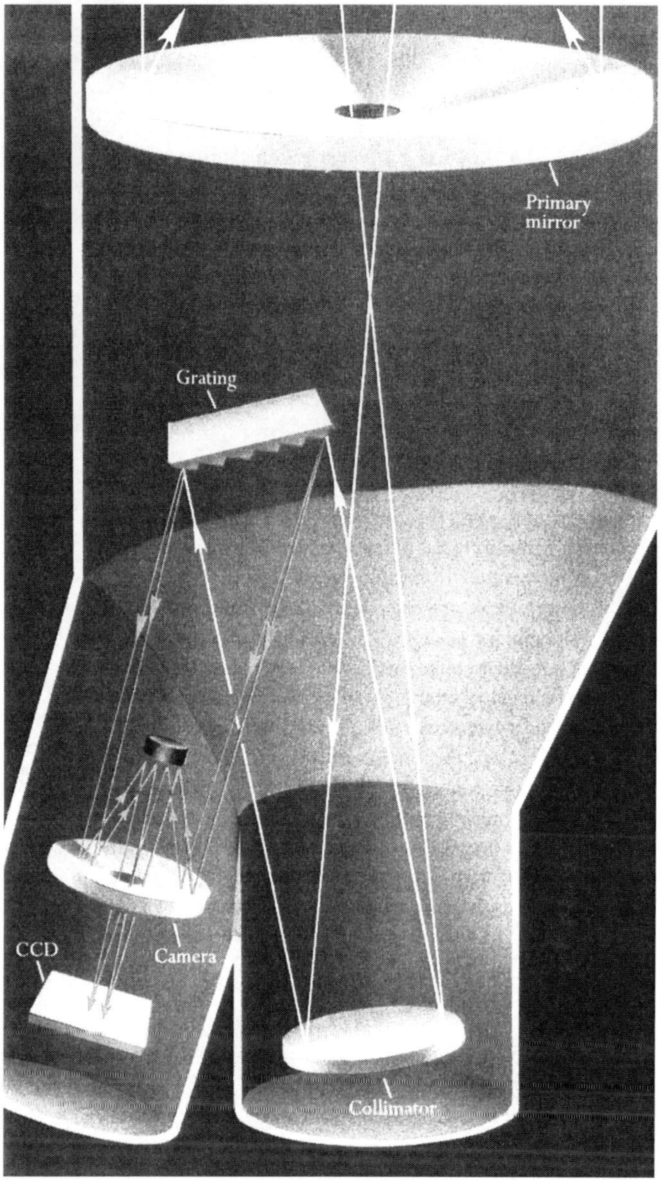

A spectrograph records stellar spectra. The light is first directed to a collimator that renders the rays parallel, then to a finely ruled grating for dispersal into the spectrum. A camera focuses the spectrum for imaging on the surface of a CCD.

Moon, which can light the sky so brightly that faint objects are unobservable. Imaging of weak sources, for example, is generally reserved for the "dark run," the two-week interval centered around the new Moon, while bright objects are observed during the "bright run," the fortnight centered on the full Moon. In the dim ages when astronomy began, people were fascinated by the comings and goings of our satellite. Even the most modern of astronomers still pay homage to it.

RADIO TELESCOPES

The electromagnetic spectrum is enormously wide. Different physical processes radiate energy in different wavelength domains, and if we confine ourselves only to the narrow optical band of the spectrum, we severely restrict our view of the Universe. The first nonoptical spectral region was opened to the sky in 1933 by a Bell Laboratories engineer named Karl Jansky, who had been assigned to track down sources of radio interference. He found that the radio noise background swelled and fell with a sidereal period of 23^{h} $S6^{m}$, a sure tipoff that at least one of the sources was extraterrestrial. It turned out to be the Milky Way. Jansky had unwittingly built the first radio telescope.

In its simplest form, a radio telescope is only a directional antenna—a long wire or a TV dipole is an example—held up to the sky. Radio waves from space excite an electrical current in the antenna, which is then amplified and measured. However, a bare antenna cannot gather much energy: Its light-gathering power is low and its directional capability poor. In order to collect a large amount of energy, the antenna is commonly placed at the focus of a paraboloidal reflector. The radio telescope is then the exact analogue of an optical instrument in which the detector, instead of being a photographic plate, a photomultiplier, or a CCD, is an antenna coupled to an amplifying system.

As in an optical instrument, this paraboloid produces a diffraction pattern. The size of any interference pattern depends directly on wavelength. The longer the wavelength, the farther away the first bright fringe and the larger the central disk. As a result, simple radio telescopes have notoriously poor resolution and in compensation must be made very large. One operating at a fairly typical wavelength of 10 cm would have to be nearly a mile across just to achieve the resolving power of the unaided human eye! Such a device is impossible to engineer and construct. The largest single steerable radio dish ever built is "only" 100 meters across, and operating it is something akin to balancing an athletic stadium on twin axes and moving it across the sky. The champion is a fixed 1000-foot-wide parabola built into a valley in Puerto Rico.

Fortunately, the problem of limited resolution is easily overcome. Imagine a mile-wide telescope. Cut out the entire middle and replace it with two separate telescopes that are a mile apart and mix or correlate their output signals. As a radiating source is tracked across the sky, the joint telescopes act like a double slit that "sees" a changing pattern of interference fringes. The interference pattern can be mathematically inverted to reconstruct the field of view. This simple "interferometer" necessarily will lose information, as it is constructed in only one dimension. But if we fill in the whole surface of our hypothetical mile-wide instrument with an array of individual telescopes, each of which is easy to build, we can accurately synthesize the effect of having observed with the big one. The largest of these, the Very Large Array, or VLA, is assembled on the plains of New Mexico. It consists of 27 25-meter antennas mounted on railroad tracks and can synthesize a single telescope 36-km wide, the size of the Washington D.C. beltway! With it, astronomers can achieve resolving powers that are actually better than those of optical instruments, since radio waves are largely unaffected by the Earth's disturbing blanket of air.

The VLA is the practical limit for electronically linked telescopes. The lengths of the electrical cables that tie each one to the central correlator cannot be accurately matched for interferometers much larger. To get around this difficulty we disconnect the telescopes and synchronize their observations with precisely matched atomic clocks. The data are merged and the interference fringes recreated in the computer. There is then no limit on size. The Very Long Baseline Array (VLBA) has receiving elements that stretch from New England to Hawaii, thus effectively creating a telescope that is 8000 km across, allowing a resolution of better than a hundred-thousandth of a second of arc! Even longer baselines have been established between Europe, the Soviet Union, and the United States, creating a telescope almost as big as the Earth. Someday, we will have an element of an interferometer array on the Moon.

A radio telescope can be used as an imaging device. The instrument can scan across the sky over a source of radiation and build up the spatial pattern of its emission. Then the appearance of the source can be reconstructed in the computer so that the image looks just like a photograph, one seen with radio eyes. Radio telescopes are also outfitted with spectrographs and can be tuned much like a home radio—which is in fact a form of spectrograph—across the radio spectrum to measure changes in the intensity of the signal.

INTO SPACE

Despite the sophistication of modern telescopes, they are severely limited by their attachment to the Earth's surface. Our atmosphere transmits radiation only in limited portions of the electromagnetic spectrum. The infrared is heavily absorbed by carbon dioxide and water vapor, and there are only a few fairly narrow bands through which it is possible to look into space. The ultraviolet is taken out by ozone absorption. Only a small portion of the near ultraviolet, that just shortward of the optical, sneaks through. For life on Earth, these absorptions are critical. Absorption in the infrared acts as an insulating blanket that keeps us warm (the famed greenhouse effect) and that in the ultraviolet keeps the burning—and lethal—solar ultraviolet at bay (although not entirely, as you can find out at the beach). X-ray radiation cannot penetrate either, and the electrified layers of the Earth's upper atmosphere block several radio bands (standard "short-wave" communication in fact relies on bouncing signals off these layers from below). For all these reasons, beneficial as they may be in other respects, we are unable to study a huge variety of high and low-energy processes from the ground. Even the optical region of the spectrum is compromised by some absorption and by the ever-present twinkling.

If we want further advances in our study of the Universe, we have no choice but to go above the atmosphere into space, where we can get an unobstructed view. Many rocket flights carrying diverse instruments have been made since the 1950s. A rocket payload in a ballistic trajectory affords only a few minutes' look, however. High-flying balloons have been used as observing platforms, too, but obviously these must swim in some residual atmosphere. The only serious way of observing from space is by satellite. The first truly astronomical satellite (as opposed to those that studied geophysical processes) was launched in the 1960s. Many more followed, and the instruments aboard them have opened up most of the hidden electromagnetic spectrum.

A few examples show the variety. The remarkable, unsung mechanized hero of the 1970s and 1980s was a 45-cm telescope called the *International Ultraviolet Explorer* (*IUE*). It was launched in 1978 with a three-year design lifetime to observe all manner of celestial bodies in the ultraviolet below the atmospheric cutoff, in the region between 1000 and 3000 Å. After 19 productive years, the *IUE* finally was shut down. The public knew little about it because it was designed only to observe ultraviolet spectra and could not produce spectacular images. Without any question, however, it generated a revolution in astronomy, allowing us to witness physical processes at work that simply cannot be seen from the ground. Such ultraviolet imaging has been impressively done with *ASTRO*, an observatory that flew for a week aboard the space shuttle *Columbia* in May 1991.

X-rays reveal the locations of very high temperature gases and the sites of other extremely energetic processes. The first effective x-ray telescope, *Uhuru* (Swahili for "freedom"), flew in 1970 and immediately began finding sources of this energetic radiation. It was surpassed by *HEAO2* (*High Energy Astronomical Observatory 2*, dubbed *Einstein*), which flew in 1978 and could image the sky. That satellite has now been superseded by *ROSAT* (*Röntgen-Satellite*), which is operated by the European Space Agency and NASA, and the *Compton Gamma Ray Observatory* is taking us to even higher energies. In the other wavelength direction, the *Infrared Astronomical Satellite* (*IRAS*) flew in 1983. It imaged the entire sky three times in four wavelength bands, compiled infrared spectra, and has provided invaluable information on cool stars and the chilling depths of space. Its work was taken up in 1997 by the *Infrared Space Observatory*.

We are now immersed in discoveries by the *Hubble Space Telescope*, the *HST*. Begun in 1979 and finally launched in April 1990, it is returning to Earth stunning views of the heavens. Although far from the largest telescope in the world, it is surely the most sophisticated, and without any question it is in the best observing site. The *HST* (designed in the Ritchey-Chrétien configuration) carries a variety of imaging, photometric, and spectroscopic instruments. Its wide-field planetary camera has a theoretical resolution of 0.1 seconds of arc, several times better than can ordinarily be achieved from Earth, and the faint-object camera a phenomenal 0.043 seconds. The much-publicized problem with its 2.4-meter mirror (mistakenly ground to the wrong figure) initially compromised the instrument's resolution and efficiency, but the repair mission that took place in 1994 and the subsequent upgrade that added new instruments made the *Hubble* a spectacularly successful telescope, as shown in the pages to follow.

MEANWHILE, BACK ON THE GROUND...

Space astronomy does not supplant ground-based research but supplements it; the two work in superb cooperation. The developments taking place on Earth are as exciting as those hundreds or thousands of kilometers above us. For example, look at the Keck 10-meters or the battery 8-meters, that will have vastly greater light-gathering power than does the *HST*. More exciting perhaps is our growing ability to overcome the deleterious effects of our shimmering atmosphere to produce remarkably high resolution. Before long, telescopes on the ground will be able to produce images the like of—or even better than—*HST*, although they obviously will never rival its ultraviolet capability (unless we destroy *all* the ozone, a horrifying thought).

There are several ways to achieve that great goal of astronomy—high resolution—from the Earth. The history of interferometry goes back to 1913. In its simplest application, two slits are used to observe the diffraction pattern of a star. Since a star is not really a point, we get in effect a smeared pattern of fringes that allow the determination of the angular diameter. In the first such experiment, A. A. Michelson placed a 10-foot steel beam with mirrors at each end across the top of the 100-inch telescope on Mount Wilson. Two more mirrors reflected the light onto the objective, where the rays were mixed and the fringe pattern observed by eye. Michelson and others reached resolving powers of a few hundredths of a second of arc, allowing them to determine the angular diameters of some stars. Modern systems, which use photometry and computer processing, do even better and permit the observation of much fainter sources. Another method, called speckle interferometry, uses ultrashort exposures of stars and can freeze atmospheric seeing patterns, enabling us to record hundreds of tiny stellar images within the big seeing disk. Computer processing of several such images also allows the measurement of angular diameters. The technique is especially useful for the separation of stars that are so close together that they cannot be resolved by ordinary observations. Even more successful are optical versions of radio interferometers run by different groups that can achieve resolutions in the thousandths of seconds of arc.

The most exciting method uses a technique called adaptive optics. We equip the telescope mirror with small motors that can deform its surface. The twinkling of the stars is monitored through the telescope and the shape of the mirror surface quickly changed to compensate and to detwinkle the incoming light waves. In principle, we can then achieve the theoretical resolving power of the telescope. Resolutions of a few tenths of a second of arc—not much greater than that produced by the *HST*—have been achieved, and that is only the start. The goal of these techniques is to image the stars themselves, to see what is on them, even to look for their planets. Success should not be too long in coming.

JOIN IN

Astronomy is a remarkably accessible science—all you need do is go out and look. Furthermore, the chief astronomical instrument, the telescope, is an extremely simple affair that anyone can own and work with (not many amateur physicists have particle accelerators). There are many stars and few professionals, and it takes only some practice and a little dedication for the true amateur to make a real contribution to the celestial science.

The amateur can work at many different levels of interest and sophistication. There are groups of devoted observers who use only their eyes to track meteor showers across the sky. (Meteors are generally caused by fluffy dust thrown from comets that enters and burns up in the atmosphere. If the Earth passes close to a comet's orbit we see a brief shower as thousands of the little bits of matter plunge into the upper air.)

The first level of optical aid is a pair of binoculars. It is remarkable what can be seen with them, especially if you use a tripod to hold them steady. Binoculars are described by a pair of numbers, for example, 7 × 35. The first gives the power, the second the aperture in millimeters. For nighttime use, 7 × 50 is excellent, and bigger ones like 11 × 80 will provide a stunning wide-angle view of the heavens. Jupiter's little moons, the larger craters on the Moon, and dozens of star clusters will jump out at you.

To see the sights at their best, though, you must get a telescope. The problem is the huge choice. Although modern refractors deliver superb images, for the price you are always better off with a reflector. The minimum usable mirror diameter is about 75 mm. If constructed in the classic Newtonian design, the costs are modest and the views nearly inexhaustible. Obviously, the bigger the objective the better. Simple and reasonably inexpensive instruments are available up into the 60-cm range. At a more sophisticated, and expensive level, you might opt for your own Schmidt (or "catadioptric") system. These use a Cassegrain design, in which a long effective focal length, affording high power, is folded into a short tube. They are therefore easy to transport into a dark site, a must for deep-sky viewing. All amateur instruments come with a variety of eyepieces, those of longer focal length for a wide field of view, those of shorter to see fine detail. Contrary to expectation, the better views are usually obtained with lower powers. The highest you can ordinarily use is about 50 power per inch of aperture, and even that is good only on exquisite nights.

Getting started requires more than just purchasing a telescope. Astronomical observing, at any level, is like any other endeavor: It is a learned skill that requires practice. Use your new telescope first in the daytime to look at distant trees and to learn how to use (and to align) the low-power finder. You will then become aware of one of the astronomical telescope's most confounding properties: The view is inverted. You move the instrument to the right, and everything appears to move to the right instead of the left. Just get used to it. Before long the Moon will look abnormal if presented to you right side up.

When you take your telescope out at night you will discover another problem. It magnifies not only the angular size of an object but also the apparent speed of the Earth's (or the sky's) rotation. You have your first object centered, and before you know it, it is gone. To follow it you have to move the telescope constantly, but in the "wrong" direction! Patience, you do get used to it. If you have paid enough, you may have a sidereal clock drive that will follow a star for you. But only, of course, if you have remembered to align the telescope's polar axis *on the pole*. That means a latitude adjustment and another at night to point the axis north (or south if you live in the southern hemisphere). A great deal of help—and fun—can be had from your local amateur astronomy club.

If you want to drive from Pocatello, Idaho, to Mobile, Alabama, you do not just jump into the car—you get a map. Similarly, maps help you find your way around the sky. Like terrestrial charts, sky maps come in all sizes and varieties. At first, you want one that will show you the bright stars and constellations so that you can get a good overview of the heavens. Then, when you are familiar with the basic layout, move up to more detailed charts that show all the naked-eye stars along with a variety of interesting celestial objects, or even to ones that show stars much fainter than the eye can see. You can then star-hop from a bright, easily locatable star to fainter ones visible only in your telescope and on to, say, a faint galaxy.

Numerous publications are available that can help enormously in guiding you around the sky. Many carry ads that will aid in finding the instrument of your dreams and the charts to go with them, and some list astronomy clubs that will help you find fellow enthusiasts. You might also discover that you wish to enter the ranks of research. Hundreds of comet-hunters scan the skies every evening and morning for these elusive balls of fuzz. The large majority of comets are found by amateurs who wander the open sky, not by the professionals who sit in darkened domes with only video screens for company. Thousands more amateurs monitor the myriad stars that vary in their light, working under the auspices of national groups (such as the American Association of Variable Star Observers), and for generations have provided valuable data to the world's professionals.

Whatever the level chosen, now we are prepared to go outside and look, to roam through the universe with our eyes and our telescopes, and to *learn*.

CHAPTER 4
WHAT ARE THE PROPERTIES OF THE STARS?

You are deceived when you look at the nighttime sky. Some stars are bright, mostly because they are close; others that look dim are incredibly luminous, but happen to be very fax away. How do we sort them out and learn their true natures? The story involves the measurement of distance, the detection of motion, and the analysis of stellar light. We begin with the distance to the Sun, then take a giant leap into interstellar space.

DISTANCES AND MOTIONS IN THE SOLAR SYSTEM

The problem of the solar distance, a fundamental quantity now called the astronomical unit, or AU, was attacked by the classical Greek astronomers. By Ptolemy's time there was at least an estimate, however erroneous, of 20 times the distance to the Moon and 600 times the diameter of the Earth. There the matter lay for 1700 years, until Copernicus (his proof supplied later by Galileo) established the concept that the Sun was indeed at the center of the solar system. With that knowledge, he could measure the relative sizes of the planets' orbits from their positions in the sky.

Copernicus' theory had a major flaw. Because he assumed circular orbits, as had the Greeks before him, the planets were never quite where predicted. Something had to be wrong. The revelation of the real structure of the solar system is the product of the awesome labor of two men, Tycho Brahe (1546–1601) and Johannes Kepler (1571–1630). Tycho had established an observatory on the Danish island of Hveen, and with meticulously graduated circles made careful measurements of the positions of stars and planets. Kepler worked with him in Prague in the last year of his life, and inherited the precious data. He adopted a revolutionary procedure. He made no presumption about orbital shapes, but used the data to find their natures. The result was a set of three laws that govern motions within the solar system.

In 1609, after eight years of crushing calculations in which he compared his orbital predictions with Tycho's positions of Mars, Kepler produced his first two laws. The first was astonishing: A planetary orbit is not a circle, but an *ellipse* with the Sun at one focus. The size of an ellipse is characterized by half the major axis, the semimajor axis, *a*. The astronomical unit is properly defined as the semimajor axis of the terrestrial orbit.

The second law governs the speed at which the planets

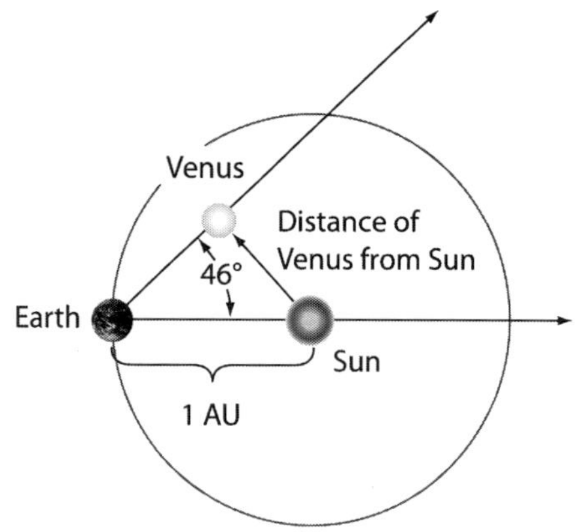

Venus's maximum angle from the Sun is 46°. From the right triangle, Copernicus found its distance from the Sun to be 0.7 times that of the Earth (the astronomical unit).

move along their paths. Kepler found that the line connecting the planet to the Sun, called the radius vector, sweeps out equal areas in equal times, which means that a planet's orbital speed changes with its distance from the Sun.

The third law took another 10 years to uncover. This one, his "harmonic law," relates the orbits of the planets to the earth and to one another: The squares of the planetary periods in years are equal to the cubes of their semimajor axes in all, or $P^2 = a^3$. Jupiter, for example, takes 12 years to go around its orbital path. Twelve squared is 144, the cube root of which is 5.2 AU, the planet's distance from the Sun.

Kepler's three laws are empirical, derived from observation and calculation. Kepler had no idea why they worked. They were explained in the late 1600s by Isaac Newton, the extraordinary mathematician and thinker who established the now-familiar laws of motion:

1. A body in motion stays in motion unless acted upon by an outside force;
2. Force equals mass times acceleration ($F = MA$);
3. For every action there is an equal and opposite reaction.

The second law is the most important to us here. A force is anything that can cause an acceleration, or a change in velocity, where velocity is defined as a body's speed and direction. Mass is commonly thought of as the amount of matter within a body, but since $F = MA$, it can also be defined and measured as F/A, the degree of force needed

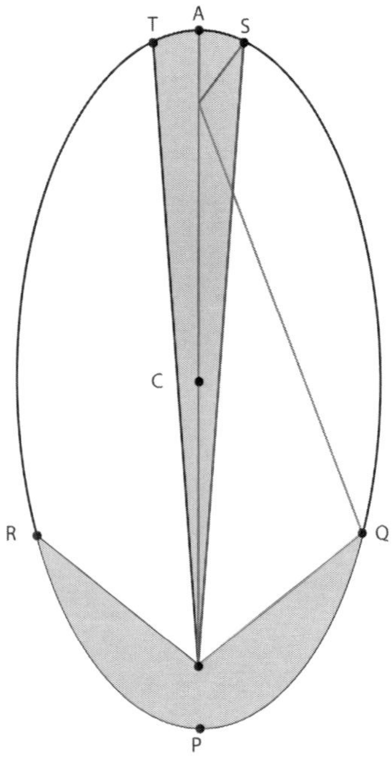

The sum of the distances of all points on an ellipse to each of the two foci, F1 and F2, is a constant: F1Q + F2= F2S + F2S, etc. The size of the ellipse is defined by half the major axis (the semimajor axis), CP, also called a, and the shape is given by its eccentricity, e, CF1/a. In a planetary orbit, the Sun is at one of the foci. As the planet moves, the radius vector that connects the planet and Sun must sweep out equal areas in equal times, so the planet moves from Q to R in the same interval of time in which it moves from S to T. The planet moves fastest at perihelion, P, and slowest at aphelion, A. (Planetary orbits are actually much more circular than represented here.)

to accomplish a particular acceleration (or the acceleration acquired by application of a specific force). It is measured in kilograms, kg.

Newton followed these with his famous law of gravity, which states that the same force that makes the apocryphal apple fall to the ground makes the Moon go around the Earth. The Moon is actually falling toward us but cannot reach us because it is also moving in a direction perpendicular to the line connecting it to the Earth. The acceleration of a falling body, g, is proportional to the mass of the Earth divided by the distance to the Earth's center squared (since the Earth attracts as if all the mass were concentrated at a central point), or

$$g = GM_{\text{Earth}}/R^2$$

where G, the gravitational constant, is determined in the laboratory.

The acceleration can be generalized to any pair of attracting bodies, and since F = MA, the force between them, F, is written as

$$F = GM_1M_2/R^2$$

where M_1 and M_2 are the two masses and R is the separation of their centers.

These laws, together with his invention (simultaneously with Leibniz in Germany) of calculus, allowed Newton to derive Kepler's laws from theory in an elegant and generalized form of which Kepler's empirical versions were special cases. He found first that orbits involving two mutually gravitating bodies can be *any* of the conic sections, not only ellipses. An orbit may also be a parabola or a hyperbola, along which the orbiting body will escape. Kepler's second law is generalized by a concept called the conservation of angular momentum. Tie a rock to a string and whirl it about your head. The angular momentum is defined as the mass of the rock times its velocity times the length of the string. Once established, angular momentum will not change unless an outside force is applied: If you shorten the string, the rock speeds up. A skater bringing in his arms almost magically whirls faster. The conservation of angular momentum makes planets speed up as they approach the Sun on their elliptical orbits, and slow down as they recede.

Since orbits are determined by gravity, and gravity involves mass, Kepler's third law must somehow include mass as well as period and distance. In his generalization, Newton derived one of the most important relations in all of astronomy:

$$P^2 = \frac{4\pi^2}{G} \frac{a^3}{(M_{\text{Sun}} + M_{\text{planet}})}$$

The orbital parameters are now divorced from those of the Earth: The period (P) is measured in seconds, the semimajor axis (a) in meters. Kepler's original law lies buried within. Because the masses of the planets are so much less than the mass of the Sun, the sum ($M_{\text{sun}} + M_{\text{planet}}$) is effectively a constant (at least to the accuracy of Tycho's naked-eye observations). If the equation as applied to any planet is divided by that for the Earth, the constants disappear, and $P^2 = a^3$, where P is again in years and a is in AU—just what was found by Kepler.

Newton's and Kepler's laws have far-reaching ramifications. Immediately, they allowed astronomers to determine accurately the relative separations between the various bodies of the solar system at any time in terms of the astronomical unit. A knowledge of the distance in kilometers to any of the planets serves to calibrate the AU and yields the distance to the Sun. The distances in kilometers to all the planets are then automatically known as well. The first such accurate measure was made more than 300 years ago by Giovanni Cassini, using his own observations and those made by the French astronomer Jean Richer. The two men took measurements from Paris and French Guiana respectively and determined the distance of the planet Mars from its terrestrial parallax (in this context, the shift in position as viewed from opposite sides of the Earth). During the nineteenth and twentieth centuries, the measure of solar distance was improved by observing the parallaxes and distances of close-passing asteroids, small bodies that ordinarily lie between the orbits of Mars and Jupiter. The modern value of the AU (1.459×10^8 km) is found by observing Venus with radar. We simply send a radio signal to the planet, find the interval of time between transmission and reception of the returned signal (allowing for slowing by refraction in the terrestrial and Venusian atmospheres),

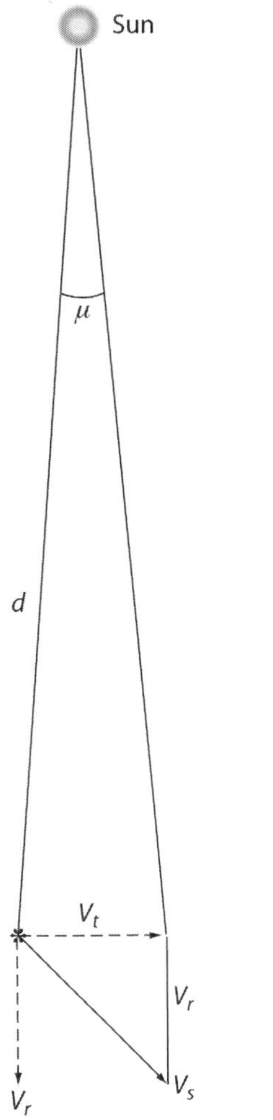

A star moves relative to the Sun at a space velocity, v_S. The motion is resolved into two mutually perpendicular components, the radial velocity, v_T, and the tangential velocity, v_t. Over a given interval of time, the star will move through a proper motion, μ, which depends on both v_t and distance, d.

multiply by the speed of light, and divide by two. The result of all this effort, begun so long ago, is the distance to one very important star.

THE DISTANCES AND MOTIONS OF THE STARS

The Galaxy is a churning, rotating mass. All the stars are in orbit about the galactic center, and all the orbits are at least slightly different from one another, some wildly so. From our home, the Sun, we would expect to see the stars constantly moving past one another. Think of a body, say an automobile, passing by you as you stand still. The vehicle has a speed (read on the speedometer) and a direction of motion that combine into its space velocity, v_s. The car is most likely moving at an angle relative to your line of sight. Its speed can then be resolved into two components, the radial velocity, v_r along the line of sight and the tangential velocity, v_t perpendicular to it. The tangential movement causes the car to appear to move across your line of sight at a certain angular rate of so many degrees per second, the "proper motion." The proper motion depends on both the tangential velocity and the distance: A car moving along the street where you stand whizzes by, whereas one seen on a distant highway seems to crawl along the horizon.

Instead of cars, think of stars: The principles are exactly the same. Measurement of proper motion, μ, and distance, d, allows the calculation of v_t from simple trigonometry. The radial velocity, v_r can be found by the Doppler formula. With these quantities known, we can then determine the actual direction and speed of the star, the space velocity, where $v_s^2 = v_r^2 + v_t^2$ from the Pythagorean theorem.

The first of these quantities to be observed were the

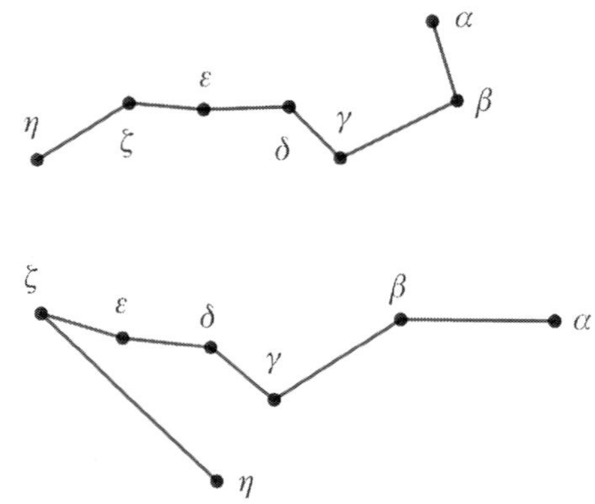

A quarter-million years from now, the Big Dipper (top) will be seen as the figure on the bottom.

proper motions. They cause changes in right ascension and declination (apart from those produced by precession), a discovery made in 1718 by Edmund Halley, who did far more than invent a constellation and get his name attached to a comet. They are now known for tens of thousands of stars. Because of the great distances involved, most of the movements are small, only a few tenths, hundredths, or even thousandths of a second of arc per year, and are sensible only with careful telescopic measurements. Nevertheless, given enough time they will cause great changes in constellation patterns.

Aristotle, Copernicus, and others all knew that if the earth indeed orbits the Sun, the stars must exhibit annual shifts or parallaxes. Moreover, since the sizes of the shifts must be inversely proportional to the distances, the distances can be measured. The stars are so far away, however, and the parallaxes so tiny, that the quest for stellar distance was not successful until 1838, when Friedrich Wilhelm Bessel turned his telescope and his trained eye to the star 61 Cygni. In his time this star, actually a double, had the highest known proper motion, a whopping 5 seconds of arc per year (it is now number two to 10th-magnitude Barnard's Star in Ophiuchus, which is clipping along at double that rate). It therefore had a good chance of being one of the closest stars, one with a relatively high parallax, and worthy of careful examination.

After allowing for the effects of proper motion, Bessel found that over the course of the year the components of 61 Cyg shifted back and forth through an angle of ⅔ of a second of arc. This small angle, 1/5400 of a degree, is the angular size of the width of your thumb at a distance of 6 km. The astronomical parallax, p, is defined as ½ the shift angle, and so for 61 Cyg it is ⅓". The distance unit used in professional astronomy is the parsec, pc, defined simply as $1/p$, which renders the star 3 pc away (the improved modern value is 3.4 pc). A parsec is the distance at which the Earth's orbital radius would subtend an angle of 1 second of arc, which is 206,265 AU (the number of seconds of arc per radian, an almost magic number in astronomy). The

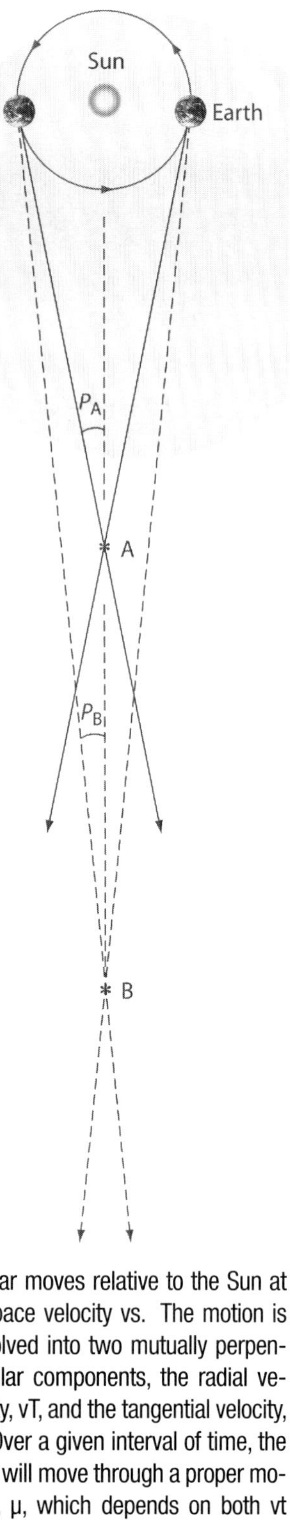

A star moves relative to the Sun at a space velocity vs. The motion is resolved into two mutually perpendicular components, the radial velocity, vT, and the tangential velocity, vt. Over a given interval of time, the star will move through a proper motion, μ, which depends on both vt and distance, d.

star nearest to the Earth (other than the Sun), however, is not 61 Cyg, but a faint, reddish 11th-magnitude companion to *a* Centauri known as Proxima Cen, which has a huge parallax of 0.76 seconds of arc and a distance of 1.3 pc, or 270,000 AU. If the Earth's orbit is the size of a golf ball, *a* Cen is some 5 km away, and the Sun a mere 0.1 mm across—no wonder that stars do not collide! Once we have the distances, transverse velocities can easily be calculated from proper motions. Now we can know something of the relative speeds of the stars, and we find them to be typically in the neighborhood of a few tens of kilometers per second.

The more famous unit, the light-year, is quite useful for the comprehension of stellar distances. The light year, 1-y, the distance that a ray of light travels in one year at a speed of 300,000 km/sec, is equal to 10^{13} km or 3.26 pc. The light you see from *a* Cen left the star *four years ago*. Of course, this star just begins our journey into space. Most of those you see in the nighttime sky are dozens, hundreds, even thousands of parsecs and light-years away. Their assembly, the galaxy, is well over 30,000 parsecs—100,000 light-years—across. And we can take images of individual stars in other galaxies that lie *millions* of parsecs away.

Parallaxes are very difficult to measure, and their determinations require years of observation. The minimum detectable shift has long been around 0.01 seconds of arc, limiting the accurate application of the technique to distances closer than about 50 to 75 pc. However, the frontier is being pushed back both from the ground and from space. The *Hipparcos* satellite has measured parallaxes and proper motions for hundreds of thousands of stars with errors of 0.001 seconds or better, and good distances up to 200 pc or so. Even better measures will surely follow.

TRUE BRIGHTNESS

We have broken through the celestial sphere. Now that we know that stars reside in the three-dimensional domain and we can plot their locations, we must expand the definition of magnitude. That measured from the Earth on the basis of direct observation by eye or telescope is the apparent magnitude, *m*. We now need a magnitude that removes the effect of distance and is a measure of true luminosity or energy output. This quantity is the absolute magnitude, *M*, defined as the apparent magnitude the star would have were it at an arbitrary standard distance of 10 pc, 32.6 1-y. Absolute magnitudes are easy to calculate once the difficult matter of distance is solved. The brightness of a light source depends on the inverse square of its distance; halve the distance and the source is four times as bright. Select a sixth-magnitude star 100 pc away. If we could move the star in to 10 pc, it would become 100 times, or five magnitudes, brighter, so that *M* would be +1. Any apparent magnitude can be converted to absolute through a simple formula, the famed magnitude equation,

$$M = m + 5 - 5 \log d$$

where *d* is in parsecs. Mark this little equation well; it will play a powerful role as we proceed.

If we could line up all the stars at 10 pc the sky would look very different. Start with the Sun, whose distance from us is 1 AU or 1/206,265 pc. The Sun's apparent mag-

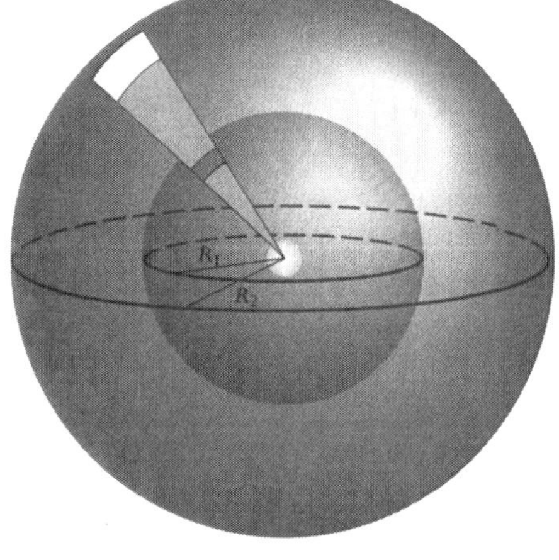

The light generated by a star spreads itself out over a sphere of radius R_1. As the size of the sphere increases, the amount of radiation per unit area becomes diluted by $1/R_2$. Double the distance to R_2, and the source looks a quarter as bright.

nitude is -26.74, and when the numbers are entered into the equation, the absolute magnitude is +4.83, so faint that it could not be seen under full moonlight. Then look at Vega. At a distance of 7.5 pc, it still has *m* equal to 0 and is nearly 50 times more luminous than the Sun. In stellar terms, however, Vega is not extraordinary. Just east of it, in Cygnus, lies the great star Deneb, *m* = 1.25, distance 500 pc. Bring it in to 10 pc and *M* jumps to -7.2, rendering it nearly 1400 times brighter than Vega and 65,000 *times the visual solar luminosity*! If the Earth were in orbit about such a beast, we would have to be 250 AU away, six times the distance of Pluto from the Sun, for humanity to survive. And even that is not the record. A handful of stars can be found that are as bright as *M* = -10, another 13 times brighter than Deneb.

From these heights, plunge to the depths. Recall 11th-magnitude Proxima Centauri, the closest star, only 1.3 pc away, yet visible in a modest telescope. Take it out to 10 pc and the brightness drops nearly 60 times to magnitude 15.5, that is, 10.6 magnitudes or 18,000 times *fainter* than the Sun. If we wished our day to be comparably bright under Proxima's feeble glow, we would have to cuddle in a distance of only 0.008 AU, *1.5 times the radius of the Sun.*

Amazingly, Proxima is downright bright compared with a faint star in Cepheus called LHS 2924, absolute magnitude 20, which is visually another 70 times fainter. In order even to see this meager star with the naked eye, it would have to be a mere 0.02 pc, 4000 AU, away, about 100 times farther than Pluto. If it were at Pluto's distance, LHS 2924 would appear only as bright to us as Venus; compare it with Deneb, already first magnitude at a distance of 500 pc. Absolute stellar brightnesses range over 30 magnitudes, an astounding factor of a *trillion*, from roughly a million times brighter than our Sun to a million times fainter.

COLOR AND TEMPERATURE

People are often astonished to find that stars show colors, some quite vivid, evident in photographs and indeed simply by eye on a clear night: Betelgeuse and Antares are rather red, Arc turns orange, the Sun and Capella yellowish, Vega and Sirius quite white, and a few are even somewhat bluish. But color represents more than just a handsome sight: It is a response to stellar temperature.

To a reasonable approximation, a star behaves something like an ideal radiator called a blackbody, a surface that absorbs all the radiation that falls upon it. It reflects none, hence the term black. Absorption of radiant energy would by itself increase the temperature of the body. In order to maintain equilibrium, it must *emit* just as much energy as it receives, and consequently can be very bright to the eye. The spectrum emitted by a blackbody is a con-

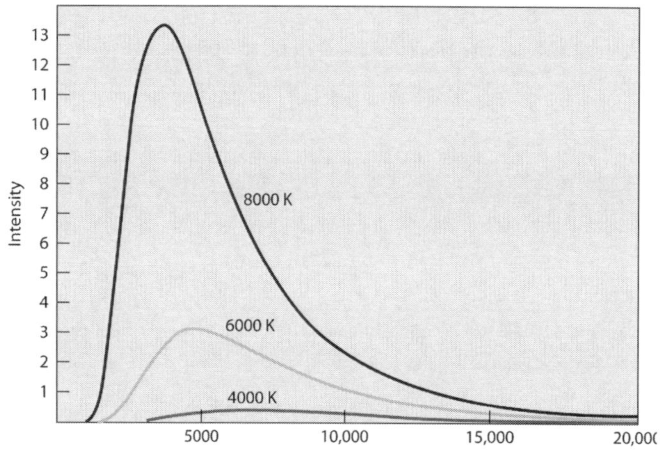

The spectra of blackbodies display a slow rise to a maximum as the wavelength decreases, followed by a rapid fall. Here are the spectral curves of three such bodies (of the same size), radiating at 4000 K, 6000 K, and 8000 K. The peak of the emission, Amax, shifts to shorter wavelengths as temperature is increased, making the hotter body bluer. In addition, the amount of radiation at every wavelength is increased. The total luminosity, the integrated area under the curve, is proportional to the temperature raised to the fourth power.

tinuum in which the different colors and wavelengths run together with no breaks or gaps. The brightness of the spectrum varies with wavelength in a very characteristic way. As λ decreases from high values the radiated energy steadily increases to a maximum, then drops suddenly to near zero. As the temperature, *T*, increases, the blackbody curve keeps the same relative shape but increases in height and expands toward lower wavelengths. At only a few degrees above absolute zero (-273°C), it radiates only in the radio spectrum. In the hundreds of degrees Kelvin (or Kelvins, K, centigrade degrees above absolute zero), it emits in the infrared as well, but is still invisible to the eye. But as *T* tops a thousand or so degrees K it begins to glow in the optical (visible) spectrum, steadily increasing the amount of blue light relative to yellow and yellow light relative to red. The effect is easily seen in a toaster or oven element, which changes color from red to orange as it heats. Theory shows that the wavelength at which the emission is at a maximum is given by the Wien law:

$$\lambda_{max}\ (\text{Å}) = 2.9 \times 10^7 / T$$

Measurement of the spectrum allows determination of the temperature. The amount of radiation released by a blackbody per square meter of surface per second, a quantity called the total flux, *F*, is proportional to the fourth power of the temperature. That is, in units of watts per square meter per second ($m^{-2}\ sec^{-1}$),

$$F = aT^4$$

(an equation called the Stefan-Boltzmann law), where a is a constant of proportionality determinable in the laboratory. One body twice as hot as another will be 16 times as bright per unit area of surface. The gross energy output of the blackbody, called its luminosity, L, must then be equal to the body's surface area times the flux. If the blackbody is a sphere, whether a star or a bowling ball, the surface area is $4\pi R^2$, and $L = 4\pi R^2 aT^4$, certainly one of the most important relations in astrophysics. This enormously useful equation allows the derivation of L, R, or T if the other two are known.

Now the origin of the stellar colors is clear. They relate to different surface temperatures through the Wien blackbody law: red stars are relatively cool, near 3000 K; yellow ones are closer to 6000 K; white are about 10,000 K and the blue stars are hotter yet, perhaps 20,000 K or greater. Moreover, color is a quantifiable stellar property that allows the derivation or at least the estimation, of temperature.

Earlier we expanded the definition of magnitude into space to allow for distance; now we expand the definition into different parts of the spectrum. The apparent magnitude of a star, m, (or as mimicked with photoelectric photometers or CCDs), is actually called the *visual* apparent magnitude, m_V, or just V, and it corresponds principally to the yellow part of the spectrum (technically at 5480 Å) where our eyesight is most sensitive. The magnitudes in Appendix 3 are all V. Astronomers also measure magnitudes in the blue part of the spectrum (where the first photographic plates were most sensitive, roughly corresponding to 4500 Å), called B.

The difference between the two magnitudes, $B - V$, is the color index. The color indices of white stars are near zero, those of blue stars are slightly negative (because the star is brighter at B and has a lower magnitude number), and those of red stars can be quite positive. At B, brilliant Betelgeuse is third magnitude! Magnitude systems can be elaborate; in addition to V and B, astronomers have added an ultraviolet magnitude U (3500 Å), which gives another color index, and several more that extend into the red and infrared.

Measurement of the color index gives the ratio of the luminosities at two wavelengths on the blackbody (or stellar-energy distribution) curve, which then tells the temperature. However, B and V by themselves provide limited information, because they are appropriate to specific wavelength bands. More physically meaningful is a quantity called the bolometric magnitude, m_{bol}, which corresponds to *all* the radiation of the star, that integrated over all wavelengths. It can be calculated if we know the stellar temperature, which gives us the form of the blackbody curve (or the true distribution of stellar energy), and the amount of radiation not included in V. For convenience,

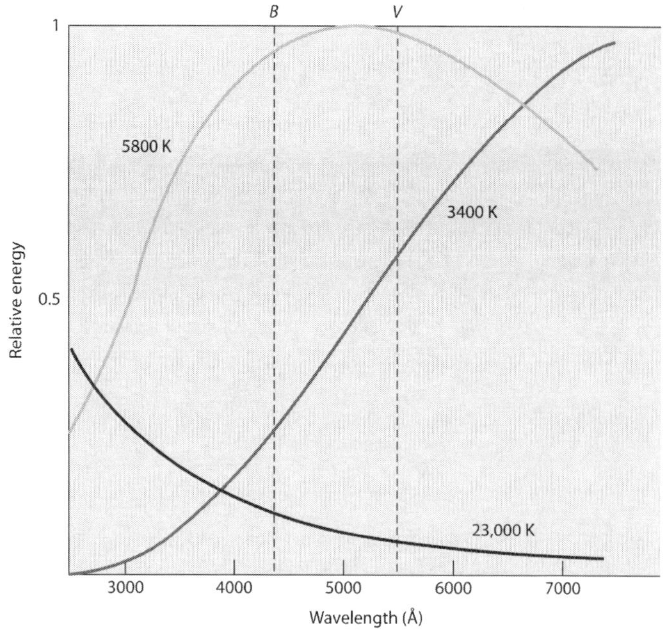

Blackbodies of different temperature are scaled to the same relative heights. The hottest (23,000 K) is brighter at B than at V, the coolest (3400 K) brighter at V. Since the magnitude goes down as the brightness goes up, the color index $B - V$ must be negative for the hot blackbody and positive for the cool one. It is only slightly positive for the Sun at 5800 K.

bolometric magnitudes are scaled to V for yellow-white stars. Consequently, for medium-temperature stars like the Sun, Vega, or even Arcturus, which radiate much of their energy in the visual spectrum, the bolometric magnitude does not differ much from V. For extremely hot or cool stars, which radiate profusely in the ultraviolet or infrared, the differences can be several magnitudes and are critically important for the determination of physical stellar properties.

Now combine the effects of distance and temperature. Absolute magnitudes can be expressed in any color. Apply the magnitude equation to V and you get absolute visual magnitude, M_v; apply it to B to find M_B, or to m_{bol} for M_{bol}. The color index remains the same whether using apparent or absolute magnitudes.

As useful as they are, however, absolute magnitudes tell us nothing about the energy output of a star in physical units.

Return to the Sun.

It is a fairly simple matter to measure the solar constant, the rate at which the Sun heats the surface of the Earth, that is, the rate at which solar energy falls per square meter per second. Observation by Earth satellite, from which we can account for energy at all wavelengths, gives a rate of 1,368 watts m^{-2}. Each square meter in a sphere with a radius R equal to 1 AU, or 1.5×10^{11} m, receives energy at this rate, so we multiply the solar constant by $4\pi R^2$ (the area of the sphere around the Sun with a radius of 1 AU) to find a total solar power of 3.8×10^{26} watts! Since a watt corresponds to a flow of 1 joule of energy per second, the Sun produces a phenomenal 4×10^{26} joules s^{-1}. And since a typical optical photon carries only 4×10^{-19} joules, something like 10^{45} photons emanate from the solar surface in that tiny fraction of time!

It is hard to have an appreciation for these numbers. Let your local power company run the Sun for a second and send you the bill. Typical energy rates are about \$0.06 per kilowatt hour, which is \$$1.7 \times 10^{-8}$ per watt second. In 1 second you will spend $4 \times 10^{26} \times 1.7 \times 10^{-8} = \$7 \times 10^{18}$. All you need do is write a check for the amount of the gross national product of the United States (\$$10^{12}$) for the next *7 million years*!

We are now in a position to calculate the solar temperature with precision. If $L = 4\pi R^2 a T^4$, then T is the fourth root of $L/4\pi R^2 a$. We already have L. From the Sun's distance and its measured angular diameter of 32 minutes of arc, its radius is easily found to be 7.0×10^8 m, or 108 times the radius of the Earth. The equation yields a temperature of 5780 K. Such a value is called an effective temperature, since it is the temperature that a blackbody of solar size would need to produce the observed solar luminosity. The surfaces of stars are not solid but semi-transparent and gaseous, and the temperature increases with depth. The result is that stellar spectra generally deviate (some considerably) from ideal blackbodies. Effective temperatures consequently provide an excellent means of intercomparison.

CHAPTER 5
WHAT ARE THE STARS MADE OF?

STELLAR SIGNATURES

Magnitudes and colors are important, but without spectra you cannot begin to understand the nature of the stars. Isaac Newton discovered that sunlight could be broken into its spectral colors. Then, in 1802, the English scientist William Wollaston found that the solar spectrum was cut by several dark gaps now called absorption lines. By 1815 Joseph von Fraunhofer had catalogued the wavelengths of more than 300 of them, assigning roman letters to the most prominent. Within a few decades laboratory investigations showed that individual lines were caused by specific kinds of atoms.

For example, those with wavelengths of 6563 Å and 4861 Å were associated with hydrogen, and a close pair in the yellow at 5801 Å with sodium.

When the nineteenth-century astronomers first turned their primitive visual spectroscopes on the stars, they were mightily puzzled by the spectral variety they saw. Only a few had absorption spectra like the Sun's. Some spectra were much simpler, but others appeared quite complicated, and the complexity correlated remarkably with color. By the late nineteenth century, hot white stars like Vega and Sirius were known to be dominated by a few powerful lines of hydrogen. Some blue ones had hydrogen and what was later determined to be helium. The orange-red stars showed little or no hydrogen but had strong bands now known to be caused by titanium oxide molecules, while the deep red ones displayed different kinds of bands produced by carbon compounds. Moreover, it looked very much as if stars of different temperatures had different chemical compositions.

The first step toward understanding in any new science is the classification of the data. The work was begun by several astronomers in the 1800s and culminated in a system developed around 1890 that is still in use today. E. C. Pickering, the director of the Harvard College Observatory, simply assigned letters to stars according to the strengths of the hydrogen lines, A for the strongest, B the next, down to O, the letters more or less correlating with color from white through yellow to red. The only glaring exception was class O, which followed red M and N even though its stars were

SOME PROMINENT SPECTRAL LINES

Wavelength Å	Fraunhofer Name[1]	Origin[1]
6563	C	hydrogen (H)
5893	D	neutral sodium (Na I)
5876	——	neutral helium (He I); hot stars
5270	E	neutral iron (Fe I)
5167, 5173, 5184	b	neutral magnesium (Mg I)
4955	——	titanium oxide (TiO); cool stars
4861	F	hydrogen (H)
4686	——	ionized helium (He II); very hot stars
4384	d	Fe I
4300	G	CH molecule
4340	——	hydrogen (H)
4227	g	neutral calcium (Ca I)
4101	h	hydrogen (H)
3968	H	ionized calcium (Ca II)
3934	K	ionized calcium (Ca II)

1 These letters were used by Joseph von Fraunhofer in the early 1800s to designate spectrum lines before they were chemically identified. The lines A and B, not listed, are produced by the Earth's atmosphere.

2 "I" denotes the spectrum of a neutral atom, "II" that of the singly ionized form, and so on.

blue-white. At about the same time, Pickering and his coworkers, notably Williamina P. Fleming, Antonia Maury, and Annie Jump Cannon, began the classification of stars in both celestial hemispheres with the aid of an objective-prism spectrograph. This simple device uses a low-angle prism placed over the lens of a telescope, so that instead of recording the images of stars, we record those of the stars' spectra, allowing rapid classification of many at a time. It quickly became apparent that several of the original letters were either unneeded or were the result of flawed exposures, so C, D, E, and some other letters vanished. With the accumulation of a large number of spectra, Maury and Cannon realized that the gradations in the strengths of the spectrum lines were continuous only if class B were placed before A and O *before* B, resolving the color issue.

The final result is the spectral sequence of the seven basic stellar types, OBAFGKM. As the observations improved, Cannon found that the seven divisions were simply too coarse, so she decimalized them, making the A stars run from A0 at the hotter end to A9 at the cooler, followed by F0, and so on. In this system our Sun is classified as G2. The spectral classes of the 40 brightest stars are given in Appendix 3. By 1920, Cannon had almost singlehandedly classified 225,000 stars in the bible of early spectroscopy, the *Henry Draper Catalogue* (stars are commonly named by their HD numbers), and by 1940, 359,000 were available. It is not surprising that her name is one of the most revered in all astronomy.

DECODING THE MESSAGE

The different stellar spectra are caused by atoms in differing states of ionization and electronic excitation. The full explanation had to wait until the revelations of the 1920s showed us how the atom works. The basic model is a cloud of one or more negatively charged electrons surrounding a nucleus that consists of some combination of positively charged protons and neutral neutrons. The charges on the proton and electron are equal in amount, the mass of the proton is about equal to the mass of the neutron, and that of the electron is some 1800 times smaller.

Opposite charges attract, allowing the electrons to be tied to the nucleus. However, like charges repel. How then can the protons stick together? There are four fundamental forces of nature, invisible fields that act over a distance and control the universe. By far the weakest, surprisingly enough, is gravity. It feels powerful only because there are so many atoms in the earth to attract us. Stronger is the electromagnetic force, which controls electricity, magnetism, the attraction and repulsion of the different charges, and the creation of photons. Both these forces behave according to inverse square laws and operate over all space, the other two only over the short range of the nucleus itself. The "weak force" involves nuclear reactions; by far the strongest of the four is the attractive "strong" or nuclear force. If protons are separated, they will indeed fly apart, but if they are close enough, the strong force overwhelms the electromagnetic repulsion, and they stay locked together.

A specific chemical element is determined solely by its number of nuclear protons, a quantity called the atomic number: for example, hydrogen (H) has 1, helium (He) 2, carbon (C) 6, gold (Au) 79. A neutral atom has as many protons as electrons, but the electrons are tied fairly loosely, and if enough energy is supplied to the atom, one or more can be lost to produce positively charged ions. Carbon with one electron gone is C+ and oxygen with three missing is O^{+3}. The sum of the number of neutrons and

THE SPECTRAL CLASSES

Class	Characteristic Spectra	Color	Color Index	Effective Temperature (K)	Examples
O	He II; He I	blue	-0.3	28,000–50,000	x Per, € Ori
B	He I; H	blue-white	-0.2	9900–28,000	Rigel, Spica
A	H white	white	0.0	7400–9900	Vega, Sirius
F	metals; H	yellow-white	0.3	6000–7400	Procyon
G	Ca II; metals	yellow	0.7	4900–6000	Sun, *a* Cen A
K	Ca II; Ca I; other molecules	orange	1.2	3500–4900	Arcturus
M	TiO; other molecules; Ca I	orange-red	1.4	2000–3500	Betelgeuse
R[1]	CN; C_2	orange-red	1.7	3500–5400	———
S[2]	ZrO; other molecules	orange-red	1.7	2000–3500	R Cyg
N[1]	c_2	red	>2	1900–3500	R Lep

1 Carbon stars
2 Mild carbon stars

protons in the nucleus is the atomic mass, written as a superscript to the chemical symbol. Common helium, ^{4}He, has two of each, and carbon, ^{12}C, has six of each. As the atomic number goes up, the neutron count tends to climb faster than the proton number, so that uranium, ^{238}U, has 92 protons but 146 neutrons. For any element the number of neutrons varies, producing one or more isotopes. Helium also comes with one neutron, called ^{3}He, and carbon comes in ^{12}C, ^{13}C, and ^{14}C varieties. Usually one isotope dominates over the others (there is only 1 atom of ^{3}He for every 100,000 of ^{4}He). The most common form is given in the periodic table of the elements.

Absorption lines are produced when an atom's (or ion's) electrons absorb energy from the continuum. Electrons bound to an atom are constrained to move in "orbits" of specific radii. (The analogy to planetary orbits is quite loose; in reality, the electrons behave as clouds of charge surrounding the nucleus.) The larger the radius, the higher the electron's energy. An electron can jump from a lower to a higher orbit if it absorbs a photon whose energy is exactly equal to the difference in energy between the two. The energy of a photon is equal to hν (Planck's constant times frequency), which is also equal to hc/λ. Consequently, the energy difference between orbits defines the wavelength of the absorbed photon. Conversely, an electron will jump *down* between these two states by *emitting* a photon of the same wavelength, which produces a bright-line or emission spectrum.

You will see an absorption spectrum when a cooler, lower density gas is placed in front of a hotter blackbody (or some similar source of continuous radiation). The emission spectrum is seen when a low density gas is viewed by itself. Stellar absorption lines are created in a cooler, lower-density atmosphere at the star's surface that partially blocks the continuum emanating from the hotter, denser gases below.

To understand how the spectral sequence is created, look at simple hydrogen. The optical absorption (Balmer) lines arise from the second orbit: The orbit 2-to-3 transition is called Hα (either in absorption or emission), 2-to-4 is H/ß, and so on. The presence of the Balmer lines requires a sufficient number of electrons in the second orbit. But nature always seeks the lowest energy state, so the vast majority of atoms have their electrons in the lowest orbit. They are raised into orbit number two largely by collisions among atomic neighbors, but that requires the gas temperature, and the particle speeds to be sufficiently high (even then, only a tiny number make it). In the cool atmosphere of an M star, the collisions are ineffectual, no electrons are premised, and there are no hydrogen lines, even though the gas is 90 percent H. We first see the hydrogen lines in class K at about 4000 K, and they rapidly become stronger up about to type A at 10,000 K as more and more electrons are excited.

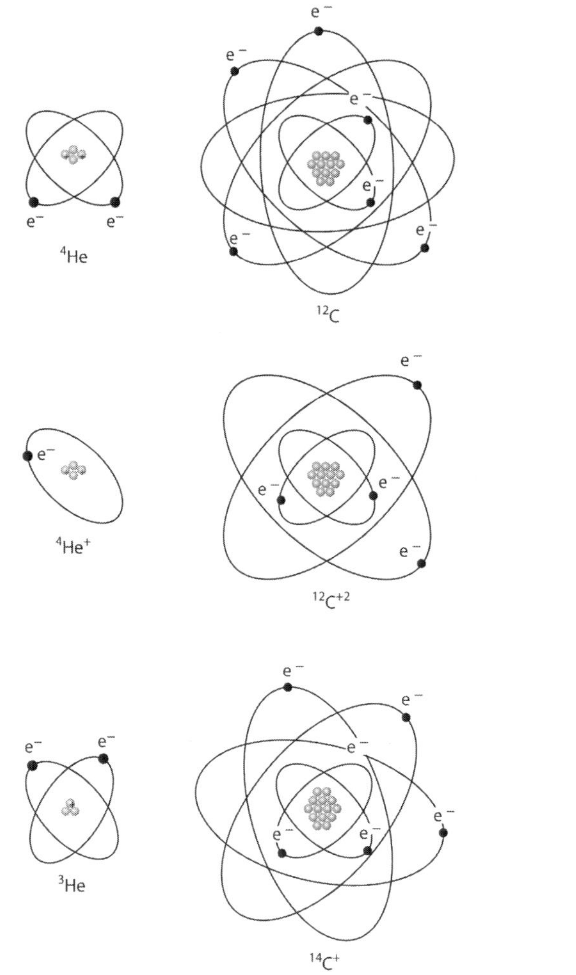

A helium atom (top left) has two electrons orbiting a nucleus that consists of two protons and two neutrons, and is therefore called ^{4}He (the superscript—the atomic mass—counts the total number of nuclear particles). Middle left: One electron has been stripped from the atom, leaving a positively charged ion He+—or more properly ^{4}He+. Bottom left is the only other possible helium isotope, a rare form with only one neutron, ^{3}He. At right are three carbon atoms, from top, ^{12}C, ^{12}C+2 (doubly ionized carbon, which has two electrons removed), and rare singly ionized ^{14}C+ ^{13}C is relatively common. Sizes of nuclei (in reality they are about 10-5 the diameter of the electron orbits) are enormously exaggerated.

Then comes a surprise. Any hotter, and the collisions become so vigorous that some of the hydrogen is ionized. The result is that there are fewer neutral atoms and the hydrogen lines weaken, the effect exaggerated by the increasing opacity of the stellar gases, which makes the absorbing layers thinner. The energy levels that produce helium are much harder to excite than are those of hydrogen. Consequently, in order for them to be seen the temperature must be much higher. They are not found until class B, and ionized helium does not appear until nearly class O.

More complex atoms, the "metals"—the term is used here in its astronomical sense to mean any element heavier than helium—have more electrons and much more complex spectra. Nevertheless, they behave in similar ways. Neutral cal-

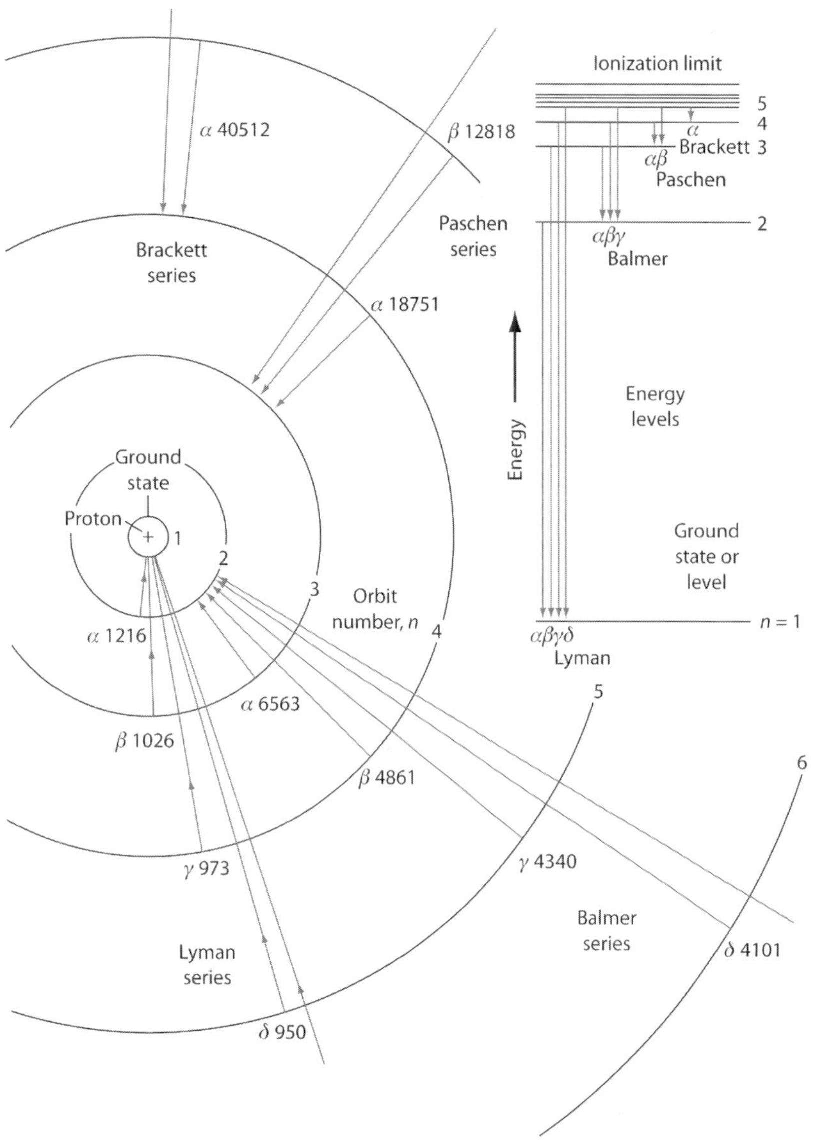

A hydrogen atom's lone electron revolves around its single proton (+) and has an infinite number of possible orbits, of which six are shown here. Each is tightly restricted to a specific energy, shown in the simple graph at the upper right. An electron is mobile between orbits and can be knocked up or down by collision, can jump upward by absorption of a photon of exactly the right energy (equal to the energy difference between the orbits), or jump down with the emission of a photon of that same energy (or wavelength). Upward jumps produce absorption spectra; downward jumps, emission spectra. Transitions that land on or arise from the Balmer series (second orbit), observable in the optical spectrum, are referred to as Ha (6563 Å), H⊠ (4861 Å), etc. The Lyman series is found in the ultraviolet; the Paschen and higher-order series are low-energy infrared and even radio.

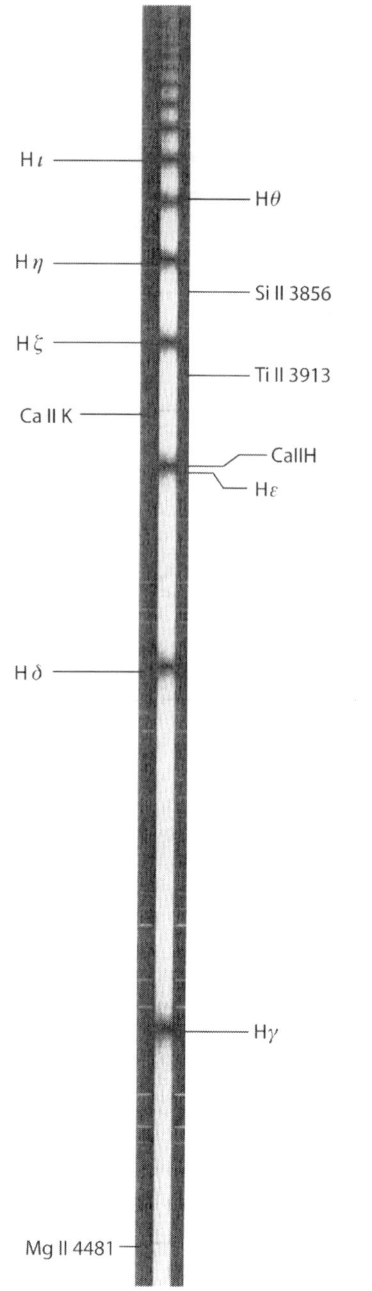

The spectrum of the star Vega, spectral class A0, for which the largest number of electrons are in the second orbit and the hydrogen lines are at their maximum strengths.

cium is powerful at the low temperatures of the M stars. But calcium is much easier to ionize than hydrogen, so that as the temperature increases to only 5000 K or so, the neutral state gives way to the singly ionized form (with one electron stripped away). Furthermore, the ionized lines in the optical, Fraunhofer H and K, arise from the lowest orbit, so that no pre-excitation is needed. Therefore the solar Ca II lines (II denotes the spectrum of singly ionized calcium, Ca+, III that of Ca^{+2}, and so on) dominate hydrogen, even though there is only one calcium atom to every 10^5 hydrogen atoms. Eventually Ca II weakens, as the temperature becomes great enough to knock yet another electron off to produce the doubly ionized form, which is followed by triple ionization. The lines of the highly ionized heavier elements tend to be in the ultraviolet, and so the optical spectra of the hotter stars become simpler, dominated by H and He. By contrast, at the lowest temperatures atoms can actually combine to form molecules, which produce extremely complex banded spectra. There is a considerable variety. TiO dominates the M-star spectra, and we find CH and CN, among others, in the solar spectrum.

Once astronomers and physicists understood the processes that produce stellar spectra and were able to formulate the rules that control excitation and ionization, they could use the strengths of the absorption lines to calculate the physical conditions within the stellar gases and the all-important chemical compositions. The scientists found a pleasing uniformity of nature. Of the 92 natural elements, 74 are observed in stars. The rest are certainly there, but their abundances are too low to render their lines visible. In gross terms, the majority of stars have the same compositions, hydrogen dominating at 90 percent by number of atoms, followed by helium, the next simplest atom, at 10 percent. All the other chemical elements are found in the remaining 0.1 percent or so. (The bizarre nature of the earth, with little H and He, can now be appreciated.) Generally, the more complex an atom, the less there is of it. Oxygen is the next after helium at roughly 5 $\times 10^{-4}$ times the abundance of hydrogen, followed by carbon, neon, and nitrogen. The heavier metals are down by one or more factors of ten, with iron dominant.

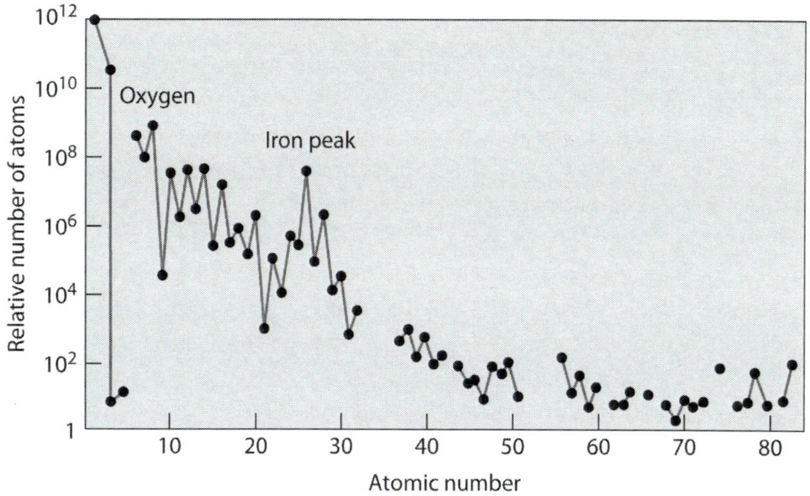

The abundances of the elements in stars decrease more or less steadily with increasing atomic number except for a great deficiency between helium and carbon and a sharp rise around iron.

The deep red stars are different, however, providing a portent of fascinating theoretical discoveries to come. The N stars, not a part of the standard spectral sequence, differ from class M not by temperature but by *composition*. Their carbon-to-oxygen ratios are reversed, the result of interior stellar processes. The R stars are warmer carbon-rich versions and correspond in temperature to classes K and G. In the 1930s yet another odd kind of star was discovered, labeled class S, which is intermediate in carbon composition between M and N, and at the same time has its TiO replaced by zirconium oxide, ZrO. All the standard spectral classes are now in place.

Stellar spectra provide the promised method of measuring radial velocities. The precise wavelengths of tens of thousands of the absorption lines of the observed elements and molecules have been measured in the laboratory. If the star is moving along the line of sight, the wavelengths of the absorptions will be shifted by the Doppler effect. The shifts can be measured with a calibration device in the spectrograph that produces emission lines of gaseous iron or another element (these can frequently be seen in the illustrations flanking stellar spectra). The radial velocity can then be determined by applying the Doppler formula. We typically find shifts of a fraction of an Ångstrom and speeds again in the neighborhood of a few tens of kilometers per second, consistent with the transverse velocities.

THE LUMINOSITY CLASSES

Class	Type of Stars[1]	Example[1]
0	Extreme, luminous supergiants; hypergiants	p Cas[1]; S Dor
Ia	Luminous supergiants	Betelgeuse, Deneb
Ib	Less luminous supergiants	Antares, Canopus
II	Bright giants	Polaris, Lyrae
III	Giants	Aldebaran, Arcturus, Capella
IV	Subgiants	Procyon
V	Main sequence (dwarfs)	Sun, *a* Cen, Sirius, Vega, 61 Cyg
sd	Subdwarfs	———[2]
D	White dwarfs	Sirius B, Procyon B, 40 Eri B

1 0-Ia

2 All faint and obscure

THE HR DIAGRAM

Since the luminosity of a radiating body depends so strongly on temperature, we would expect to see a clear correlation between absolute magnitude and spectral class—and we do. But there is a delightful surprise. The first investigations were made early in this century by the Danish astronomer Ejnar Hertzsprung and independently by Henry Norris Russell, the dean of twentieth-century American astronomy. In 1913, Russell plotted absolute visual magnitude against spectral class for stars of known distance. His graph shows the anticipated effect, the stars tending to lie along a broad band that climbs up and to the left, visual magnitude dropping (and luminosity increasing) with increasing temperature.

But look! There is another band that climbs up and to the *right*, the brightness increasing with *decreasing* temperature. The only way that can happen along this locus is if the stars become larger as temperature drops to offset the effect of the rapidly decreasing stellar flux. Russell, following Hertzsprung's earlier suggestion, discriminated between these two bands by calling their stars dwarfs and giants, thereby ensuring a wonderful stellar nomenclature. The Sun occupies the main band, and in spite of its tremendous size is still a G2 dwarf. The giants, however, are of *truly* impressive dimensions. Their diameters can easily be estimated from the blackbody equation, $L^* = 4\pi R_*^2 a T_*^4$, where * symbolizes a star. If we place everything in solar (Ø) units, the constants disappear, and $R_*/R_{\emptyset}$ is just $\sqrt{(L_*/L_{\emptyset})}/(T_*/5780)^2$. The effective temperature comes from the spectral type (or the color), L_* from M_{bol}, and we instantly have the radius. Look at the sky's archetypal giants, orange Arcturus (Kl) and Aldebaran (K5). With an absolute bolometric magnitude of -0.3, Arcturus is 105 times more luminous than the Sun. Its great brightness and its relatively cool temperature of 4400 K yield a radius 18 times solar; and Aldebaran is more than twice as big as that. Mira, a cool M7 giant, is *300* times the solar size and would roughly fill the orbit of Mars. Only the giant branch contains the carbon-rich N and S stars, important evidence to be applied later to the mystery of stellar life and death.

Some 15 years earlier, Antonia Maury, one of Pickering's group at Harvard, had noted that some of the hotter stars had broad hydrogen lines, whereas those of others were quite narrow. In 1907, Hertzsprung found that narrow-line stars have considerably smaller proper motions than the broad-line stars, implying that they are much farther away and brighter. These rare and amazing bodies turned out to be ten to a hundred times more luminous and up to ten times larger than even the giants. These supergiants, if plotted, would sprinkle across the top of Russell's graph, which long ago became known as the Hertzsprung-Russell or HR diagram.

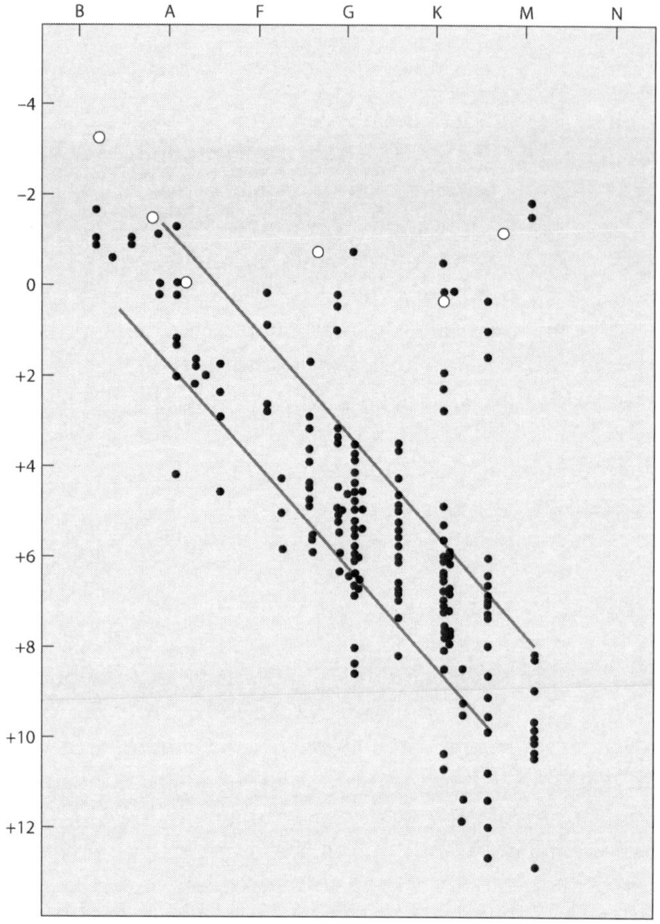

Henry Norris Russell's original diagram, in which he plotted absolute visual magnitudes of stars against their spectral classes. The dwarf sequence (now called the main sequence), which includes the Sun, goes from lower right to upper left. The giant branch scatters out toward the upper right. One white dwarf sits at the bottom.

The red supergiants are the largest known stars. Apply the blackbody equation to Betelgeuse, the star that marks Orion's right shoulder. Visually, it is 11.8 magnitudes, 54,000 times brighter than the Sun; yet at 3300 K each square meter produces only 1/10 the solar flux. Bolometrically it is another 1.4 magnitudes brighter, and the blackbody equation yields a star with a radius of almost 7 AU—over 1300 times solar, over four times the dimension of Mira, large enough to overfill the orbit of Jupiter by 30 percent. And that is not the limit! A fourth-magnitude red jewel in Cepheus, μ Cephei—the "garnet star"—shines 1.2 absolute magnitudes brighter than Betelgeuse and is 1.7 times larger, 11.8 AU in radius, comparable to the orbit of Saturn. The volume of such a behemoth could contain over *one billion suns*. The angular diameters of supergiants can be several hundredths of a second of arc, easily measured by interferometry.

Return now to the dwarf band, which has since become known as the main sequence. Along it, the stars brighten more than would be expected from temperature alone, indicating that their radii increase as well. Vega, for example, is 2.5 times the size of the Sun. At the top of the sequence we

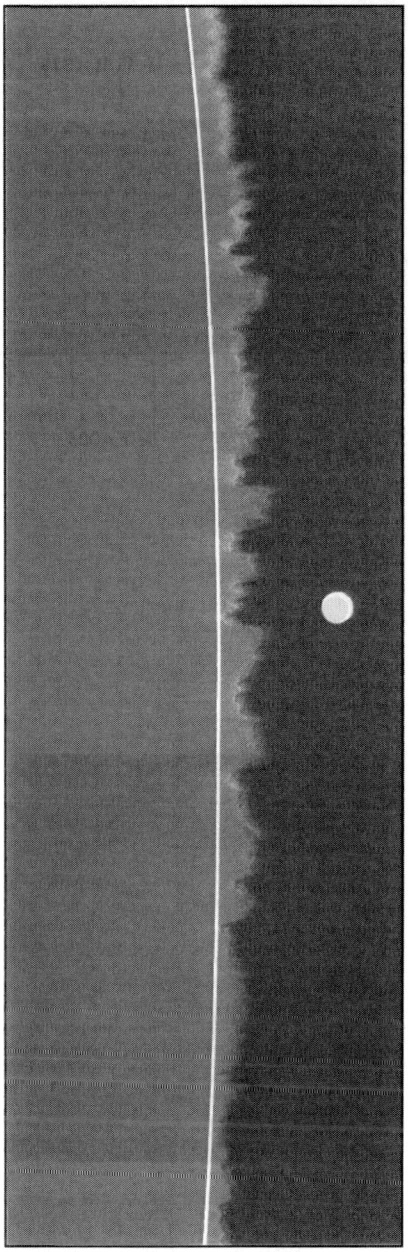

The Sun compared to the surface of Mu Cephei, which on this scale is some 8 meters across.

find HD 93129A, V = -7 and M_{bol} = - 11.3, 2.6 million times the solar luminosity. Only part of this extraordinary brightness derives from the star's extreme temperature of 48,000 K. The rest is the result of a dimension 25 times solar, comparable with the smaller giants (yet because of its placement on the HR diagram, HD 93129A is still termed a dwarf).

At the other end of the main sequence are dwarfs that live up to their names. Proxima Centauri's reddish color and complex molecular spectrum indicate an effective temperature of some 3300 K. The bolometric luminosity is 1/1800 that of the Sun, giving a radius only 0.07 solar. Then drop down near the end of the main sequence, to LHS 2924, which has a temperature of only 2600 K and a bolometric luminosity of 1/36,000 solar; its radius is down another factor of 2.5, making it a mere three times the size of Earth. At the distance of the Sun, Proxima Centauri would be just barely discernible as a disk, and LHS 2924 would appear merely as a brilliant star.

Now for something even odder. Down at the bottom of Russell's original diagram is one lonely white star called 40 Eridani B. It is both hot and faint, and as the exact opposite of the giants, it must be very small. This is the first white dwarf, a star roughly the size of Earth. A close look at brilliant Sirius of the northern winter sky, magnitude -1.5, reveals another. This star has a dim companion called Sirius B that is 10 magnitudes fainter, about 2.5 times as hot, and only three-quarters the size of the Earth. As more white dwarfs were discovered, they were seen to string out along the bottom of the graph into the realm of the blue and even the red stars. The name "white dwarf" stuck, though, in spite of the variety of colors.

These various categories of stars define luminosity classes, indicated by a set of roman numerals. Main sequence stars are V, the giants III, and the supergiants I (divided by brightness into Ia and Ib). In between the supergiants and the giants are the bright giants, class II, and between the giants and the dwarfs are the subgiants, number IV. Then to the left of the lower main sequence on the HR diagram is an odd set of stars called subdwarfs. These are low-metal versions of the ordinary dwarfs, their metal deficiency making them appear a bit bluer than normal. The white dwarfs are simply referred to as D. Note that an upper main sequence dwarf can be just as bright as a supergiant. The terms and roman numerals refer to the gross characteristics of the *classes*, not to those of the individual stars. This two-dimensional (temperature and luminosity) classification scheme was developed by W. W. Morgan, P. C. Keenan, and E. Kellman in the 1940s, and is referred to as the MKK or MK system. The Sun is a G2 V star. At last its classification is complete.

INTO THE GALAXY

The HR diagram provides a powerful means of determining stellar distances. We know the luminosity class and consequently the absolute magnitude of an A star merely by looking at its line width. Measurement of apparent magnitude and application of the magnitude equation then tells us how far away it is. Because of their extreme sizes, the supergiants have much lower atmospheric densities than do the dwarfs. If atoms are close together, they disturb one another, slightly broadening the allowed radii of the electron orbits and smearing the absorption lines. As a result, the dwarfs have broader lines. Each spectral class has such identifying criteria: For example, in class K the giants and supergiants have stronger cyanogen (the CN molecule) absorptions. These standards are all laid out nicely for us by Morgan, Keenan, and Kellman, allowing us simply to look at the spectrum of a star to find its gross luminosity class and thus obtain a spectroscopic distance. We are now in a position to work our way across the Galaxy, and even into other galaxies—providing that we can observe stellar spectra. The distances of all stars listed in Appendix 3 that are greater than about 400 parsecs were obtained in this way.

Parallaxes unfortunately still have a limited range. Within the grasp of this distance method there are no luminous supergiants or O stars, and few B stars, so how do we know their absolute magnitudes in the first place? How do we calibrate the entire HR diagram?

We make use of the profound property of stars to group together into clusters, which abound throughout space.

better than 8 percent, allowing us to construct the cluster's HR diagram. We can join this value with the Hyades' parallax and add the parallax stars that are not in clusters, the combination providing a powerful fundamental reference.

Unfortunately, the Hyades has no really bright stars, nor any super-giants or even bright giants. But the Pleiades, with a bevy of bright B dwarfs, *does.* To find its distance, we construct an HR diagram with *apparent* magnitudes instead of absolute. We then place the diagram of the Pleiades atop that of the Hyades, aligning the lower axes. Magnitudes are logarithmic, so we can slide the Pleiades diagram vertically until the two main sequences fit, allowing us to determine the difference *m* — *M*, and find the distance by using the magnitude equation. The Pleiades, at a distance of 150 pc, is currently within parallax range, and combination of the two methods and the distance to the Hyades will produce a superb reference base. The next step is to pick clusters that have O stars and supergiants, apply the same method, and there it is: a complete HR diagram.

Once we are able to measure distances (and there are other methods in addition to that based on spectra), we are able to look at the numbers of different kinds of stars. As we climb the main sequence, the so-called luminosity function (the number of stars per unit volume of space per absolute magnitude division) goes down dramatically. It is more

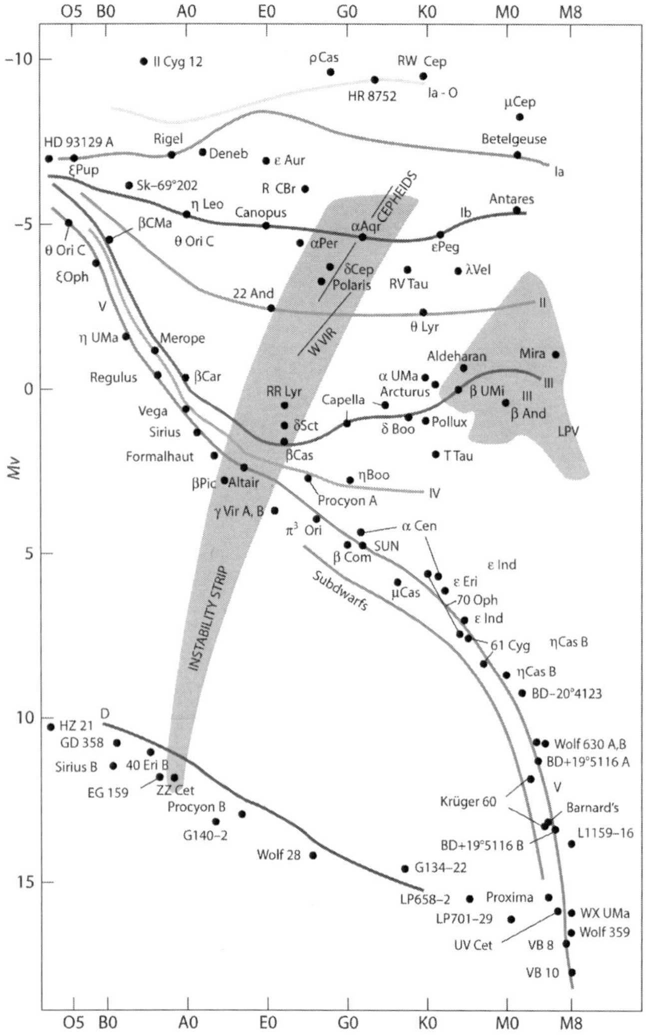

A modern HR diagram shows distribution of well-known stars and loci of all recognized branches. In addition to the main sequence (V) and giant branch (III), we see supergiants across the top (Ia and Ib); extreme supergiants or hvpergiants (Ia-O) above them; subgiants between the giants and dwarfs (IV); subdwarfs; and a whole sequence of white dwarfs (D). (Along the main sequence, over two-thirds of stars actually fall into class M.)

Some are among the most loved objects in the sky: the Pleiades, or seven sisters, in Taurus up and to the right of Orion; the Hyades, which form Taurus' head and surround Aldebaran; the jewel Box (κ Crucis), next to the Southern Cross.

The Hyades cluster is very close to us, so near that we can not only measure accurate proper motions, we can actually see the convergence of the proper motion vectors, telling us where in space the cluster is moving. Once we know the angle between where the cluster is now and where it is going to be after an arbitrarily distant time in the future, we are able to calculate the cluster's gross tangential velocity from its observed radial velocity. Combination of the tangential velocity and the average proper motion yields a distance of 47 pc (consistent with the parallaxes of its stars), accurate to

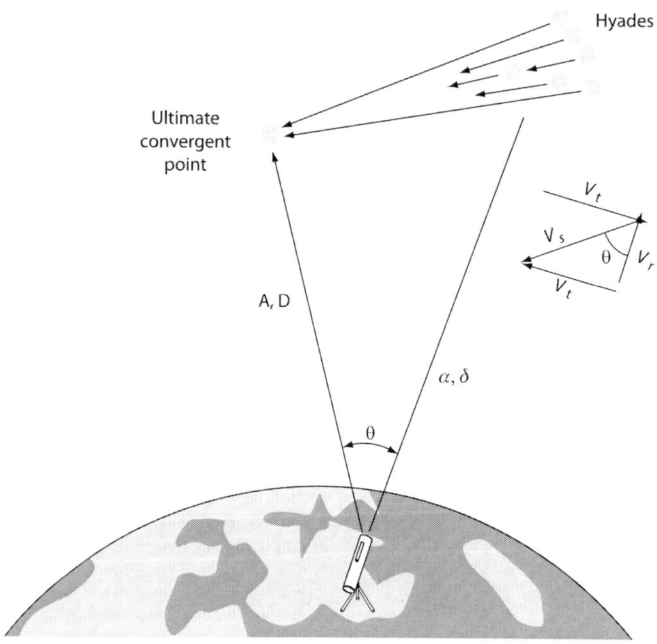

We look out to the Hyades at right ascension and decimation (a, 8). The proper motion vectors of the individual stars converge to a point in space with coordinates (A, D) that define the angle between the line of sight and the direction in which the stars are actually going; this is also the angle between the space motion vector and the radial velocity vector (see the diagram to the right). We can then calculate the space velocity (v_s) and the tangential velocity (v_t) from the radial velocity (v_r). The distance of 45 pc follows.

meaningful to express it here in terms of spectral class. Over 70 percent of the main sequence stars fall into class M, with masses under one-half solar, and only 9 percent are class G like the Sun. The most luminous stars are remarkably rare: A mere 0.1 percent are class B, and—a stunning statistic—-only 0.00004 percent fall into class O! We see a similar phenomenon among the stars off the main sequence. Dim white dwarfs are quite common, whereas the giants are rare compared with the number of stars on the main sequence, and the supergiants are exceedingly few and far between.

These counts are all out of proportion to our naked-eye experience; we see a sky dominated by giants and main sequence stars of classes B and A. This apparent anomaly is a fine example of observational selection, a harrowing problem in astronomy. The dim red dwarfs are everywhere, swarming around us like invisible fireflies, none at all visible to the naked eye. Though the intrinsically bright stars are fairly rare, we can see them for great distances, and so they seem to overpopulate the sky.

MASS

Whether for people, rocks, or stars, no distinguishing characteristic is quite so significant as mass. The mass of a star controls nearly everything, from its luminosity to its surface temperature to the length of its life and the way it will die. How can we determine the masses of bodies that we cannot touch and weigh? Again, begin with the Earth. You certainly cannot put it on a scale. All you have to do, however, is to drop a ball and measure its acceleration in the Earth's gravitational field. Since $g = GM_{Earth}/R^2$, and R and G known, the mass of the Earth is easily found to be 6 X 10^{24} kg.

Simple enough, but how do you apply this concept to the Sun? Again, just drop something, in this case the Earth itself. Remember that an orbiting body is no more than a falling body. To derive the solar mass, use Kepler's third law as generalized by Newton, which requires the Earth's orbital period and semimajor axis. The result, 2 X 10^{30} kg, is the sum of the masses of the Earth and Sun, in which the Earth's is inconsequential.

This formula is general enough to apply anywhere, to Jupiter's satellites (from which we find that the giant planet is 318 times the mass of the Earth) and to double (or binary) stars. Remember 61 Cygni, *a* Centauri, Sirius and its companion? They are not rare. At least half of all stars are doubles, triples, or higher-order multiples. If a pair is close enough, we can watch the stars move and map out the orbit. The Alpha Centauri pair completes a circuit in 81 years, Sirius in SO, others in far fewer. *P* is easy to find, and *a* is known if we have the distance. Then follows the sum of the masses.

It is simpler to think of the more massive of two bodies as being at the focus of the orbit. In fact, however, both bodies must revolve about each other, with a point in between them—the center of mass—at a common orbital focus. The position of the center of mass depends on the ratio of the masses. For example, the Moon can be said to have an orbit about the Earth with a semimajor axis *a* equal to 384,000 km. But the Moon's accelerating mass is 1/81 that of the Earth, so the center of mass is 1/81 of the way from the terrestrial center to the lunar center. The Earth's orbital semi-

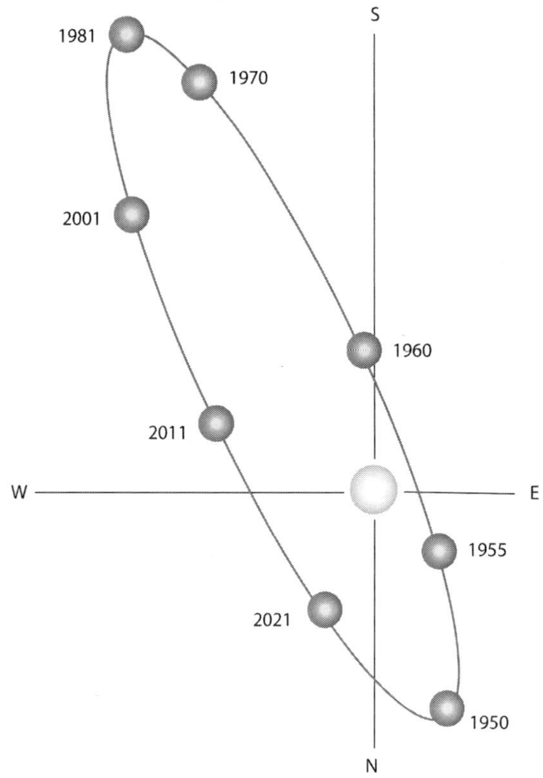

Alpha Centauri B, a Kl dwarf, orbits *a* Cen A (G2 V); although the orbit is an ellipse, *a* Cen A is not at a focus. This apparent violation of Kepler's first law results from the orbit's tilt to the line of sight. From the position of *a* Cen A relative to the focus, the orbit can be rectfied, allowing us to find the true path and the sum of the masses from Kepler's third law as generalized by Newton. In fact, both stars have elliptical orbits about a common center of mass.

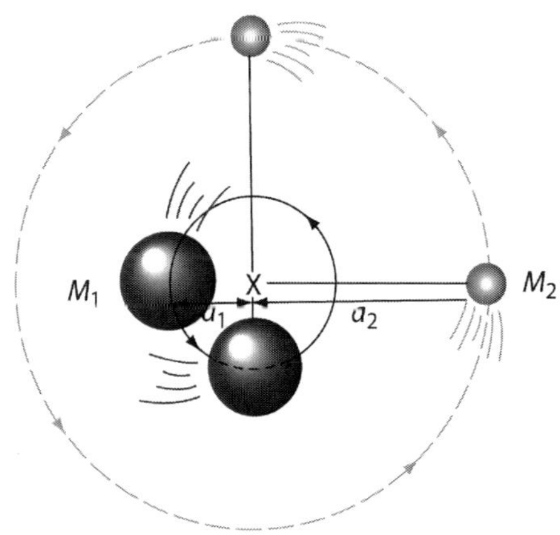

A and G dwarfs orbit one another: Mx/M2 — a2/a1

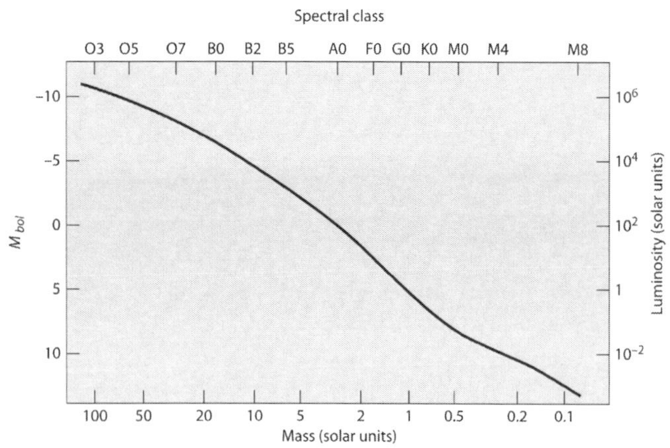

One of the two orbiting components of the star U Sagittae is large, the other small and bright. The latter can hide behind its large companion, producing a primary eclipse every 3.3 days. When the small one passes in front of the large star it occults only a portion of its surface, leading to a much smaller secondary eclipse halfway between the primary eclipses. The shape of this light curve provides orbital parameters that allow masses to be deduced from radial velocities. The duration of the primary eclipse yields the size of the big component, and the time it takes for the small star to go into eclipse yields its diameter.

major axis is then 4600 km and that of the Moon 379,000 km. If through careful measurements we can locate the center of mass of a double star, we will be able to determine the mass ratio, which, when combined with the mass sum from Kepler's third law, yields the individual values.

A large number of binaries are unresolved: that is, they are so near one another that we cannot see the separation between the components. If they are very close they must move very quickly in order to maintain orbit. The variations in their radial velocities are then observable, and though we see only one star, we might see two sets of stellar spectra that Doppler-shift back and forth as the stars swing around each other, changing their directions. If we are looking at the orbit edge-on, the radial velocities give the orbital velocities, which with the period give the semimajor axes and the masses. Unfortunately, the orbital orientation is unknown, but at least we *can* get mass ratios and some statistically valid information.

The real treat comes when an orbital plane is in the line of sight. Then the two stars can get in the way of each other and produce eclipses that are sometimes very dramatic as the apparently single star dims by one or even several magnitudes. The celestial hero Perseus slays the monster by showing him the gorgon's head, represented in the constellation by ß Persei, Algol, *al ghoul*, the "demon star." Every 2.9 days it drops from second magnitude to third as a small but bright star is partially eclipsed by a huge companion. There are enormous numbers of eclipsers in the sky, and they are a treasury of information. Since we know the orbital orientations from the eclipses, we can get actual velocities from the Doppler shifts, orbital sizes, and stellar masses. From the durations of various phases of the eclipses and the known orbital velocities, we can also get another measure of the physical diameters of the stars. Some eclipses last but a few hours, showing that the stars are small VV Cephei, just visible to the naked eye, is in eclipse for over a year, revealing a star comparable in size with the monster Mu Cephei.

After all the labor, the thousands upon thousands of measurements, we find a major key to stellar astrophysics, a strict relation between luminosity and mass for main sequence stars. Bolometric luminosity climbs roughly as the mass raised to about the 3.5 power (the exponent is actually a function of mass). A star like Sirius A, 2.3 times the solar mass, is some 20 times as bright! The main sequence is really a *mass* sequence.

The Sun is in the middle of the mass range. The observed masses run from a minimum of about 8 percent that of the Sun (the smallest that can actually make a self-sustaining star) to several dozen solar masses. The top of the range is very difficult to define. The rarity of these stars ensures statistically that they are all very far away, and binary orbits cannot easily be studied. But by extending the mass-luminosity relation with a little help from theory, we estimate that at the top, near main sequence spectral class 03, stars can reach 120 times the solar mass, allowing them to blaze forth at bolometric magnitudes near -12 and to be easily visible across intergalactic distances. Low-mass stars are the preferred variety and are everywhere. At the other extreme, there may be only one or two stars near the 120-solar-mass limit in our whole Galaxy of 200 billion. Nature simply does not *like* to make massive stars.

The giants and the supergiants also show something of a mass-luminosity relation, although not so well defined. The giants have masses typical of the middle main sequence. of perhaps 1 to 6 solar masses, and the supergiants possess masses tens of times greater than that of the Sun. The white dwarfs are perhaps the most astonishing. The mass of Sirius B is easy to find from its 50-year orbit about Sirius A, and the two stars are near enough so that the center of mass is accurately located. Although the dim white dwarf is smaller than the Earth, *its mass is about equal to that of the Sun.* (This star is heavy; most white dwarfs hover around 0.6 solar.) Its creation would require the Sun to be shrunk into a sphere with less than one-millionth its current volume, thus raising the density by over a million. The Sun's average density is roughly 1 gram per cubic centimeter, about that of water. A cubic centimeter sampled from the interior of Sirius B would weigh a metric ton, 100,000 times that of a similar volume of lead!

CHAPTER 6
DO THE STARS EVER CHANGE?

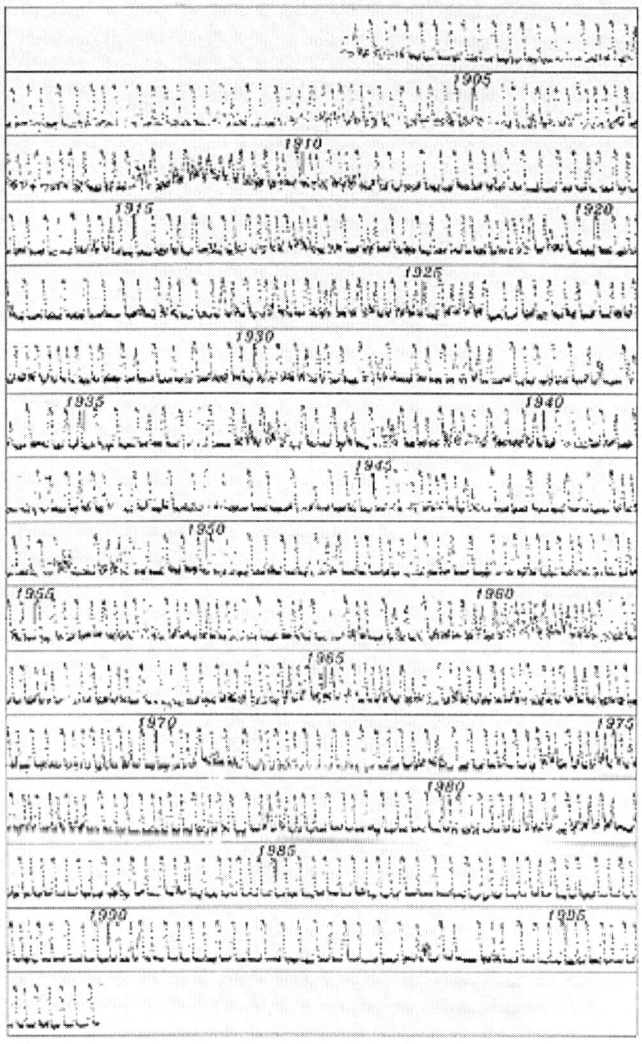

100-year light curve of AAVSO observations of the variable star SS Cygni.

INTRODUCTION

Variable stars are, quite simply, stars that change brightness. More than 30,000 are known and catalogued, and many thousands more are suspected variables. Their light variations are a vital clue to their nature, but professional astronomers do not have the time or the telescopes to monitor the brightness of thousands of variables night after night; their efforts are directed toward understanding specific aspects of the stars using such sophisticated instruments as spectrographs, photometers, radio telescopes, and satellites. Such instruments enable us to probe many different regions of the spectrum, such as x-rays, ultraviolet, infrared, and radio wavelengths. Yet it is crucial to measure the stars in ordinary visual light (the *optical* region of the spectrum) as well. The data obtained with special instruments and in shorter or longer wavelengths of light are compared with optical measurements, which serve as a benchmark. By observing variable stars, a serious observer (such as an amateur astronomer or student) can make a significant contribution to astronomy. Also, to understand and create theories about why and how stars vary, astronomers need to know the long-term history of the stars; hence it is essential that we have long-term observations. To date, the vast majority of long-term data has been provided by amateur observers.

Observations of variable stars are plotted on a graph called a *light curve* as the apparent brightness (magnitude) versus time, usually in Julian Date (JD). The light curve is the single most important graph in variable star astronomy. The light curve allows astronomers to unlock some of the secrets of stars and decode the messages hidden within the starlight. Information about the periodic behavior, the orbital period of eclipsing binaries, or the regularity (or lack) of stellar eruptions can be directly determined from the light curve. More detailed analysis of the light curve allows astronomers to calculate such information as the masses or sizes of stars. Several years' worth of observational data can reveal the changing period of a star, which is a signal of a change in the structure of the star. In the same way that a histogram is a useful tool to inspect the precision of a set of measurements, we can visualize the nature of a star's variation more easily by plotting a light curve of apparent magnitude versus time.

Light curves show that many variable stars are periodic. Periodic phenomena repeat in a regular way that is predictable. However, the line between periodic phenomena and non-periodic phenomena is not always a sharp one. Sometimes a process may seem to be periodic from a large-scale perspective, while investigation on a fine scale reveals non-periodic variations. Consider the change in apparent altitude of the Sun. If you measure the highest apparent altitude of the Sun each day for a week, you will not notice much change. The Sun seems to reach the same maximum altitude each day. Over a period of a month, however, changes will become obvious. But if we take measurements each day from December 31st until the summer solstice around June 21st, the apparent daily altitude of the Sun gets higher and higher. This is still non-periodic behavior. If we take our measurements for an entire year, from December 31st to December 31st, we will see a definite pattern. When we examine the change in the apparent altitude of the Sun over several years, we see that in addition to monthly variations, it follows a regular pattern with a period of one year.

Many variable stars do exhibit simple periodic behavior. Some semiregular variables have time intervals of periodic behavior in between non-periodic intervals. Some have two separate periodic cycles superimposed over each other, affecting the overall result. The only way to determine the periodicity is by plotting and analyzing light curves. The emerging patterns will then tell their story.

There are four main classes of variable stars: *pulsating* and *eruptive* variables whose variability is *intrinsic*—due to physical changes in the star or stellar system; and *eclipsing binary* and *rotating* stars whose variability is *extrinsic*—due to an eclipse of one star by another or the effect of stellar rotation. Below is a brief description of several different types of variable stars with their corresponding light curves. There is also more detailed information about some of the variables mentioned below throughout the *Hands-On Astrophysics* manual.

I. INTRINSIC VARIABLE STARS

A. Pulsating Variable Stars

Pulsating variables change brightness because they change their size and/or shape; the whole star is actually "vibrating." Most of them simply expand and contract repeatedly, swelling and shrinking in a continuing cycle of size changes; this is known as the *fundamental mode of pulsation*. Others change not only their size, but the internal arrangement of material within the star as well. Still others change shape, alternately flattening and elongating, or showing even more complex shape changes. And to make things yet more complicated, some stars vibrate in more than one way (more than one *mode*) at the same time. Like most vibrating systems, pulsating variables repeat their changes; they tend to be *periodic*.

One important class of pulsating variables is the *Cepheid variables*. These large yellow stars pulsate with periods from 1 to 70 days, with an amplitude of light variation up to 2 magnitudes. They are intrinsically very bright (high luminosity). The greater the absolute magnitude (luminosity) of a Cepheid, the longer its period. In fact, there is a strict relationship between a Cepheid's period and its luminosity, called the *period-luminosity relationship*. We can determine the period of a Cepheid from its light curve; then, using the period of the Cepheid, we can apply the period-luminosity relationship to compute the Cepheid's luminosity. Knowing its magnitude (how bright it appears to observers on Earth) and its luminosity (how bright it actually is) allows us to compute its distance from us, thereby enabling us to use Cepheid variables as distance markers. For example, just as astronomers in the first half of this century used Cepheid variables within the Milky Way itself to measure distances within our own galaxy, we can measure the distance to any given galaxy by computing the distance to its Cepheid variables.

The light curve for the Cepheid variable X Cygni is shown below; each dot represents one observation. **(All observations of X Cyg, and all subsequent light curves come from the AAVSO International Database, unless otherwise noted.)**

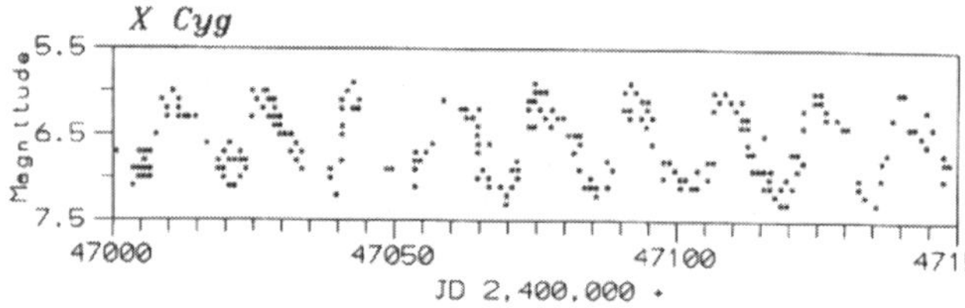

One of the largest groups of pulsating variables is the *long-period variables*, or *LPVs*. They are further divided into two major subclasses, the *Mira-type* and the *semiregular* variables. The visual light curves of Mira-type variables show well-defined periods ranging from 80 to nearly 1000 days, with amplitudes of 2.5 magnitudes or more. Mira-type variables are red giant stars, often of enormous size. Many of them are slowly ejecting a steady stream of matter into the surrounding space; this mass loss can have very dramatic consequences for their future evolution. This class of variable is named after the star *Mira* (also known as omicron Ceti), the first such star discovered; its light curve is shown below.

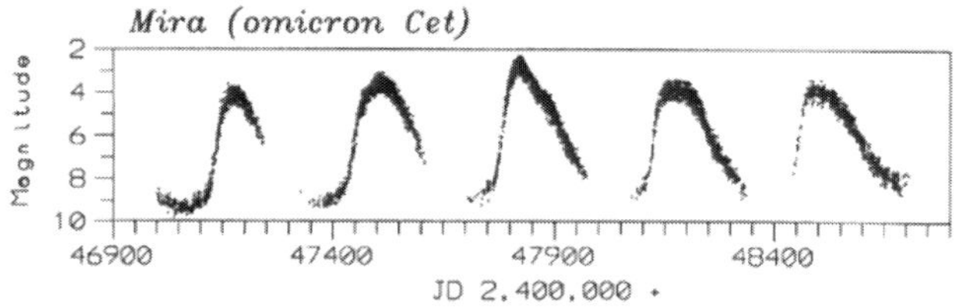

Yet another group of pulsating stars is the *semiregular variables*. These giants and supergiants show appreciable periodicity accompanied by intervals of irregular light variation. The periods range from 30 to 1000 days, with light amplitudes of not more than one to two magnitudes. An example is Z Ursae Majoris.

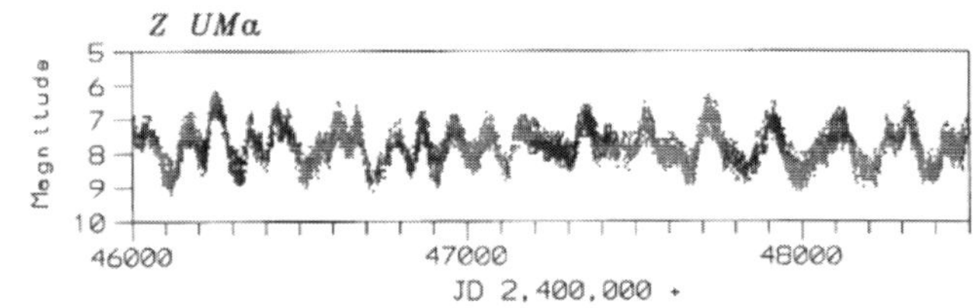

RV Tauri stars are also variables that pulsate. These yellow supergiants have a characteristic light variation with alternating deep and shallow minima. Their periods, defined as the interval between two deep minima, range from 30 to 150 days. The light amplitude may vary as much as four magnitudes. Many of these variables show long-term cyclic variations in brightness from hundreds to thousands of days. An example of an RV Tauri variable star light curve is that of R Scuti.

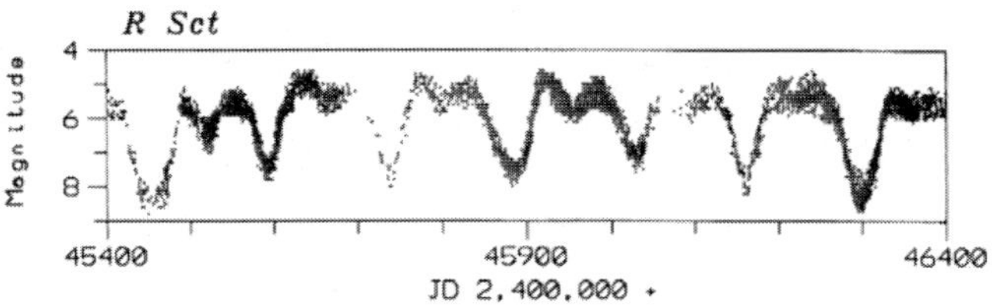

B. Eruptive (Cataclysmic) Variable Stars

Another major category of variables is the eruptive variables. These are not one-star systems, but two stars orbiting very close to each other, one of them a normal Sun-like or giant star, the other a white dwarf star. White dwarf stars are very compressed: they may pack the mass of our entire Sun into a size no bigger than the earth. Some of the outermost material from the larger star is pulled away by the white dwarf's gravity, but this material does not fall directly onto the white dwarf. Instead, it builds up in a disk called an *accretion disk*, which orbits the white dwarf. The combination of normal or giant star, white dwarf, and accretion disk can lead to some very spectacular celestial fireworks. That is why, instead of varying smoothly like most pulsating variables, eruptive variables exhibit *outbursts* of activity, usually brightening by a large amount. The changes in their light curves are usually very unpredictable, and tend to be sudden and dramatic; that is why these stars are also called "cataclysmic variables."

Novae explosions (eruptions) occur in close binary systems consisting of a white dwarf orbiting a larger and cooler star. A layer of hydrogen-rich material is slowly accreted from the cooler star onto the compact white dwarf. The accreted material provides the fuel for the nova explosion—a thermonuclear fusion reaction similar to the detonation of a hydrogen bomb. The system increases in brightness by 7 to 16 magnitudes in a matter of one to several hundred days. After the eruption, the light slowly fades back to its original brightness over several years or decades, as shown in the light curve of V1500 Cygni at right.

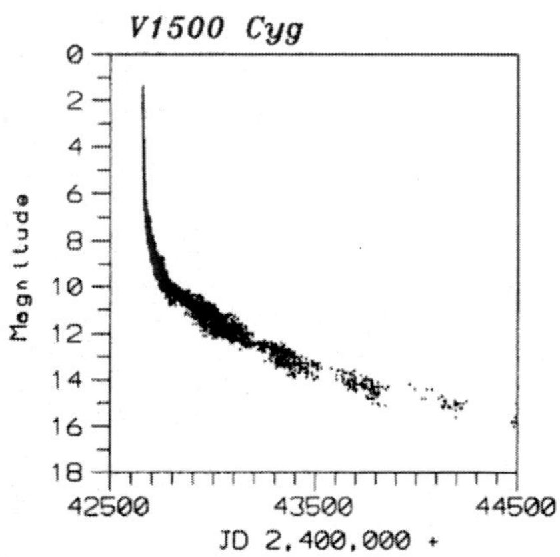

Systems which have undergone two or more nova-like eruptions during their recorded history are referred to as *recurrent novae*. Such recurrent eruptions have a slightly smaller amplitude than that which is observed in novae. An example of a recurrent nova is RS Ophiuchi.

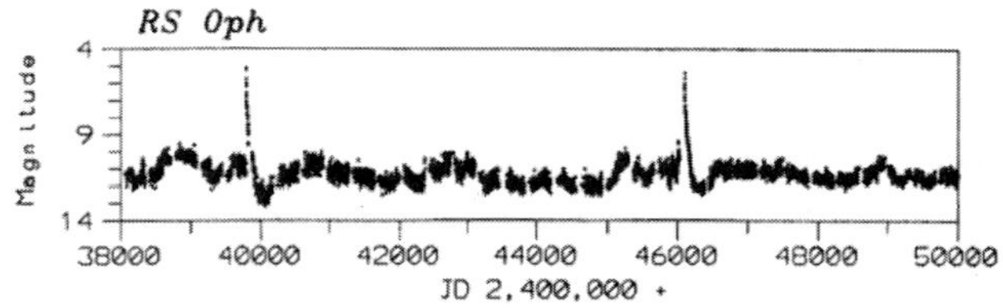

Yet another class of cataclysmic variable stars is called *dwarf novae*. These are close binary systems made up of a Sun-like star, a white dwarf, and an accretion disk surrounding the white dwarf. As matter accumulates in the accretion disk, the disk becomes unstable. Eventually, matter from the unstable disk will fall onto the white dwarf, leading to an outburst. There are three sub-classes of dwarf novae.

Dwarf novae of the *U Geminorum* sub-class (also called SS Cygni sub-class) are also in close binary systems, with orbital periods on the order of a few hours. After intervals of well-defined quiescence at minimum magnitude, they brighten suddenly within 1-2 days. Depending on the star, the eruptions occur at intervals of 30 to 500 days. The light amplitude of the outbursts ranges from two to six magnitudes with a duration of 5 to 20 days. An example is SS Cygni.

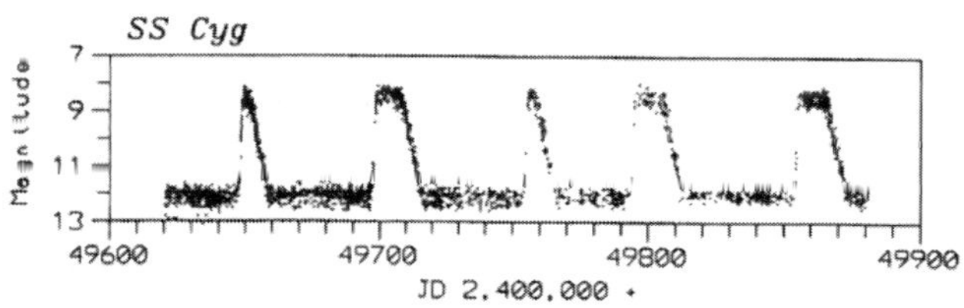

Z Camelopardalis stars, the second dwarf novae sub-class, are physically similar to U Geminorum stars. Their eruptions occur at intervals of 10 to 30 days or more; however, the quiescent level of Z Camelopardalis stars is not well-defined. In addition, the cyclic variations are interrupted by intervals of constant brightness called *standstills* (also called *stillstands*). The standstills last the equivalent of several cycles (sometimes years), with the star "stuck" at a brightness approximately a third of the way from maximum to minimum. In general, at the end of a standstill, the star continues to fade to its quiescent level, as in the light curve of the prototype star Z Camelopardalis, below.

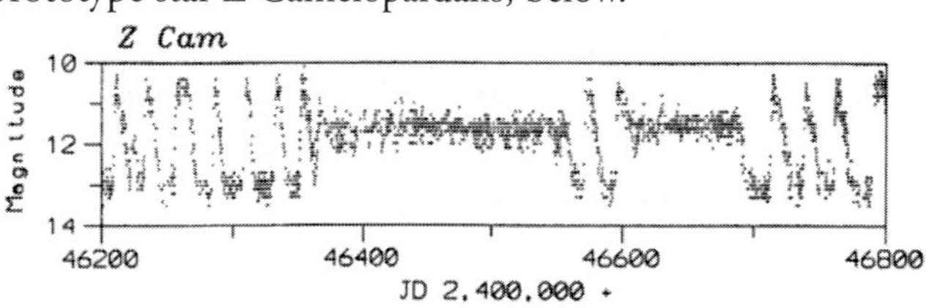

SU Ursae Majoris stars, the third dwarf novae sub-class, are also physically similar to U Geminorum stars. They have short orbital periods of less than two hours. In addition, they have two distinct kinds of outbursts: one is short, with a duration of one to several days, faint and more frequent; the other (*superoutburst*) is long, with a duration of 10 to 20 days, bright and less frequent. During superoutbursts, small-amplitude, periodic modulations (*superhumps*) appear with periods two to three percent longer than the orbital period of the system. The light curve of SU Ursae Majoris appears below; can you find the superoutburst?

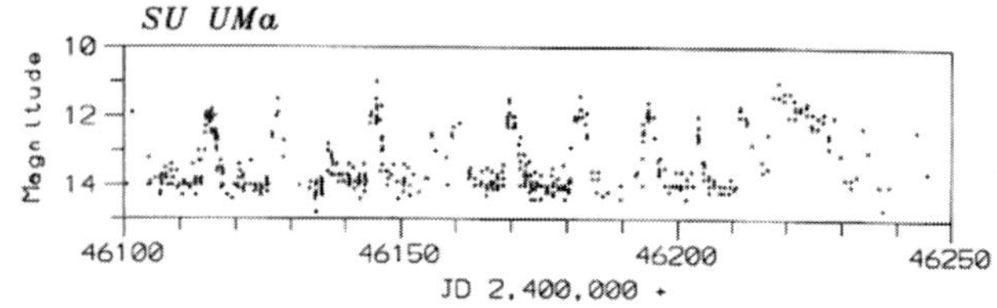

Another group of eruptive variables is the *symbiotic stars*, which display semiperiodic nova-like outbursts of up to four magnitudes. These are close binary systems with one component a red giant and the other a hot blue star. They are embedded in *nebulosity*, a region of gas and dust. An example is Z Andromedae.

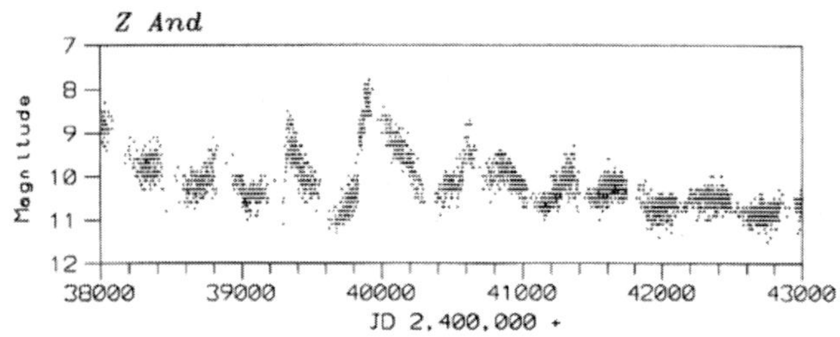

R Coronae Borealis variables are high-luminosity stars (supergiants) which go into "outburst" not by brightening, but by fading! They spend most of their time at maximum magnitude, and at irregular intervals fade by one to nine magnitudes, slowly recovering to their maximum magnitude after a few months or years. The drop in brightness is believed to be caused by the formation of soot in the star's atmosphere, which is abnormally rich in carbon. When the carbon veil is blown away, the star returns to maximum magnitude. The light curve (one-day means) of R. Coronae Borealis is shown below.

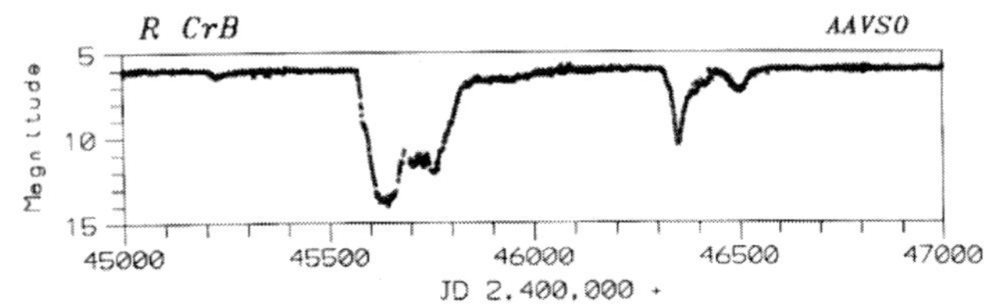

Perhaps the most spectacular of all variables is the *supernova*. A supernova is the violent end of a massive star during which the star quite literally explodes, blasting most of its material out into space in seconds. As a result, it can brighten by 20 magnitudes or more; a large supernova may briefly outshine the entire rest of the galaxy. The tremendous explosion will leave behind a fragment of very dense material at its center. This may take the form of a neutron star, whose material is compressed to the density of an atomic nucleus. Or it may be so compressed that it forms a black hole, a region where the gravity is so strong that nothing, not even light, can escape.

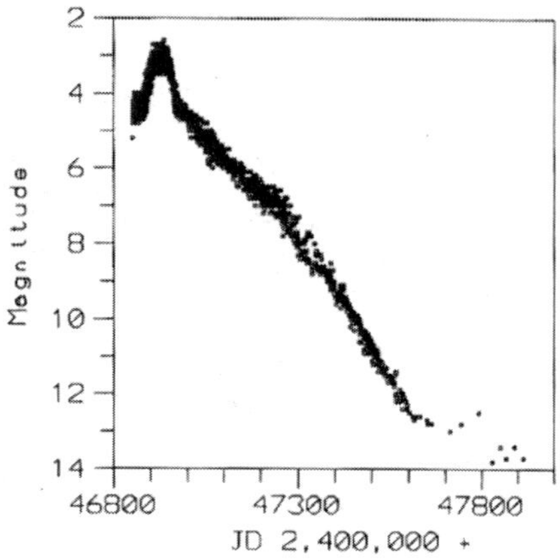

The light curve of the spectacular Supernova 1987A, which appeared in 1987 in the Large Magellanic Cloud (a dwarf galaxy near our own Milky Way), is shown above.

The material blown off in a supernova explosion is rich in complex elements such as carbon, nitrogen, oxygen, and iron. Eons later, this material can condense as part of a newly-forming star system. The enrichment of our galaxy by supernova explosions has provided the necessary complex elements to form Earth-like planets, and even life. It is no exaggeration to say that we are stardust: many atoms in our bodies, including carbon, calcium, iron, and magnesium, were synthesized within the core of a star, and probably ejected by a supernova explosion.

A supernova happens, on average, about every 300 years in a galaxy. The last supernova in our own galaxy was in the year 1054, over 900 years ago. Are we overdue?

Stars vary when they are in the stages of birth, old age, or death, all of which are periods of extreme instability in the life of a star. Many pulsating variables, for example, are near the end of their lives, and running out of their main fuel, hydrogen. Thermonuclear fusion does not simply turn on at birth and quietly turn off when the hydrogen is exhausted: fuels other than hydrogen can be burned, causing a wide variety of nuclear reactions and changes to the star's structure. The changes in the fuel economy of a star can upset the delicate balance between gravity and radiation pressure which has kept the star stable for most of its lifetime. Through the variations in their brightness, these stars "talk" to us—they try to tell us their story. Thus by studying variable stars, astronomers try to put together the puzzle of how stars evolve and die.

II. EXTRINSIC VARIABLE STARS

A. Eclipsing Binaries

Eclipsing binaries are binary star systems whose members eclipse each other, blocking one another's light, thereby causing the system to look fainter to observers on Earth.

The light curve of an eclipsing binary depends on the sizes and brightnesses of the stars, their separation from each other, and the geometry of our view from Earth. For example, the first star may be larger, and totally eclipse the second, while the second only partially eclipses the first. The time between primary eclipses is the orbital period of the binary system. Detailed analysis of the light curve shape yields much information about the size, mass, and shape of the stars, and the shape of their orbits.

Beta Persei (Algol), whose folded light curve is shown below, is an eclipsing binary that can be seen with the unaided eye. (Folded light curves, or phase diagrams, have two or more cycles superimposed on each other.

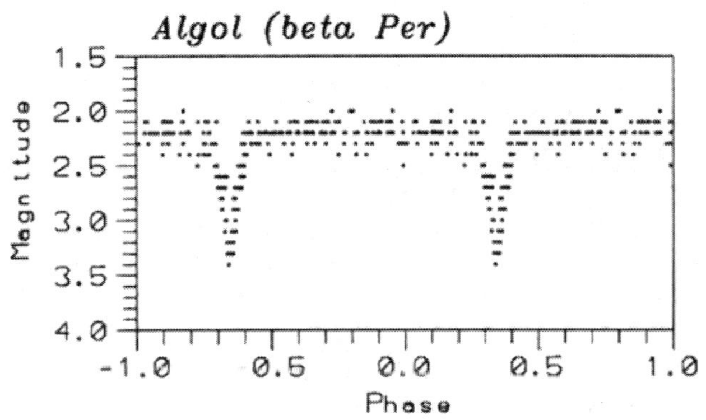

B. Rotating Variables

Rotating variables are rapidly rotating stars, often in binary systems, that undergo small-amplitude changes in light which may be due to dark or bright spots on the star's surface, similar to sunspots on our own Sun.

CHAPTER 7
HOW ARE THE STARS DISTRIBUTED?

DYNAMICS AND DISTRIBUTION

A look outside at the Milky Way will quickly tell you where most of the stars are. But our Milky Way Galaxy is far more complex than the simple disk introduced earlier. Now that we have the all-important distances, we can begin to pick the Galaxy apart to see where the different kinds of stars are located and how they are moving.

In 1944, Walter Baade, a staff astronomer at Mount Wilson Observatory, was engaged in studying the magnificent Andromeda Galaxy, M 31, with the 100-inch reflector, then the greatest telescope on Earth. Even by then the lights of Los Angeles had largely destroyed the once-superb seeing conditions, but during World War II the city was occasionally blacked out as a defense measure, rendering the skies beautiful once again. Baade had photographed the graceful spiral system through red and blue filters, and to his surprise he found that stars of different colors were differently distributed. Bright blue ones appeared sprinkled throughout the disk and its spiral arms, while red ones sparkled in the thick central region. This central portion is seen from photographs of other galaxies to be a great bulge in the disk, looking like a huge, vaguely spherical blister. Different kinds of stars occupy different positions. To discriminate between them, Baade called the blue disk stars Population I and the red ones of the bulge Population II.

We had long seen a similar effect in our own Galaxy. The hot blue O and B stars adhere tightly to the disk of the Milky Way, in Gould's Belt, named for the American astronomer B. A. Gould (1824–1896). Within the disk, they tend strongly to cling together into what are called OB associations that occupy our own spiral arms. Such a grouping can immediately be seen in the constellation Orion, where practically all of the stars are blue. Scorpius is similar. Although the red supergiants are much rarer than the B dwarfs, they lie in the galactic plane as well, and even occupy some of the associations, clearly showing a link between these two apparently disparate groups of stars. With known distances, we can get an accurate picture of the stars' three-dimensional distribution, and we find something quite astounding: The main sequence O and B stars and the supergiants are mostly found in a disk that is only 120 pc or so thick, despite being over 30,000 parsecs across.

Now look at the positions of stars elsewhere on the HR diagram. Other classes define thicker disks. The G stars—those like the Sun—and the K and M dwarfs spread over some 700 pc. The M giants are more extreme and form a disk 2000 pc thick. Moreover, they spread far beyond such confinement into a great spherical halo that completely surrounds the galactic plane and concentrates into a central bulge, giving it a reddish color. We then see M 31 reflected in ourselves and our own Populations I and II. We also see from their galactic distributions that the supergiants are not just luminous extensions of the giants, but are profoundly different kinds of stars.

The same effects are evident if we look only at velocities, those derived so arduously from measurements of radial velocity and proper motion. The O and B dwarfs and the supergiants have low velocities relative to the Earth. They, together with countless dwarfs, including the sun, are all going around the galactic center on roughly circular orbits at hundreds of kilometers per second, and there is little differential among them, a mere 15 or so km/sec. Moreover, the components of their speeds perpendicular to the galactic plane are low, only about 6 km/s, consistent with the thin disk they occupy. But down the main sequence, and in the realm of the giants, the velocity dispersions and the perpendicular speeds are some three times greater, allowing these numerous, cool, reddish stars to spread far away from the plane of the Milky Way and into the halo.

The Population I disk displays a wonderful variety of sights. Not only do we find the loose OB associations, but thousands of much more tightly bound "open" (sometimes called galactic) clusters, those like the Pleiades and the Hyades. A small telescope or even binoculars show that the Milky Way is filled with them. They are sprinkled in among gaseous nebulae, great clouds of gas that are illuminated by the hot O and B stars.

As we concentrate our vision to the central bulge of the Galaxy in Sagittarius, Ophiuchus, and Scorpius, and to the galactic halo—into the realm of Population II—another kind of object takes center stage, the globular cluster. These magnificent stellar gatherings can contain up to half a million stars, packed into a space only two dozen or so parsecs across. They are rare—only about 150 are known in the entire Galaxy—and are generally quite distant. Yet a trio, one in northern Hercules (M 13) and two in the southern hemisphere (ω Centauri and 47 Tucane), are so bright as to be visible even to the naked eye. The telescopic view of ω Cen is extraordinarily beautiful, great masses of stars con-

A schematic view of our Galaxy shows a thin disk of bright blue stars (plus a few red supergiants), decidedly clumped together into associations. The white and yellow stars spread farther from the disk, and the red giants extend out into a gigantic halo that coalesces into the central bulge of the disk. Within the disk are thousands of small open clusters of stars, but the halo is dominated by huge, rich globular clusters.

centrated so tightly at the center that it is impossible to separate them. Over half the globulars are packed within 60° of the galactic center in Sagittarius. The other half extends an amazing 50,000 pc outward.

The biggest surprise, however, is revealed by the HR diagrams of the globular clusters, which look very different from those of the open clusters or that of the general stellar field of the disk. Most of the main sequence is missing. Instead, there is a prominent horizontal branch that we have not seen before at about absolute magnitude zero that begins at the giant branch and extends leftward toward hotter stars, in many instances as far as class B. All globulars show some sort of horizontal branch, but none of the open clusters does. The two kinds of clusters are entirely distinct from one another; there is no overlap. Some open clusters are so rich that at first glance they may look like poor globulars, but their HR diagrams will quickly reveal them as imposters. They differ spectroscopically, too. A large number (though not all) of the globulars are quite deficient in metals, which relates them to the subdwarfs. The main sequence stars of the metal-poor globulars *are* subdwarfs.

Where the instability strip hits the horizontal branch, it produces a kind of metal-deficient Cepheid called an RR Lyrae star. These vary by up to a magnitude in less than a day. The gap in the horizontal branch of the globular M 5 shows the realm of these pulsators. M 5 has about 100; other globulars have none. All are about the same absolute magnitude and consequently are also excellent distance indicators.

These differences extend to the whole general collections of stars in Populations I and II. Population II by and large has a lower metal content, and its HR diagram has a horizontal branch. It also contains RR Lyrae stars and a form of Cepheid (with its own period-luminosity relation) called a W Virginis star. The lower main sequence comprises subdwarfs, and the upper main sequence is missing. Population I has no horizontal branch, but contains the brilliant massive stars instead. The reason is age and evolution—not just of the stars, but of the Galaxy itself: subjects that we will now begin to explore.

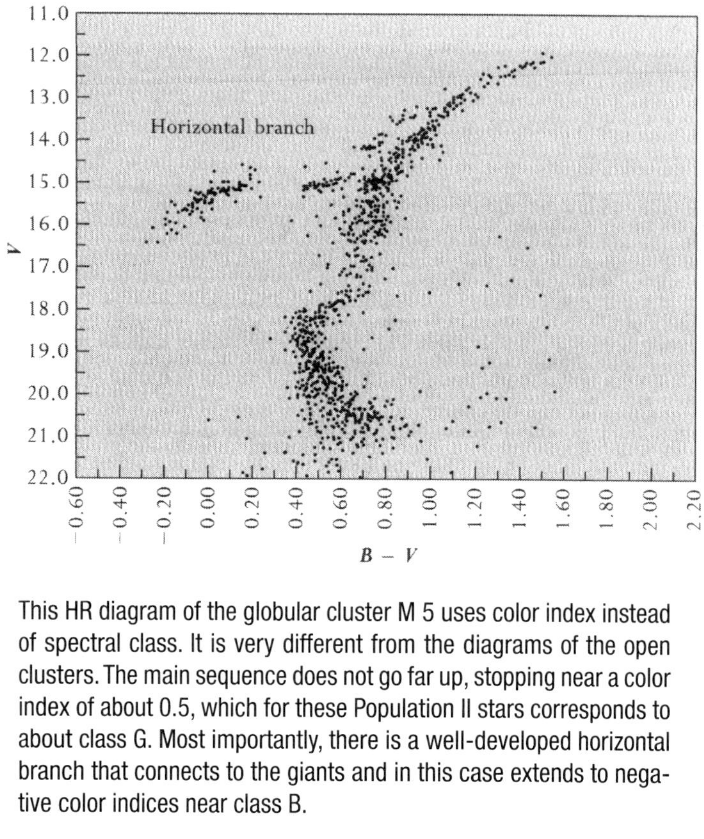

This HR diagram of the globular cluster M 5 uses color index instead of spectral class. It is very different from the diagrams of the open clusters. The main sequence does not go far up, stopping near a color index of about 0.5, which for these Population II stars corresponds to about class G. Most importantly, there is a well-developed horizontal branch that connects to the giants and in this case extends to negative color indices near class B.

CHAPTER 8
WHICH STAR IS BEST STUDIED?

THE INNER AND OUTER WORKINGS OF THE SUN

Bubbling, churning, ringing, alive with energy, the brilliant yellow Sun passes across our sky every day, eliciting little notice: We simply expect it to be there, lighting our day, warming our gardens. We are unaware of the madness going on within and without, oblivious of the intense nuclear furnace that utterly dwarfs all energy resources on Earth, the flaming gases that erupt from its surface, the mighty explosions that hurl great sheets of particles toward our tiny planet. So far, we have seen stars only from the outside, as colorful moving points in a boundless firmament. Now we close in on them, take them apart, and see how they are made, beginning with the most typical of them all, the one that belongs to us.

Look at a summary of its remarkable properties: a diameter of 1.39×10^6 km, 108 times that of Earth; a mass of 2.00×10^{30} kg, enough to balance 332,800 Earths on the scale pans, 745 times the mass of all the planets put together; a surface temperature of 5780 K; and a luminosity of 3.83×10^{26} watts, which relates to an absolute visual magnitude of 4.83, sufficiently bright to be seen with the naked eye at a distance of 20 parsecs.

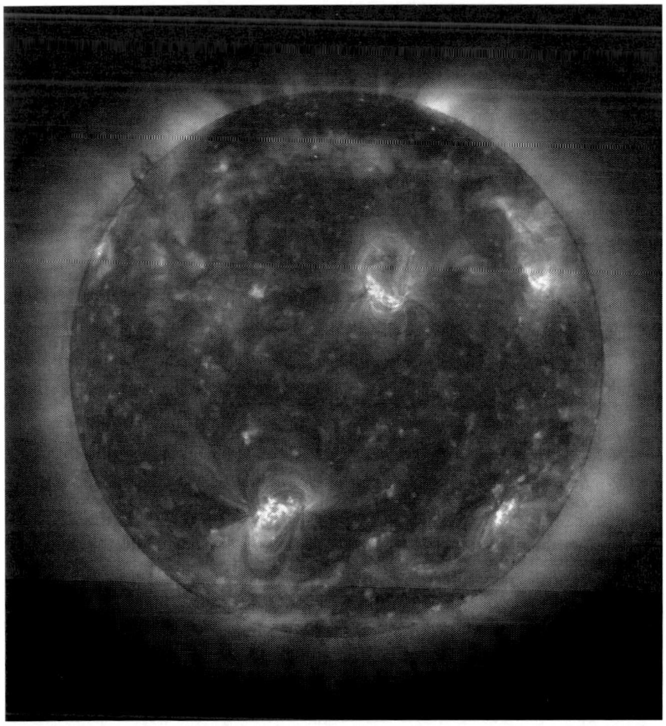

The brilliant solar corona, full of holes and loops, seen in X-radiation. [courtesy of SOHO (ESA & NASA)]

Perhaps the most astonishing feature is the Sun's durability. It is a relatively simple matter to measure the age of a planet. Most atoms have unstable isotopes that break up into lighter ones with the emission of radiation and are hence called radioactive. Beyond bismuth in the list of elements (atomic number 83), all isotopes are radioactive—some, like radium, dangerously so. All rocks contain some portion, however tiny, of these radioactive elements. Consider the nucleus of the common isotope of uranium, ^{238}U. It will eventually decay, with many intermediate products, into an atom of lead (^{206}Pb) and several helium atoms. Statistically, the decay rate is constant and is specified by the element's half-life, the time it takes for half the atoms of a given quantity to disintegrate: A kilogram of ^{238}U will reduce itself to half a kg in 4.5×10^9 years. When a rock solidifies from its molten state, it seals in the ambient elemental and isotopic abundance ratios. As time passes, the ratio of ^{206}Pb to ^{238}U will steadily increase, a sure measure of the age if we just know the half-life. The oldest rocks on Earth, as determined by several isotopic ratios, are about 3.8 billion years old. The Moon, which is smaller and solidified earlier, gives us 4.3 billion, and the meteorites, rock and iron shards of small smashed bodies from between the orbits

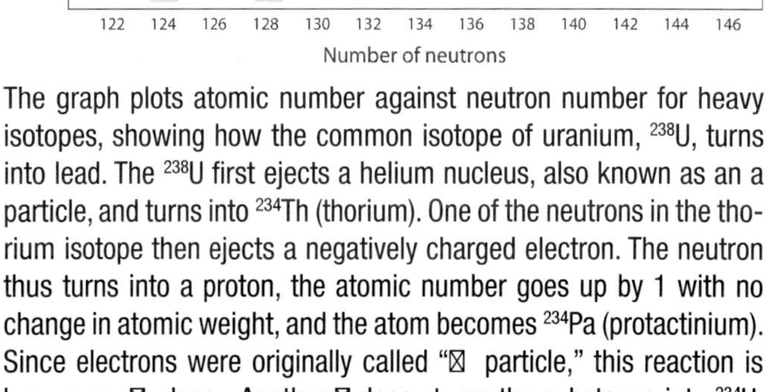

The graph plots atomic number against neutron number for heavy isotopes, showing how the common isotope of uranium, ^{238}U, turns into lead. The ^{238}U first ejects a helium nucleus, also known as an a particle, and turns into ^{234}Th (thorium). One of the neutrons in the thorium isotope then ejects a negatively charged electron. The neutron thus turns into a proton, the atomic number goes up by 1 with no change in atomic weight, and the atom becomes ^{234}Pa (protactinium). Since electrons were originally called "⊠ particle," this reaction is known as ⊠ decay. Another ⊠ decay turns the substance into ^{234}U. After successive a particle ejections and ⊠ decays, the chain arrives at lead.

of Jupiter and Mars, hold the record at 4.5 billion. The great regularity of the solar system, its planets orbiting in the solar equatorial plane in the same direction as the solar rotation, clearly implies that everything formed nearly at once and that the Sun must also be 4.5 billion years old. The fossil record shows that life formed on Earth somewhere around 3.5 billion years ago, indicating that the Sun has undergone no really gross changes since then and that it has been radiating at a rate of something close to 10^{26} watts *for all that time,* emitting enough energy to run our entire modern world for 10^{20} years.

What is it that gives the Sun and the stars such awesome power? Let us begin to probe inside to see, beginning with the solar surface layers, whose properties will, with suitable application of laboratory data and theory, tell us how the interior functions and what the source of power must be.

PICKING THE SUN APART: THE SURFACE LAYERS

No attempt should ever be made to look at the Sun without proper filtration; even a short glance can cause permanent blindness. On a misty morning, though, you can sometimes watch it as it rises red above the horizon, the brilliant harsh rays absorbed and scattered by our thick hazy atmosphere. It looks like a solid ball, the edge, or limb, sharp as a razor, an impression reinforced by telescopic examination. Yet, remarkably, the Sun is gaseous throughout, the apparently hard surface an illusion caused by gases of extremely high opacity. The radiation we see as Sunlight has worked its way out slowly from the dense interior of the Sun a fraction of a centimeter at a time as it is successively absorbed and reemitted by atoms along the way. It is finally released from a highly opaque boundary layer called the photosphere—literally, the "sphere of light"—only a few hundred kilometers thick, which at the distance of the Sun corresponds to but a fraction of a second of arc, small enough to be unresolvable. All stars have photospheres, although they are not all so thin as that of the Sun. The effective temperature of a star is always that of the photosphere, the zone from which the colorful visible starlight originates.

The spectrum gives indisputable evidence about the character of the photosphere, since the absorption lines, which give us its chemical composition, can be formed only in a gas under relatively low pressure. Even a simple white-light (unfiltered) photograph provides a clue. Remember that the Sun is not a solid disk but a gaseous sphere: As we look toward the edge, our line of sight penetrates at an angle to the shallower, cooler layers that radiate at a lower rate. Thus the apparent disk is not uniformly illuminated, but is darkened toward the limb. All stars would show such darkening if they could be seen close up. We can then use blackbody laws to show how temperature climbs with depth. The rate, or temperature gradient (at the surface about 3 K per km), provides data on which to begin to build a mathematical solar model that can be a foundation stone for models of other stars for which such detailed information is not generally available. Since starlight comes from different gaseous strata, each with its own temperature and each obscured by overlying layers in which the opacity changes with wavelength, whole stars cannot actually radiate as black-bodies—hence the need for the contrived "effective temperature."

The Sun shows a wealth of detail, including a darkened edge, myriad sunspots (some of which are comparable to the Earth in size), bright patches called faculae, and a distinctly textured, granulated surface that demonstrates active convection. [GONG/NSF/AURA/NSO]

The Sun is not at all what it seems. On the horizon it looks smooth and perfect, but through the telescope we see great texture, a million tiny bright points called granules at most only a second or two of arc across embedded in a darker matrix. These are the tops of vast convection cells several hundred kilometers wide and deep. They sweep heat upward in hot columns of expanding, rising gas that release their energy by radiation, then cool, darken, and fall. Time-lapse movies show a churning cauldron, each bright cell lasting for but a few minutes before it is replaced by another, and high-resolution spectra display absorption lines that subtly shift in wavelength in response to ever-changing velocities. These granules are grouped into much larger and deeper convection cells called supergranules. And whole areas of granules seethe up and down in large-scale oscillations, the solar surface ringing like a bell in many overtones.

At least twice a year, our orbiting Moon passes between us and the Sun, casting its shadow toward the Earth. By wonderful coincidence, the angular diameters of the Moon and Sun are almost exactly the same, so that under favorable circumstances the shadow can just barely reach us and cut off the light of the photosphere during a total eclipse. As the brilliant surface disappears, the edge of the Moon is suddenly outlined by a narrow deep red ring of light, the "sphere of color," or chromosphere. It is in turn surrounded by a beautiful white solar "crown" or corona that extends outward to many solar radii. Neither of these outer solar layers can be seen in ordinary daylight without special instrumentation because their feeble glow is overwhelmed by the light of the blue sky. The corona is a million times fainter than the Sun itself.

The corona and chromosphere defy expectations. Photospheric limb darkening shows that temperature decreases outward. We would naively anticipate that as we journey above the photosphere and approach the apparent chill of outer space the rapidly thinning gases above the solar surface would cool. And for a while that is true, the temperature of the upper photosphere dropping to 4200 K. Then, remarkably, the temperature begins to *rise* and, in the chromosphere, stabilizes at about 7000 K over most of its 6000-km thickness. Still farther outward, where the chromosphere makes the transition to the corona, the temperature shoots up to an extraordinary million degrees, reaching an average of 2 million degrees in the corona, with hot spots approaching 5 million.

Recall that flux of radiation depends on the fourth power of the temperature: Why, then, are we not fried here on Earth? In order for a gas to be a blackbody, it must be dense and opaque to its own radiation. The photosphere is so thick that it looks like a solid surface, radiation escaping only from the top. However, the low-density corona and chromosphere are transparent, so blackbody rules do not apply. There is simply not enough gas to produce significant luminosity. The temperature of the corona applies only to the energies and velocities of the atoms in the gas, and is properly called a kinetic temperature.

Because of their very low densities and low opacities, neither the chromosphere nor the corona can create absorption lines. Instead of absorbing radiation, these layers *emit* it, and when these layers are presented without the photosphere in the background, we see the photospheric absorptions replaced by bright emission lines. The spectrum of the chromosphere is dominated by hydrogen, in particular $H\alpha$ at 6563 Å, which gives this transition zone its red color. The element helium was actually discovered in the Sun by its chromospheric emissions (the photosphere is too cool to allow helium absorptions), hence its name, from the Greek *helios* for "Sun." The corona is so intensely hot that we find emissions of iron and other metals that have been ionized 13 or more times. More remarkably, the high coronal temperature is responsible for copious production of x-rays. Indeed, x-ray imaging from above the Earth's protective atmosphere reveals nothing of the photosphere, but instead shows the whole corona. It is a far more complex structure than can be seen by eye during an eclipse, with vast loops, whorls, and holes. These outer layers are clearly electromagnetic phenomena. The tenuous coronal gases are constrained by great looping magnetic fields that emerge from the photosphere and are heated by the constant release of magnetic energy.

In spite of their tumultuous natures, all these features are collectively referred to as the quiet Sun. They are quiet in contrast with a superimposed set of erratic magnetic phenomena called the active Sun, which we must examine in order to make sense of the chromosphere and corona.

SOLAR ACTIVITY

When Galileo first turned his primitive telescope to the Sun in 1610, he discovered that, contrary to the dogma of the day, the surface was not perfect, but marred with black spots. Several can be seen in the white-light photograph (on p. 107), surrounded by bright areas called faculae (from the Latin for "torches"). The spots tend strongly to come in pairs and to cluster into groups. Individuals range in size from the just barely visible to occasional monsters that are several times the size of the Earth and can be seen with the naked eye. A spot is composed of a depressed dark center, the umbra, surrounded by a grayish halo called the penumbra that slopes upward into the bright photosphere and displays light and dark radial lines. The umbra is far from black, but its temperature of about 4500 K makes it look so in contrast to its brilliant surroundings. A particular spot or spot group will last anywhere from a few days to months depending on its size. The spots are excellent indicators of solar spin, and from them we find that the Sun is a differential rotator, the equatorial regions going around in about 25 days and the period increasing to near 30 days toward the poles.

After Galileo's discovery of the spots, little mention was made of them until the early 1700s, and it was not until 1851 that Heinrich Schwabe discovered that their appearance is cyclic, with a period now known to be about 11 years. The spots of a new solar cycle appear far from the equator at about 50° north and south solar latitude. As their numbers increase, their average position creeps toward the equator. At maximum, over a hundred may appear at a time. Finally, at minimum, only a few spots are seen, about 10° or so on each side of the equator, and the next cycle simultaneously begins at high latitudes.

The solution to the mystery of the spots began to be revealed in 1908 by George E. Hale (the man after whom the 200-inch telescope is named), when he found their spectrum lines to be split by the Zeeman effect, which is seen when radiation is created in the presence of a magnetic field. The electrons of an atom act as tiny electric currents, and their orbits and the absorption lines they produce are split by the magnetism. The degree of splitting is proportional to the field strength, and the light within the components is polarized according to the field direction. By examining line-splitting and polarization across the entire Sun, solar astronomers can generate detailed magnetic maps. These show powerful fields whose strengths can range up to 3000 times that of Earth and that loop out of one of the spots of a pair and back into the other. In a way still not understood, the intense magnetism suppresses the upward convection of warm gases and acts as a giant refrigerator. Such a field, sometimes encompassing an area larger than our planet, requires an electrical current flow of over 10^{12} amperes.

The magnetism is highly organized. The Sun has a general, weak dipole magnetic field (one with opposing positive and negative poles) that, per unit area, is only about five times that of Earth. If the north rotation pole of the Sun has a positive magnetic polarity, the leader spot of the pair in that hemisphere—the one that is ahead in the direction of rotation—will always be positive, the follower negative. In the southern hemisphere, the polarities will be reversed. This pattern will hold for the entire 11-year cycle as the spots in each hemisphere approach one another at the equator. Then, for the next 11 years, in the new cycle, *all the polarities will be exchanged,* including that of the general polar magnetic field. The spot cycle is then seen to be a magnetic cycle that is actually 22 years long.

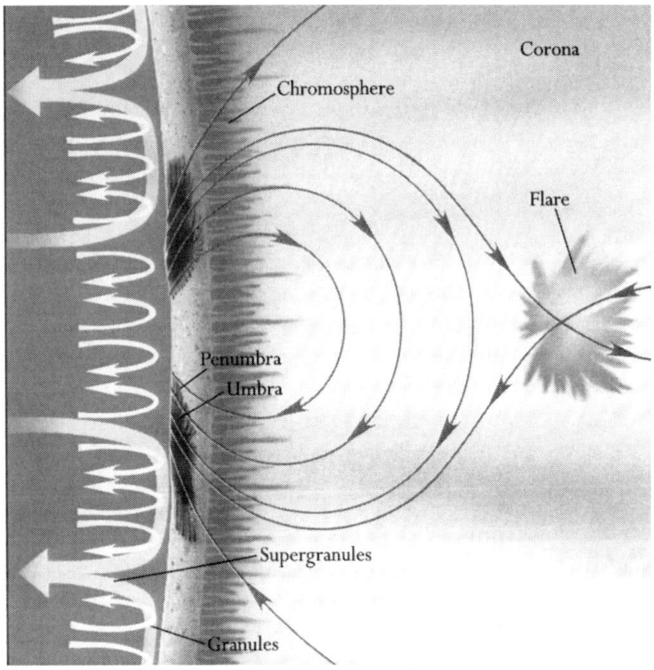

A highly schematic view of the solar surface shows a pair of depressed sunspots created by a magnetic field that loops out of the photosphere into the corona. The powerful magnetism somehow inhibits the photospheric convection. The striations in the spot penumbras [??] along the field lines. At the top of the loop, the field lines can connect to produce a flare that lights up the chromosphere below.

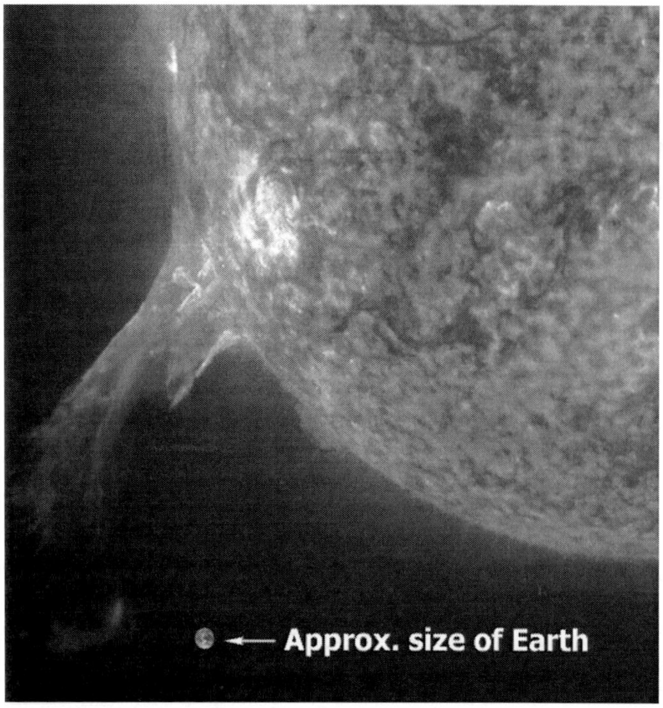

This X-ray image shows a powerful solar flare - the result of an electromagnetic explosion in the corona. The Earth is shown to scale. [courtesy of SOHO (ESA & NASA)]

The spots are but one example of solar activity, and for all the energy involved, they represent a relatively benign one. For the real action we look to the corona, as instruments aboard the orbiting *Skylab* satellite did for six months in 1973 and 1974. Several times a day near maximum in the spot cycle, Earthbound instruments detect small bright flashes of light called flares in the chromosphere above and between spot pairs. On rare occasions a flare may grow to terrestrial size and last for a [missing word] hours. Solar x-ray images show that the flares originate high in the corona above the active spotted regions in crowded, tangled magnetic fields that suddenly release their energy in a great electrical spark; in ten minutes the local temperatures can climb to an astounding *20 million degrees,* hotter than the solar center. The event can accelerate electrons to over a third the velocity of light and generate x-rays that slam into the chromosphere, producing the optically visible flares. The chromosphere then boils off some of its gas, supplying material back to the corona, in violence beyond imagination. It is probably this input of magnetic energy and the heat generated by myriad tiny flares—even when the Sun is "quiet"—that maintains the high general coronal temperature.

Nor do the wonders end there. Associated with the activity are great prominences, cool sheets of gas that condense out of the corona above the active regions. Some are quiet and hang there for days and weeks, while others rain matter down on the photosphere. The collapse of local magnetic fields can also release huge "coronal mass ejections" that violently shove blobs of coronal matter from the Sun altogether. The Sun is a true astonishment, constantly changing day to day, minute to minute.

In spite of our vast ignorance, we do know that all solar activity is ultimately caused by solar rotation. The Sun is a spinning body made of ionized—that is, electrically charged—gas. The rotation and the churning convection zone, which extends a third of the way to the center, act together, in a way that still eludes us, as a dynamo that produces the magnetic field. At the start of a new cycle, solar magnetism is ordered and relatively simple, with the field locked into the hot gases below the surface. Since the equatorial gases shear past those at higher latitudes, the field lines begin to wind up around the Sun in the direction of the rotation. They are always pushing upward through the solar surface in tiny loops that are intimately related to the granulation, and where they accumulate we see a pair of spots aligned with the equator. The greater the degree of magnetic wrapping, the more the magnetism pops through the surface and the greater the activity. It takes some 11 years for the field to become so chaotic that it breaks down and reorders itself with the opposite polarity, the solar north magnetic pole becoming south and vice versa. Then the whole process starts again, yielding the 22-year observed cycle.

It rather sounds as though we understand something of what is going on, but awareness of our ignorance deepens when we look further into the subtleties of rotation. At any one time there are latitudes on the Sun that are spinning just slightly faster or slower than average, by only a few meters per second. These zones move downward from the polar latitudes, taking 22 years to reach the equator. The spots develop about halfway down in the zones, where the velocities drop. The phenomenon seems to be related to great, longitudinal rolling tubes of gas that extend deeply into the Sun. Why, we simply do not know. But the question, and others like it, is of more than academic interest. All these events touch our Earth and our lives in many ways, some of which we are sure we have not discovered.

THE SOLAR WIND

Once every decade or two, our skies will be graced with one of nature's grandest sights, a bright comet sporting a delicate gossamer tail that can stretch from horizon to zenith. Dozens of faint ones are found every year. These fragile bodies, usually no more than a few tens of kilometers across, are made of dust set within a matrix of ices, mostly water. The comets, which may be the most primitive bodies of the solar system, reside in a pair of huge clouds far beyond the realm of the planets. A large number, however, are on highly elliptical orbits that take them into the inner part of the planetary system and toward the Sun's heat. As they pass the orbit of Jupiter they begin to warm, and the ices sublime into space in massive jets, carrying the solids along with them in a growing cloud that eventually produces a tail that can grow to tens of millions of kilometers in length.

The comet's gas tail is a creature of the solar wind, which wraps the Sun's magnetic field around it like a tight sleeve and sends it millions of kilometers away from the nucleus. (Comet LINEAR, 2004, courtesy of James McGaha, Grasslands Observatory)

It is not hard to notice that no matter what the direction of the comet's motion, the tail is always pointed away from the Sun. As early as 1950 it was noted that comet tails were likely interacting with a gas streaming from the Sun that constantly shoves the tail radially outward. Theoretical analysis a few years later demanded that the hot corona *had* to be losing matter into space. Confirmation came late in the decade with the detection by satellites of a solar wind, a flow of protons and electrons that blows vigorously outward all the way past the Earth, where some of it is waylaid by the terrestrial magnetic field, and into the far reaches of the solar system. The Sun is actually *evaporating*, albeit at a paltry rate of 10^{-14} solar masses per year (far worse things will happen to it long before 10^{14} years pass). At the Earth, we typically count about 10 or so protons and electrons per cubic centimeter at a (nonblackbody) temperature of 200,000 K howling past us at a speed of some 500–700 km/sec after a journey of about two days. In a sense we are actually *in* the outer reaches of the solar corona. The flow is complex. The corona is tightly confined by the great magnetic field loops and acts as a barrier to the streaming wind. The gases emanate primarily from the coronal holes seen in x-ray images, and the wind's strength is highly variable and dependent on solar magnetic activity.

The wind also highly distorts the Earth's magnetic field,

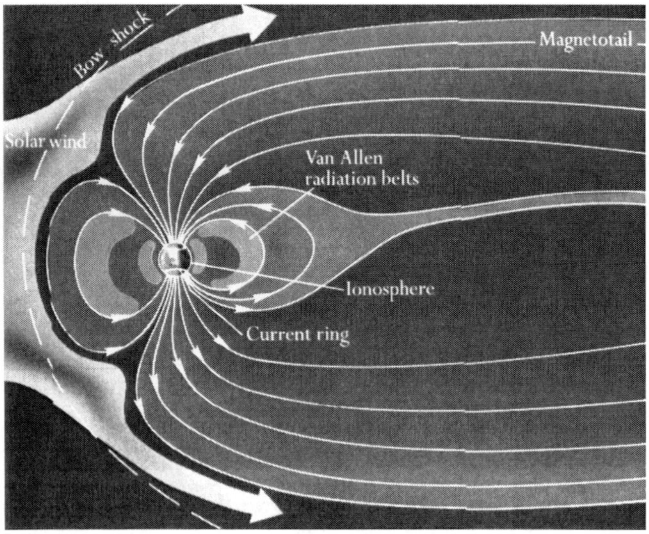

As the solar wind slams into the Earth's magnetic field it creates a powerful bow shock and a long magnetotail, as it does with Jupiter, Saturn, Uranus, and Neptune. The interaction of the solar wind and the terrestrial magnetic field produces the Van Allen radiation belts and drives massive rings of electric current around the magnetic poles.

compressing it in front in a shock wave as the prow of a speeding boat compresses the water in front of it. Within this boundary, particles become magnetically trapped in two zones called the Van Allen radiation belts. Closer to Earth, the interaction between the wind and the field also sets up intense rings of electrical current that surround the magnetic poles. To the rear, the wind sends the field into a magnetotail to a length of over 100 terrestrial radii. All the magnetic planets of the solar system are affected by the solar wind, even as far as Neptune, 30 AU from the Sun. The solar influence finally comes to a halt where it slams into the surrounding interstellar gas. We do not yet know where that barrier, the heliopause, lies. Two extraordinary spacecraft, *Voyager I* and *Voyager 2*, are outbound to find it, perhaps arriving in true interstellar "outer" space about the year 2020.

SOLAR ACTIVITY AND THE EARTH

The night is dark and clear as you stand outside to admire the stars. You look to the north and see an odd red patch in the sky, perhaps the glow from a fire over the hill. But the spot grows and begins to develop streamers that arc over the heavens, the colors changing from red to green to white. Suddenly the night is a theater stage hung with great draperies, lit with pulsing spotlights that illuminate a celestial review beyond all imagination one that exalts the sky, frames the stars, and testifies to the awesome power of nature—and the Sun. You have just seen a display of the aurora (borealis in the north, australis in the south), the northern or southern lights. An erupting coronal mass ejection sends a shock wave screaming down the flow of the solar wind. About two days later it smashes into the Earth, where it sets off a disturbance that causes magnetic field lines to reconnect close by, depositing a vast amount of new energy into the polar current rings. The result is an electrical display that produces ionization of the upper air. Recombination of the electrons and their parent molecules and atoms then lights the sky. The electrical potentials during such "magnetic storms" are so great that they can cause currents to flow in long power lines that are sufficient to blow circuit breakers and produce power blackouts—making it easier to see the aurora!

Solar—terrestrial relations extend much more deeply. Between 1645 and 1715 the cycle seems to have disappeared. During this period, called the Maunder minimum after the man who first pointed it out, the world was plunged into a "little ice age": Europe was covered with deep snows, and rivers in the southern United States froze. One data point does not make a correlation, of course, but there is additional evidence from the geologic record of radioactive ^{14}C. Its manufacture in the atmosphere by high-energy particles (cosmic rays) from deep space is suppressed by high solar activity, allowing the demonstration that low activity coincides with cooler weather. What will happen when the activity drops again, which it surely will? Do we really want to find out? And why does it happen?

Long-term trends also relate solar activity to droughts, and recently the activity has been great enough to expand the atmosphere sufficiently to bring *Skylab* hurtling to the ground and to necessitate placing the *Hubble Space Telescope* into a higher orbit. We have really only begun to explore this vast subject, so long are the intervals over which we must look and so difficult is the theory. But the effects are profound. To know the Earth and, as we will see, the stars, we must know the Sun.

SOLAR ENERGY

For all of its marvels and mysteries, what we have seen of the Sun so far is only its outer husk. All of its external properties, from its normal radiation to its activity, are driven by what goes on inside. The most important problem is the origin, of the solar luminosity. What causes the Sun—and all the other stars—to shine at so great a rate? The first really serious discussion, initiated by the physicists Lord Kelvin and Hermann von Helmholtz in the 1860s, involved the conversion of gravitational potential energy into heat. There is no question that this process must take place. A body the mass of the Sun is held together by its own gravity. The internal squeeze would elevate the outward pressure and raise the temperature to extraordinary values, as high as 40 million degrees. Any body hotter than its surroundings must radiate away part of its internal energy; when it does, it contracts, the shrinkage producing more energy, and the process continues. Only a minute rate of contraction, about 20 meters per year, is needed to produce the Sun's observed power of 4×10^{26} watts. So far so good, but a problem remained: not one of luminosity, but of *time*. Even at this rate, the Sun would have used up all its gravitational energy in only 100 million years. This is an immensely long time and satisfied the requirements of the astronomers and physicists of the late 1800s; but it was soon found to be totally out of line with the age of the Earth. Gravity alone simply cannot do the job.

Understanding had to wait for the early part of the twentieth century, which brought the discovery of the nature of the atom, the development of the theories of relativity and quantum mechanics, and the realization that the Sun and almost all stars are mostly made of hydrogen. The first major step was taken in 1926 by Sir Arthur Eddington, the English physicist and astronomer who dominated much of his science in the early part of the twentieth century. He was one of the first to embrace the young Albert Einstein's theory of relativity, which contains the famed formula for the equivalence of matter and energy, $E = mc^2$. The import of this relation is that because of the large value of the speed of light, 3×10^8 m/s, only a tiny amount of mass is needed to create a huge amount of energy, enough to light the Sun. Eddington knew that a helium atom is slightly lighter than four hydrogens—the deficit is 0.7 percent—and he surmised that if hydrogen could be turned into helium, an enormous amount of energy could be created out of the missing mass. The solar luminosity could be produced by the conversion of 6×10^{11} kg of hydrogen per second. While this number seems huge, it is only 3×10^{-19} the mass of the Sun. At that rate, if the Sun were made of pure hydrogen, it could shine for an astounding 10^{11} years, far longer than the age of the Earth.

Such numbers alone, as satisfying as they may be, are still not enough to prove that hydrogen fusion is responsible for solar power. At the time, it was not at all clear that the Sun had enough hydrogen to do the job, and even if it did, no one could say how the transformation actually took place. Cecilia Payne-Gaposchkin solved the first difficulty in the late 1920s with the aid of the new quantum view of the atom, when she showed that hydrogen dominates the stars even though its spectrum lines may be weak or absent.

The second difficulty is more profound. In any gas, the atoms are flying about with an average speed that is given by temperature. At 40 million degrees, the temperature calculated by Eddington under the gravitational theory, the hydrogen nuclei—protons—typically smash into one another at the astounding speed of 2000 km/s. Moreover, there is a great distribution of speeds about the mean, so that a tiny fraction are going 10 or more times faster. But even at the extreme, the repulsive barrier imposed by the protons' similar charges is too great to allow the tiny particles to approach one another closely, let alone stick together. How then can they fuse into a helium nucleus?

This problem was overcome in 1928 by George Gamow, Fritz Houtermans, and Robert Atkinson, working at the University of Göttingen, with the application of quantum mechanics, specifically Heisenberg's famous uncertainty principle. As we have seen, light behaves simultaneously as a wave and as a particle in the form of the massless packet of energy called a photon. Remarkably, subatomic particles have the same kind of duality (with their wavelengths dependent upon the particles' energies). As a result, one cannot simultaneously know both the exact position and velocity of a particle such as a proton or an electron. It is not a matter of imprecision in measuring devices but a fundamental limitation of nature. The uncertainty in location times the uncertainty in momentum (mass times velocity) is of the order of h, Planck's constant. The number is so tiny that the uncertainty cannot be noticed on any large scale, but it has a powerful impact on the atoms.

The probability that a particle is at any given point depends on the amplitude of its wave, which is greatest at the particle's most likely position. However, in principle the wave extends to infinity. Although the amplitude rapidly decreases away from the most favored location, a particle can actually be found anywhere along its wave. By the uncertainty principle, the position of a particle cannot be known: It may be here, it may be there. If two protons can be brought sufficiently close together in a high-temperature gas, one might suddenly find itself somewhere else along its wave and tunnel right through the electrical barrier between them. If they can attain a separation less than about 10^{-13} cm from one another, the strong force takes over, overwhelms the electrical repulsion, and makes the particles adhere to create a new atom or isotope. Tunneling is so efficient that the temperature can be considerably lower than Eddington's calculation: a "mere" 15 million degrees.

However, even the discovery of tunneling still does not reveal how the reactions that produce helium from hydrogen actually proceed and what mechanisms produce the radiation. Several other discoveries were necessary, including that of the neutron, the deuteron (^{2}H, hydrogen with a proton and a neutron as its nucleus), the positron, and the neutrino. The first three of these were isolated in 1932. The positron is a positive electron, the first type of antimatter (matter with reversed charges) ever seen. Oddly, the strangest of the four, the neutrino, was indicated a year earlier. This presumably massless particle, which travels at the speed of light, had been suggested by Wolfgang Pauli to explain a mysterious disappearance of energy in nuclear reactions. Not actually discovered until 1956, it will play important roles in our later assessments of stars.

Eight years after these elementary particles were found, Hans Bethe and Charles Critchfield had the answer to hydrogen fusion (or hydrogen "burning," as it is frequently called) in the proton-proton (p-p) reaction, which involves a stepwise process. The first step is the creation of deuteri-

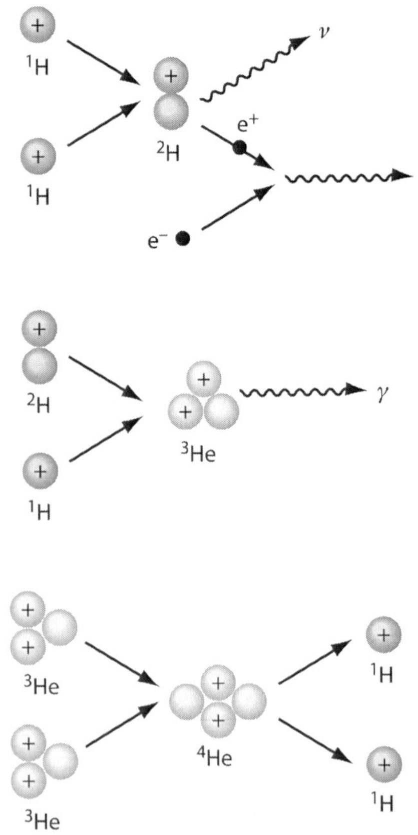

The proton-proton reaction requires the collision of two protons to form a deuteron, ^{2}H. The deuteron combines instantly with another proton to create ^{3}He, and two ^{3}He nuclei eventually combine to form ^{4}He with the ejection of two of the protons. The net result is one helium atom (four protons) and the release of energy.

um by a tunneling collision between two protons. One of the protons instantly turns into a neutron with the ejection of its positive charge in the form of a positron. The process, which is accompanied by the release of some energy in the form of a neutrino, is called inverse beta decay. It is the reverse of beta decay, in which a neutron bound within an atom ejects an electron (in the old jargon known also as a "beta particle"), thus turning itself into a proton. (Beta decays are in part responsible for the decay of uranium into lead; see the figure on p. 106). Antimatter and normal matter cannot coexist. As soon as the positron encounters a free electron, which it does almost instantly since there are so many available in the intense ionizing heat of the solar core, the two annihilate each other in a flash of energy in the form of a gamma ray.

The wonder of solar power is its stability. Once we begin a reaction in a nuclear bomb, the release of energy is almost instantaneous, but the Sun liberates its energy over billions of years. Why does it not explode too? The key to the puzzle is this first reaction. Even with tunneling, ^{1}H + ^{1}H —> ^{2}H is exceedingly slow. A typical proton waits over *10 billion years* before it attains enough energy by random collision to have sufficient speed to overcome the electrical barrier. The process works only because there are so *many* protons in the Sun that at any one time a few are actually making the transition. Once a deuteron is created, however, it reacts with blinding speed. In under a second it encounters and absorbs another proton (with the further production of a gamma ray) that does *not* turn into a neutron. The three particles, two protons and a neutron, are now helium, albeit in a light form, ^{3}He. In the final step, typically after a million years, two ^{3}He atoms collide with enough velocity to merge into normal helium, ^{4}He, in the process ejecting a pair of free protons back into the burning maelstrom. The vast bulk of solar power, about 77 percent, is produced in this way, aided by several other reactions that will have their "place in the Sun" later.

This sequence of discoveries, based on the work of a great many people, should still astonish us. By inferring processes involving atoms that we cannot see inside a body that our vision could not penetrate, astronomers and physicists had found the source of the energy that keeps us alive.

THE SOLAR INTERIOR

Let us now build a Sun. We can make a mathematical model and calculate the temperature, pressure, and density at all interior points, so that we can calculate the rate at which hydrogen is converted into helium. The construction of such a model is a demanding procedure that requires the simultaneous solution of several interlocked equations, a task that lends itself admirably to high-speed electronic computers. The basic law is that of hydrostatic equilibrium, which Eddington had used earlier to find the interior solar temperature under the assumption of pure gravitational contraction. In every layer within the Sun, the weight of the overlying gas must be equal to the outward-pushing pressure; otherwise the Sun would either expand or contract, and by simple observation it does neither. One must also adopt an equation of state, a relation among pressure, temperature, and density. Under most conditions encountered in astronomy (though not all, as we will see), the pressure is proportional to the density times temperature, a relation known as the perfect gas law ($P = n\mathrm{k}T$, where n is the number of particles per unit volume and k is Boltzmann's constant). The more atoms there are per unit volume and the faster they are going, the greater must be the outward push. Since as we proceed into the Sun, the weight on a layer imposed by gravity gets larger, the density and temperature must correspondingly increase.

Next, we must consider the rate at which energy is generated by hydrogen burning as a function of distance from the solar center; this will depend on the variation of temperature, density, and the compositional fraction of hydrogen fuel. The rate per unit volume will be greatest at the center and will drop as the temperature decreases outward. As we move away from the center and include more and more mass within successively larger shells, the total luminosity generated rapidly increases. Eventually the increase in luminosity slows and comes to a halt when, in our outbound trip, we reach a vaguely defined limit of about 7 million degrees, where the atomic velocities are just too slow to sustain reactions.

Perhaps the most difficult problem involves the way in

which energy is transported, by conduction, radiation, or convection. Conduction (direct heat transfer by contact) can immediately be dismissed, since its rate in a gas is very slow. Radiation, however, is vitally important. The rate at which energy flows by radiation depends principally upon the temperature gradient. Place yourself somewhere inside the star. Radiation is coming at you from all directions within the hot, hazy surrounding gases. Temperature always decreases outward, so that as you look in the direction of the surface (you will only see a few centimeters) you view gas that is at a slightly lower temperature than the gas you see if you look inward. Because the flux of radiation depends on temperature, there must be slightly more outbound radiation than there is inbound. As a result, there is always a net flow of energy outward. The steeper the gradient, the greater the flow.

The gradient is controlled by the gas's opacity, which is the degree to which the gas will absorb, or impede, radiation. The higher the opacity, the greater must be the gradient in order to get energy to move outward. The opacity is dependent upon the detailed chemical composition; it can be very difficult to calculate, requiring intimate knowledge of thousands of atomic absorption processes, and is thus one of the limiting factors in building stellar models. We read in the newspapers and science magazines of glamorous discoveries of galaxies, black holes, and the like, but never of the unsung heroes of a massive five-year effort to improve our knowledge of opacities, without which we cannot understand the inner workings of stars.

Eddington pointed out that a flow of radiation must produce a pressure that adds to the gas pressure and plays a role in hydrostatic equilibrium. When an atom absorbs a photon, it also absorbs the photon's momentum (even though a photon is massless, it has momentum because it carries energy). As a consequence, the atom recoils. The radiation flow, which depends on the temperature gradient and the opacity, then provides an outward push. Near the center of the Sun, the flux of radiation is so amazingly high that the radiation pressure is about 10 percent of the gas pressure and helps buoy up the Sun considerably.

All the equations that govern the Sun's physical processes must be solved at the same time. The object is to determine the internal temperature, density, pressure, interior mass, and energy-generation rate at every value of radius, so that given a solar mass and the chemical composition, the calculated solar radius, total luminosity, and effective temperature can match the observed values. We find that the temperature at the center must be about 15 million degrees and that the core of the Sun, the region in which the energy is generated, occupies about 30 percent of the solar radius and about 3 percent of the volume. Because of its astounding density, which reaches 160 grams per cubic centimeter (more than 10 times that of lead!), the core contains 40 percent of the solar mass. In spite of this huge density, because of the enormous temperature *the core is still a gas*. At the very center, about half of the hydrogen has already been fused into helium.

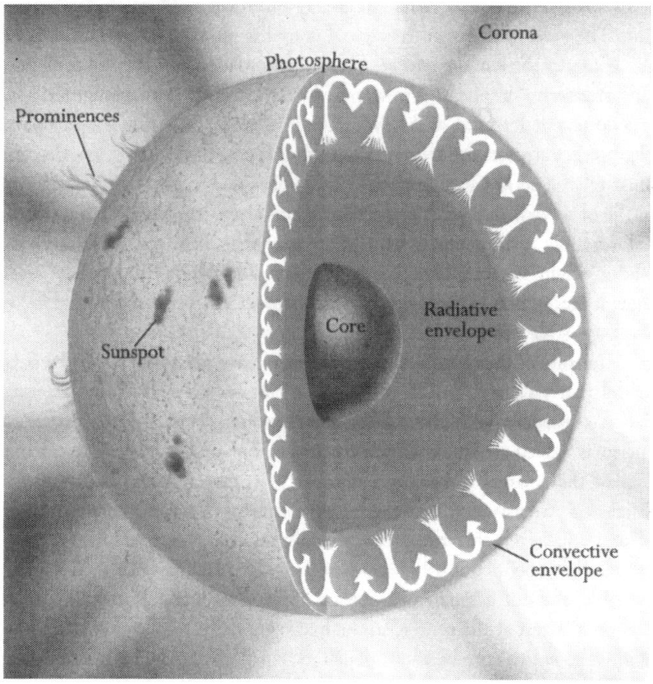

A model of the Sun shows us the size of the nuclear-burning core and its surrounding insulating nonreactive blanket, the outer part of which is in a complex state of convection.

Beyond this central zone, at a point where the temperature drops below about 7 million degrees, we enter the solar envelope, wherein the remaining 60 percent of the solar mass is spread out over about 70 percent of the radius. The envelope acts much as an insulating blanket that impedes the flow of radiation and maintains the core's pressure and temperature. It also degrades the energies of the outbound photons. Energy created in the core works its way out of the Sun by a random-walk process, aided by the very slow push of the temperature gradient. It is rather like a drunk who walks from Los Angeles to New York by bumping into lampposts and trees along the way, gently propelled by a very slight westerly wind. Because of the enormous opacity of the solar gases, even at the speed of light energy will take over a million years to arrive at the surface! Each time the energy is re-emitted, it comes *on the average* from a layer at a slightly lower temperature. Consequently, because of the blackbody law, the mean photon energy must also be lower. But since the energy cannot be destroyed, the number of photons must go up in compensation. What began as a single deadly gamma ray finally emerges from the photosphere as a thousand low-energy optical photons, the envelope thus converting the radiation of the core—the kind emitted in nuclear bomb explosions—into a spray of soft yellow light.

The complications of the opacities are bad enough, but

there is a worse problem. About 70 percent of the way from the center to the photosphere, the temperature gradient becomes such that it is more effective to transport energy not by radiation but by the third possible mode, convection. The gases then begin to roll in a complex series of layers that eventually come to the surface and are seen as granulation and supergranulation. Unfortunately, despite the power of physics and modern computers, we still do not understand well how convection works. We cannot, for example, predict a priori the sizes of the cells, the so-called mixing lengths, and must rely on ad hoc assumptions—those designed to give the "right," that is, the observed, answer. It is the convection layer, combined with differential rotation, that generates the solar magnetic field and solar activity, whose origin we can then hardly understand either. Remarkably, even though convection is confusing, we can probe the layer through analysis of the large-scale solar oscillations and observe its depth directly.

In spite of these problems, astronomers have been able to build a solar model that will both match the observed luminosity and show a lifetime that is consistent with that of the solar system, demonstrating that solar energy is indeed generated largely by the proton-proton cycle. The calculation of the solar age involves a number of considerations. Earlier in this discussion, we assumed that the entire solar mass could be converted into helium and found that the process would take 10^{11} years. However, now we see that only about 40 percent of the solar mass, or $0.4M_{\odot}$, is initially available, and our estimate of the solar life is accordingly shortened. Fusion proceeds fastest at the center, so the hydrogen there will be exhausted long before the entire core becomes depleted. Finally, we must look at how the p-p fusion rate changes as the available hydrogen diminishes. The pressure of a gas under the perfect gas law depends only upon the *number* of atoms per unit volume, not upon their kind. As four atoms of hydrogen are converted into one of helium, this number decreases, so that to maintain the internal pressure, the core has to contract under the weight of the overlying layers. As a result, the interior temperature increases, so the fusion rate and the luminosity are maintained even as the fuel supply dwindles. It is this process that gives the Sun its stability and that has allowed the time for life to develop on Earth; it is responsible for our very being.

The calculations show that the total lifetime of the fuel in the central core is about 10 billion years. When it is gone and the solar hydrogen-burning stage is effectively ended, the Sun will blossom into a brilliant red giant, a matter to be explored in the next chapter. With the age of the Earth and Sun known to be about 5 billion years, we see that the Sun is distinctly middle-aged, with a long and presumably good life ahead of it. Nature has produced a wonderfully balanced fusion machine, with the unimaginable violence in the core so tightly controlled that we will be able to go outside and enjoy a sunny day for a long time to come.

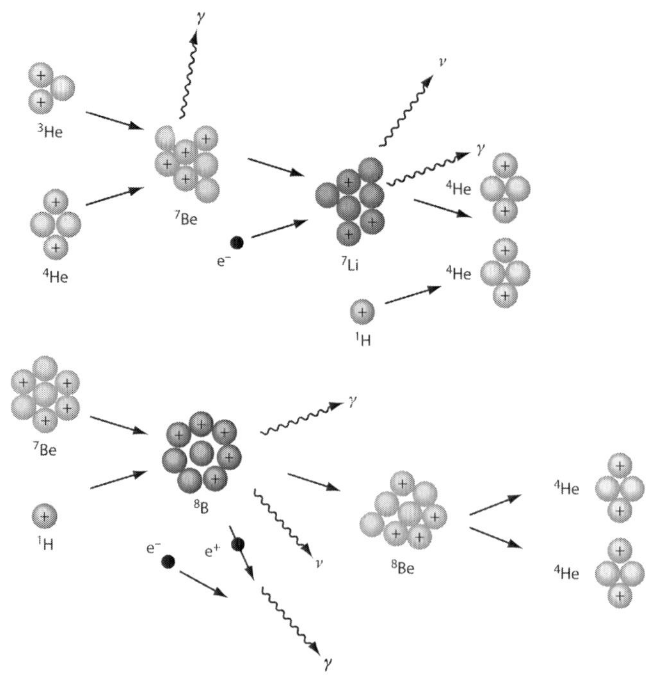

Four different reactions generate the countable neutrinos in the Homestake mine experiment, of which two are shown here. Top: ^{3}He produced in the p–p reaction can combine with ^{4}He to yield ^{7}Be and a gamma ray. The beryllium is then struck by an electron, which converts one of the protons into a neutron, creating ^{7}Li, a gamma ray, and a neutrino. Alternatively, bottom, the beryllium can capture a proton to become unstable ^{8}B, which can decay into ^{8}Be, a positron, and a very energetic neutrino. Both reactions end in the production of two helium nuclei.

NEUTRINOS

All our standard observations of the Sun are of its surface or of its surrounding halos, the chromosphere and corona. Even probing the convection layer by the analysis of solar oscillations takes us only partway into the Sun; it provides no information on the core and gives no direct proof that hydrogen fusion really runs the Sun. So far, all we have is the indirect evidence of steady solar luminosity over a very long period of time.

Neutrinos may provide such proof. These are the most noninteractive of all subatomic particles. It would typically take a *light-year of lead* to stop one. Consequently, they are effectively unimpeded by the overlying solar layers and come flying out immediately upon creation. A neutrino telescope would allow us to look directly into the solar heart. But if matter is transparent to the little pieces of energy, how do we detect them?

Matter is not *perfectly* transparent. Neutrinos can induce a variety of nuclear reactions, and even though these are all intrinsically very unlikely, there are so many neutrinos emanating from the Sun (the particle flux at the Earth is an astounding 10^{10} per square centimeter) that we should occasionally see one. The first detection experiment has been running for nearly 20 years, deep in the Homestake gold mine in South Dakota, where Ray Davis set up a tank containing 100,000 gallons of the cleaning fluid perchloroethylene (C_2Cl_4). About a quarter of the chlorine is the ^{37}Cl isotope, which has a particular propensity to capture a neutrino and turn into radioactive ^{37}Ar. Radioactive atoms are detectable when they decay. A count of the number of converted atoms along with the known capture probability and decay rate tells how many neutrinos are emerging from the Sun.

The experiment unfortunately does not count the neutrinos that are produced by the p—p process itself, which have energies that are too low to affect the $^{37}Cl \longrightarrow {}^{37}Ar$ reaction, but it does count the much more energetic neutrinos generated by four secondary chains. Still, the solar model tells us with considerable accuracy how many of these there are and how they relate to the more prolific main chain. What has been found over two decades of work is quite disconcerting: The experiment has counted only somewhat more than a quarter of the expected number. There are a variety of possible explanations. The experiment may be flawed, the estimate of neutrino capture rates may be wrong, the Sun's core temperature or chemical composition may be different from that expected (and thus the solar model wrong)—or we do not understand neutrinos.

The question is so important, both in terms of the Sun and the basic model of nuclear physics, that several other experiments have been mounted or planned. The next after Homestake, built in 1987 in the Kamioka zinc mine in Japan, uses a large pool of ordinary water. Neutrinos striking the atomic electrons produce a brief flash of radiation that is detected by photoelectric cells lining the walls of the tank. Unfortunately, this detector has an even larger energy threshold than does the Homestake experiment and can capture neutrinos from only two of the side chains. But the results are consistent, in that it finds only about half those expected. It is also directional, and clearly demonstrates that the neutrinos it does detect are in fact coming from the Sun. So the problem does not seem to be with the experiments themselves. Further evidence for what is happening derives from long-term plots of the Homestake data. The neutrino flux is apparently anticorrelated with the solar cycle! The Sun may indeed be responsible. But how?

Two experiments, a joint United States and Russian venture in Russia (*SAGE*) and a European experiment in Italy (*Gallex*), use the rare element gallium, and count neutrinos when they convert ^{71}Ga into radioactive germanium, ^{71}Ge. These instruments can detect the neutrinos of the main p—p reaction, and together find about half the expected number. The solar model should not be that wrong. Consequently, we look to the neutrino itself.

Complexity is piled on complexity. We have seen that there are five nuclear reaction chains that generate solar energy (and there is actually one more). Moreover, there is not one kind of neutrino but *three*, and each of them has an antineutrino version that spins in the reverse direction. The kind of neutrino produced in solar reactions is associated with the electron. But there are also heavier versions of the electron, the muon and the tau particle, that have their respective neutrinos, which are not detectable by any of the experiments (neither are their antineutrinos). If the electron neutrinos generated by the p—p cycle have a small amount of mass, then ambient solar electrons can convert them into uncountable muon and tau neutrinos. The more energetic neutrinos of the side chains are less subject to this effect, and consequently we can detect some of them, which would explain why Homestake finds a quarter what is expected and Kamioka a half. Moreover, if the neutrinos have a small amount of magnetism, the solar magnetic field might convert them into undetectable antineutrinos, which would explain the relation to the solar cycle.

When these experiments, and others soon to be developed, have been run for many years, we should be able to sort out all the effects and test the theories of the solar interior. Perhaps more importantly, astronomy and the Sun have given new insights into the nature of the atom and the subatomic particles associated with it, in a wonderful symbiosis between sciences.

CHAPTER 9
HOW DO STARS LIVE?

ON THE MAIN SEQUENCE

Now we look out into space to see myriads of other stars. Since the Sun is an archetypal dwarf, the main sequence ought to be a hydrogen-burning sequence, and if the theories that are applied to the Sun are correct, they should fit the other stars as well. Over 50 years ago Henry Norris Russell and H. Vogt demonstrated that mass and chemical composition were all that mattered in determining stellar structure and hence luminosity. The chemical compositions of stars are similar. The Russell-Vogt theorem is, then, the theoretical counterpart of the empirical mass-luminosity relation. As we proceed down the main sequence from the Sun, the mass and the gravitational pressures decline, and so do the internal temperature and the luminosity. The p-p reaction proceeds ever more slowly. At 0.08 $M_{\odot}$, core temperatures are down to 5 million degrees; any cooler and fusion can no longer be sustained. We have reached the bottom, the lower limit to stellar masses. Our own planet Jupiter, although made largely of hydrogen, is too small to be a star by a factor of 50. However, when we go upward in mass from the Sun and the interior temperature and luminosity increase, something quite different begins to take place. Just after the p-p reaction had been announced, Bethe realized that if the temperature is sufficiently high, another reaction is possible, one that uses carbon as a nuclear catalyst. In normal galactic stars, about one atom of every 2000 is carbon. At lower temperatures, below about [??] million degrees, the atomic velocities are too low for carbon to be penetrated easily by a proton. However, the probability rises rapidly as temperature climbs, so that above this limit a proton is *more* likely to combine with carbon than it is with another proton, initiating the carbon cycle.

When a ^{12}C nucleus merges with a proton (after a typical wait of 13 million years), it has both its atomic weight and number increased by one and becomes ^{13}N. This isotope of nitrogen is radioactively unstable. After about seven minutes one of the protons suffers an inverse beta decay, converting itself into a neutron with the ejection of a positron and a neutrino. The mass stays the same, but the atomic number goes down by one, and the ^{13}N becomes ^{13}C, which is as stable as ^{12}C. The positron and an electron meet, and a gamma ray is again created. After a wait of a few million years, the ^{13}C will grab another proton with the production of another gamma ray, making stable ^{14}N, the stuff we breathe. Yet another proton smashes into the ^{14}N, producing radioactive ^{15}O, which after a minute or so decays into stable ^{15}N. So far, three protons have been consumed. A fourth makes its entry by tunneling into the ^{15}N. Now, instead of the creation of stable ^{16}O, nature offers a surprise. The reaction causes the ejection of a helium nucleus (in the old jargon, an alpha particle), which re-creates the original ^{12}C. So, four atoms of H produce one of He, and the carbon remains unchanged. The Sun is hot enough to produce about 10 percent of its energy by the carbon cycle, and the process must be taken into account in the solar model. The carbon and p-p cycles generate roughly the same amount of energy at about spectral class F0 on the main sequence. Above that, the carbon reactions dominate, and the p-p chain is an inconsequential contributor among the B and O stars.

The basic model of the Sun and stars now also tells us the *upper* limiting mass on the main sequence. As mass and luminosity rise, so does radiation pressure. At 120 or so solar masses, the outward shove of a star becomes so great that gravity can no longer hold it together. Any body that were to form larger than that limit would immediately be dis-

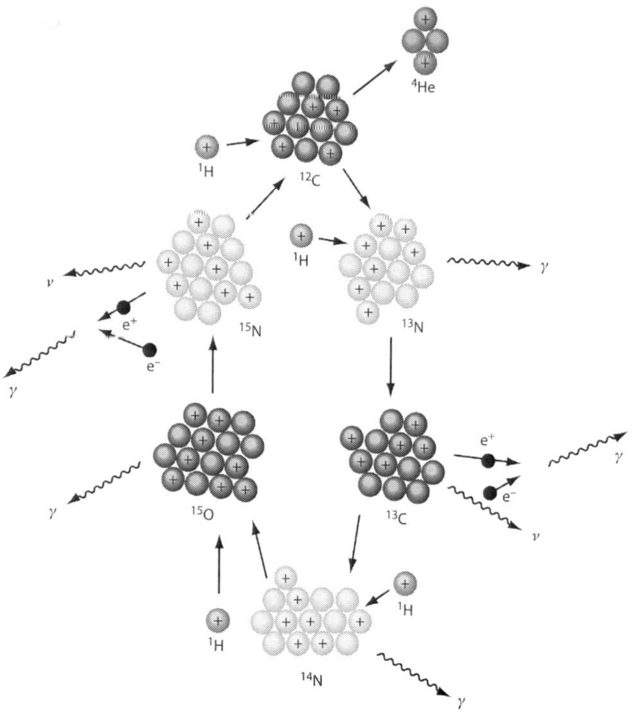

In the carbon cycle, sometimes called the carbon-nitrogen or CN cycle, carbon acts as a sort of catalyst that allows four hydrogen atoms to combine into helium. At all but the last step a gamma ray is produced either directly or by the creation of a short-lived positron.

rupted and literally cut down to size. Even at lower masses, the luminous dwarfs have enormously strong winds, as their tremendous radiation shovels quantities of mass into space in stellar winds up to nine orders of magnitude greater than the Sun's.

Theoretical models show that stars of different masses undergo great changes as we proceed up and down the main sequence. Stars more luminous than the Sun have thinner convection zones in their envelopes, and from class A upward, envelope convection disappears altogether, followed by the advent of core convection. Downward through classes G and into K and M the convection thickens. At the bottom of the main sequence it is possible—though not proved—that the convection takes over the star completely, circulating gases from all over into the hydrogen-burning core.

Since convection helps generate the magnetic fields of stars, we would expect these changes to have great effects on what we observe, and we are not disappointed.

BROWN DWARFS

The mass range of stars goes from 120 M_o to 0.08 $M_\odot$. The upper limit appears absolute; the lower limit does not. Below this value a star cannot sustain itself by thermonuclear reactions, but that does not mean that such submassive objects—called brown dwarfs—cannot form. Their existence is potentially important to theories of star formation and to the assessment of the Galaxy's mass.

Even though they do not stably burn hydrogen, the brown dwarfs should be visible, radiating dimly first as they burn their small original supply of deuterium to helium and then as they convert their gravitational energy into heat. Since the number of stars increases precipitately down the main sequence, we might think that there should be huge numbers of brown dwarfs, but the search has been daunting. After years of looking, only a small number of candidates have been uncovered. One was found by its very low surface termperature of about 1000 K. A handful are suspected from the existence of atmospheric lithium, an element easily destroyed at normal stellar temperatures. And about a dozen are inferred from the gravitational effects they have on stellar companions, from which we estimate mass, the candidates themselves unseen. In spite of their bulk popularity, the number of M dwarfs seems to drop off cooler than M5 or so, the rare brown dwarfs apparently continuing the trend.

STELLAR ACTIVITY

There is no reason to believe that the Sun is unique. Other stars should have the same sorts of bizarre characteristics: the granulation, spots, magnetic fields, activity, flares, and the like. There are probably few astronomers alive who have not dreamed of actually seeing the disks and surface features of other stars, first just for the wonder of it, and second, because the exterior properties provide powerful constraints on the internal structures. But the inscrutable stars give us little satisfaction. Even the angularly largest of them is only a few hundredths of a second of arc across, and since our turbulent atmosphere gives at best a resolution of about a half a second, the disks remain hidden from view.

Nevertheless, astronomers have learned a great deal by taking more indirect routes to knowledge. We have been able to detect and study stellar chromospheres, and coronae, have found starspots, and have even seen active stellar phenomena that mimic—and sometimes violently exceed— those of the Sun. The solar chromosphere and corona are overwhelmed by the brilliant photosphere, and in integrated light (radiation summed over the whole surface) they would be very hard to see. But careful spectrographic observation, with which we can increase the contrast, allows the detection of stellar chromospheres, which appear from class F down to and including M. In these stars, the photospheric H and K absorptions produced by ionized calcium are enormously powerful. At the centers of these lines nearly all the photospheric radiation is blocked, and the stars become extremely dim. However, the chromospheres, consisting of as under low pressure, produce *emission* lines at H and K, which can then easily be seen against the darkened background. As we proceed down the main sequence from the Sun, the ionized calcium emissions strengthen relative to the integrated stellar luminosity, showing that the chromospheres become thicker, apparently in response to the growing magnetic fields generated by the deeper and deeper convection zones.

We know that the magnetic dynamo is produced by a combination of convection and *rotation*. Remarkably, we can determine the spin rates of these tiny specks of light with considerable accuracy. An absorption line is generated from all points on the surface of a star. If a star is in rotation (and if we are not looking right down on the pole), some of it must be coming at us (relative to the overall radial velocity) and some must be going away.

Therefore part of the absorption is Doppler-shifted to shorter wavelengths relative to the average and part to longer, resulting in the broadening of the line. The profile of the absorption—its functional form relative to wavelength—takes on a very characteristic shape, from which we can deduce rotation velocity. Since the rotational axis of a star is rarely placed exactly perpendicular to the line of sight, what we actually derive is the projection of the velocity against the plane of the sky. By assuming that stellar axes are randomly oriented, we can find statistical averages for specific spectral classes. Activity in the stellar chromospheres also provides information. We can actually see the H and K line emissions change periodically as active regions move in and out of sight, allowing the rotation period to be determined directly.

We find that there is a strong correlation between main sequence spectral class and rotation speed. The G, K, and M stars all turn slowly, the speed dropping from about 5 km/s to 1 km/s or so near the end of the main sequence. But as we enter class F from below, we encounter what is called the rotation break, and suddenly the spin rates jump to 100 km/s and up to over 200 km/s in class B. The break occurs on the main sequence about where envelope convection ceases. Furthermore, rotation is closely associated with stellar age. As we will see in the next chapter, the theory of stellar evolution allows us to tell the ages of star clusters. Stars in older open clusters are always spinning more slowly. The rotation combined with the convection zone produces the magnetic fields and stellar activity, *and then the magnetic field applies a brake that actually slows the star*. As mass is lost from a star, as it is in the solar wind, it carries the magnetic field lines with it. We know this to be true from observation of the Sun. But the field lines are still anchored to the star, and slowly, over billions of years, they act like ropes that drag on the rotating ball of gas, gradually bringing it to a near halt. Nevertheless, activity stays strong in the cooler stars, showing that as long as there is a deep convection zone, not much rotation is needed to produce a significant magnetic field.

It is unfortunate that we cannot all spend a few hours at the telescope watching a few faint red dwarfs. These dim stars are generally so unassuming that we pay them little attention. A substantial number, however, exhibit flares that put those of the Sun to shame. It is rare to see a solar flare that is visible through the telescope in white light. We normally have to look in chromospheric emissions like the H and K lines or Hα to see them. Not only are flares visible from M dwarfs, they can increase the brightness of the star by one or several magnitudes. Imagine the Sun suddenly becoming two or four times brighter for a few minutes! Like the solar variety, these flares are also visible in the radio, ultraviolet, and X-ray parts of the spectrum. They are so powerful that they may involve magnetic release over the whole star. Flare activity is clearly related to rotation and age, as young clusters contain many more active flare stars than do older ones. It is a fascinating exercise to consider what our own Sun may have been like billions of years ago when life was first developing and when it was young and spinning more rapidly. What effect could enhanced activity have had on the evolution of living things?

Where there are flares there ought also to be coronae. We would not necessarily expect a stellar corona to be like that of the Sun, but it should certainly be a fascinating variation on the theme. We know that coronae exist around other stars from simple quiescent observation of their X-rays by Earth-orbiting satellites.

Activity seems to go berserk in certain kinds of binary stars. The RS Canum Venaticorum stars are close pairs of G dwarfs or subgiants, tidally locked onto one another in much the way that the Moon keeps one face pointed toward the Earth. Quick orbital revolution has produced rapid axial spins, which together with the stars' deep convection zones produce awesome activity. They are so covered with starspots (it is the only explanation) that we can actually see the whole star vary significantly as active regions turn in and out of sight. They also display prominent coronae through X-ray radiation and occasionally pop enormous flares, solar-type activity gone mad, clearly demonstrating the role of rotation in generating the magnetic dynamo.

On a quieter level, if we look closely enough at solar-type stars, we see evidence for entire stellar magnetic cycles that are quite reminiscent of our own of 22 years. About two-thirds of the stars examined display regular variations with multiyear periods. It is tempting to relate the third that show nothing at all with the times when the solar cycle shuts down, as it did during the Maunder minimum.

Such long-term studies are some of the most difficult in astronomy. Telescope time is precious, and the pressure to publish results is great. Anyone examining the problem of stellar cycles must have enormous patience to follow a star over intervals of decades—yet the payoff is huge. Eventually, we will be able to correlate stellar activity with other observational parameters, giving us a much better understanding of the phenomenon. The connection between other stars and the Sun may then allow us to know our own star better, the one that we depend on for our very lives.

UP THE MAIN SEQUENCE

Above class F the convection disappears, and we might think that stellar surfaces would be more placid, if not even boring. Again, we are in for a delightful surprise. Zeeman splitting of spectral lines shows that A dwarfs can exhibit magnetic fields far more powerful than that found in the Sun or even in the biggest sunspots. The record, held by dim ninth-magnitude HD 215441, is a general field some 110,000 times the strength of Earth's! Regular variation in the field strengths of these Ap (A-peculiar) stars clearly shows that the magnetic axis is tilted relative to the rotational axis, an effect commonly seen in the magnetic fields of the planets. There is a dynamo of some sort at work, but as yet we do not know what it is.

These stars also display weird abundance anomalies, showing overly strong lines of silicon, chromium, strontium, and even the rare earths such as europium. The line strengths are variable along with the magnetism. Apparently, there are patches vaguely reminiscent of solar active regions that are highly enriched in particular elements and that rotate in and out of sight The best explanation is that some kind of

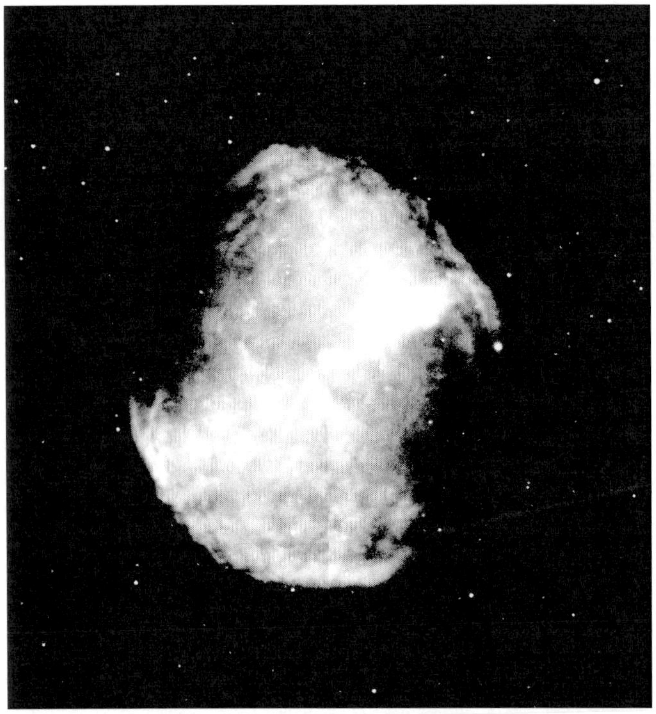

The Dumbbell Nebula. Photograph by Tom Fleming.

diffusion is at work that can cause some elements to settle downward out of the atmosphere and others to rise. The whole matter is a considerable mystery.

Some type of diffusion may also be at work in the metallic line or Am stars. These show no significant magnetic fields, are underabundant in such elements as calcium, and strongly overabundant in heavy metals like strontium, yttrium, and barium. There is also good evidence that they are in binary systems, scoring one more effect for stellar duplicity.

It seems the only time we can say that we "understand" the stars is if we do not look too closely. The wealth of detail they exhibit is overwhelming, and the closer we look the more we see to amaze us.

COMING OF AGE: THE DEVELOPMENT AND AGING OF THE SUN AND THE LOWER-MASS STARS

The old astronomers spoke of the "fixed-stars," which, unlike the wandering planets, apparently never change their course across the celestial sphere. Watch all your life, you will see no deviation. Our constellation patterns, seemingly unaltered over thousands of years, proclaim the constancy of the heavens. However, we saw that the stars *do* move and that, given long enough, the mythological beasts and beings that populate the sky will all disapper. Now we address a different kind of change. Stars live long lives, the Sun 10 billion years, but none lasts forever. Their energy supply is limited, and when it runs out they— he forms of the friendly constellations—will alson die.

Something so large and powerful as a star is not going to go quietly. Stars do not just suddenly wink out but go through death throes that can produce massive swelling, wild variations, devastating explosions, literal punctures in space, and some of the most beautiful sights of all nature. This activity, neither subtle nor hidden from view, is in the sky for all to see, for it is the brilliant giants and supergiants that are departing.

The death act is quick in a stellar context but, with some singular exceptions, it is creepingly slow in human terms. Astronomers have to look at the entire array of celestial sights and string them together with the aid of theory to see how the quiet dwarfs of the main sequence transform themselves into the wonders of the celestial sphere.

STELLAR LIFETIMES AND THE REALMS OF THE MAIN SEQUENCE

The main sequence is the broad band of stellar youth and middle age where stars lounge while burning their initial supplies of hydrogen into helium. Earlier, we saw that stellar temperatures and luminosities are controlled bv mass. We will now see that mass commands even how long the stars live and how they will mature and finally expire.

Given two piles of wood and a match, you would naturally expect the bigger pile to burn the longer. Apply this reasoning naively to stars and you will be utterly wrong. More massive stars clearly have larger energy supplies; whether the fuel is gravity or nuclear power is at the moment irrelevant. However, in 1926, Eddington showed that in order for a star to maintain equilibrium, the luminosity must be roughly proportional to M^3, in rough agreement with the mass-luminosity relation derived from binary stars. This relationship allows an examination of stellar lifetimes. Look at a simple argument that ignores all the subtleties. Crudely, the lifetime of a star, τ, must be proportional (α) to its energy supply divided by the rate at which it is used. Since energy supply α mass and the rate α luminosity, $\tau \; \alpha \; M/L$. But since $L \; \alpha \; M^{3.5}$ (the average relation from binaries), $\tau \; \alpha \; M/M^{3.5}$, or $\tau \; \alpha \; 1/M^{2.5}$. On the basis of this simple relation, Vega, a typical A dwarf 2.5 times the solar mass, will live $(1/2.5)^{2.5} = 1/10$ as long as the Sun, and a B dwarf of 10 solar masses will expire in only 1/300 of that time. At a tenth of a solar mass, down near the bottom of the main sequence among the dim red M dwarfs, stars use their meager fuel supply so parsimoniously that they will endure 300 times longer than the Sun. More precise values can be calculated by factoring in the actual value of the exponent In the mass-luminosity relation, which changes some with the mass.

Absolute stellar ages, as we saw with the Sun, depend on the nature of the fuel. Nuclear fusion allows the Sun to live on the main sequence for 10 billion years. If we combine this number with the above proportionalities (using the variable exponent in the mass-luminosity relation), we find that a rare O dwarf at the top of the main sequence will survive for a mere 3 or 4 million years—here today, gone tomorrow. *Some stars have come and gone since our human race has walked the Earth.* Down in the depths of the main sequence, a cool M dwarf will last for an astounding *3 trillion years.*

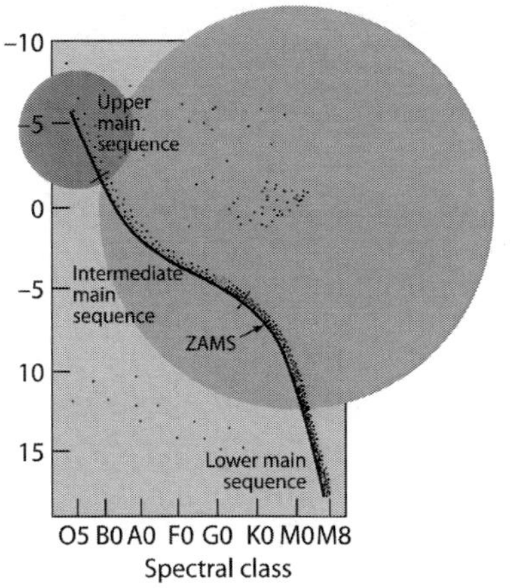

Stars on the lower main sequence—61 Cyg, Barnard's Star, Proxima Centauri—have never had time to evolve. An intermediate-mass star like the Sun or the A dwarfs Sirius and Vega will, in a Jew billion years, evolve first into a giant, then a white dwarf. The young massive stars of the upper main sequence, which mark the Galaxy's spiral arms, are supernova candidates. The left-hand edge of the distribution is the zero-age main sequence, or ZAMS, where stars begin fusing a fresh supply of hydrogen. The density of points is schematically proportional to the number of stars of each kind per unit volume. The giant branch begins at class G on the main sequence and goes up to the right. Typical main sequence and giant diameters are superimposed.

The main sequence ends at 0.08 $M_\odot$ (class M8), below which we find the fleeting brown dwarfs, "nonstars" that are only now being discovered. Below 0.8 $M_\odot$ (class G8), lifetimes are longer than the age of the Galaxy, which we believe to be about 13 billion years (how this number is known is a principal subject of this chapter). None of the stars between these limits, those within the lower main sequence, has yet had time to perish—every one that ever was *is still there.* They consequently play no further role in any meaningful discussion of evolution. Remarkably, because the number of stars rapidly increases down the main sequence, this realm includes some 90 percent of all dwarfs.

The intermediate main sequence, though far less populated, is much more interesting. Its range is between 0.8 and about 8 $M_\odot$ (spectral class B3), and thus includes our Sun. The upper limit is poorly known: It might be 6 $M_\odot$(B5) or it might be 10 $M_\odot$ (B2). This domain includes some 10 percent of the dwarfs and is the origin of all the giants like Arcturus, Aldebaran, and Mira, and of the white dwarfs such as 40 Eridani B and Sirius B. These main sequence stars, though they undergo some spectacular changes, die with quiet dignity.

Upper main sequence stars, those between about 8 $M_\odot$ and the maximum allowable value of about 120 $M_\odot$, are rare, constituting less than 1 percent of the total population. These are ephemeral beings, lasting at most a few tens of millions of years. Since they are the sources of the supergiants, their enormous luminosities and evolutionary effects make them famous out of all proportion to their numbers. Apart from the difference in lifetimes, the great distinction between these stars and those of the intermediate main sequence is that they are too heavy to produce white dwarfs. Some may expire quietly, but the majority (perhaps all) do not. Instead, they are wiped out in sudden blasts called supernovae, in the process creating some of nature's most bizarre objects. We will wander through the wonders of the upper domain. Though scarce, the great beasts who live there have a profound impact on the Galaxy and on the intermediate mass stars. Here we will look at the intermediate zone—at stars like the Sun—to learn our own fate.

EVOLUTION ON THE MAIN SEQUENCE

All of stellar evolution can be understood from one simple rule: A star is always under the influence of its own gravity and is trying to make itself as small as possible. For many reasons it will be prevented from doing so for various periods of time. All the stages along the way can be identified as either pauses in the shrinkage or as the acts of the shrinkage itself.

A star begins by collapsing out of the turbulent mixture of interstellar gas and dust. The first stop in its life history is by far the longest, the main sequence. Brand new stars lie along the left-hand edge of the band, on a locus called the zero-age main sequence, or ZAMS. Here the temperatures in the cores of the contracting bodies become high enough to sustain thermonuclear fusion. The burning of hydrogen into helium creates so much internal pressure that the great squeeze is temporarily halted, and the star stabilizes.

Much like human beings, however, stars are always aging and evolving, even on the main sequence. It is a matter of degree, of the rate at which transformations occur. In early adulthood we change relatively slowly, but old age moves in quickly, producing ever-accelerating alterations. From the time it is born, a star begins using its finite fuel supply, and therefore must change. The close balance in the core between the diminishing fuel supply and the increasing burning rate maintains stellar luminosity. Without this stability stars would change quickly, and there would be no main sequence. Good as it is, however, the balance is not perfect. The ever-increasing temperature actually wins out a bit over the diminishing fuel, and the outer part of the core expands slightly into the envelope, where fresh hydrogen awaits. The result is that the luminosity of the star slowly *increases* with time. The surface of the Sun now glows with an effective temperature of 5780 K, but 5 billion years ago, when it was born and had a full complement of hydrogen (90 percent H, 10 percent He), it was somewhat redder and cooler, closer to 5500 K. It was also smaller and produced only about 70 percent of its present radiation. We have now used about half of the core hydrogen. Over the next 5 billion years, the rate of change will accelerate.

When the hydrogen finally runs out, our Sun will be twice as bright as it is now and 75 percent larger. Slow changes such as these help give the main sequence its breadth, making it a band on the HR diagram and not a line. It is further broadened by differences in composition that shift the subdwarfs of Population II to the left.

What do these changes mean to life on Earth? A drop of 30 percent in the luminosity of the Sun would be devastating to us—water would freeze and the planet would become more like Mars—yet the fossil record shows that the conditions for life here have remained hospitable over the ages. Perhaps our atmospheric blanket was different earlier, containing more carbon dioxide that trapped the lower solar heat. The Sun must also have been spinning considerably faster when it was younger, before its magnetic brake had been applied. The magnetic dynamo had to be more powerful, so that flares and the accompanying X-ray and ultraviolet radiation should have been at higher levels. We may never fully know the reason.

Our Sun's future, though, is predictable and inevitable. It will become much brighter. Within 2 billion years, or perhaps far fewer, there will be little in the way of winter. Higher temperatures will first cause greater evaporation of the oceans and increased atmospheric opacity. The temperature will climb still further, leading to a runaway greenhouse effect and conditions more like those found on Venus. Solar ultraviolet radiation will then break down the water molecules, and their hydrogen will escape forever. Even if our atmosphere were to stay stable, which it almost certainly cannot, by the time the main sequence phase is over the Earth will be so hot that life will be long gone. But such events are far in the future. Considering that humanity has been on the planet for a period measured in only millions of years, and that we have written records for only thousands, the thought should not be too depressing. We have a long way to go.

CHAPTER 10
HOW DO STARS DIE?

CASE 1

GIANTS IN THE SKY

Main sequence evolution is but a prelude, its stars remaining relatively stable until nearly the entire central fuel supply is gone. Then the *real* action begins. Look first at the more populous stars, those like the Sun. When nuclear reactions finally cease deep in the core, there is no longer any support against the inexorable force of gravity. The star has waited 10 billion years to use its vast supply of gravitational energy, and now the time has come: The core must rapidly (at least in astronomical terms) contract. The outer part of the core, where it has been cooler, still has some hydrogen left, and nuclear reactions continue in a shell, which now grows into the envelope. The changes in internal chemical composition, energy-generation rate, and opacity begin to produce massive alterations in the star's internal and external structure. For a period equal to about 20 percent of the main sequence lifetime, luminosity stays fairly constant as the radius nearly doubles and the surface cools to under 4500 K. Then evolution speeds up. The point that represents the star on the plane of the HR diagram begins to move rapidly across its surface to climb the evolutionary giant branch, the path that takes it into the realm of the giant stars. In only a bit over half a billion years the brightness shoots to over 1000 times the current solar value. The outer envelope expands enormously, causing the effective temperature to drop to 3500 K, and there is a new M0 giant in the sky, blazing forth at an absolute bolometric magnitude as bright as -3. Somewhat more massive stars follow the same general course, becoming even brighter—another magnitude or so—in shorter times. A 2-$M_{\odot}$ star will make the transition after the cessation of core H burning in only 300 million years, and one of 3 $M_{\odot}$ in but 100 million.

The effect as viewed from an orbiting planet—could anyone survive to see it—would be awesome. Our Sun will increase its size perhaps 100-fold to an All in [is this correct?] diameter (it is now 0.01 All), pushing past the orbit of the planet Mercury. Reddish and bloated, it will appear 50° across in the Earth's sky, quadruple the angular size of Orion, and (if we ignore the inevitable slowing of the terrestrial rotation) will take over 3 hours to rise and set. The new giant will now be seen with the naked eye to a distance of over 500 parsecs (1900 light-years).

As it matures, the surface gravity goes down, and the once-quiet wind from its surface increases its strength a million times or more. The star is beginning one of its greatest acts: It starts to whittle itself away, returning its mass to interstellar space, its birthplace.

A swollen Sun—half the size of the orbit of Mercury—rises as a red giant to blister Earth.

As the Sun expands something strange begins to happen that presages the end product of evolution. Remember the uncertainty principle—that the uncertainty in momentum times uncertainty in position approximately equals h? The Planck constant, h, represents something of a minimum in dimensionality. If you cube it you have the unit volume of a six-sided space called phase space that has three dimensions in momentum and three in real space. In 1925, Wolfgang Pauli formulated the exclusion principle, which states that within a unit phase space no two identical electrons can exist. Like all subatomic particles, electrons have a property called spin whose value must be plus or minus half a unit, the different signs indicating spins of opposite directions. Therefore, within the volume h^3 there can be at most two electrons with opposite spins. You cannot squeeze the electrons together any tighter. (One result of the exclusion principle is that no two bound electrons can occupy exactly the same orbit, so multiple electrons are forced into shell structures.)

The disparity in size between the envelope and the helium core of a fully developed giant reaches astounding proportions. As the outer parts expand to interplanetary dimensions, the core contracts toward the size of Earth. The density gets so

"NGC 290, an open cluster of stars, has most of its main sequence intact."

high, in the neighborhood of a million grams per cubic centimeter, that the lower-velocity electrons fill their phase spaces. The gas is now said to be degenerate. If any electrons are added, they can exist only at high velocity, where phase space still has room. The perfect gas law breaks down, and pressure depends on density alone and no longer on temperature—that is, temperature can increase while the pressure is constant. The core now has the characteristics of a white dwarf.

When the interior temperature reaches 100 million degrees, what was the ash of the old reaction, the helium, suddenly and explosively begins to fuse and becomes a new fuel in an event called the helium flash. Helium does not burn easily. If two helium atoms collide, they produce ^{8}Be (the normal isotope of beryllium is ^{7}Be), which is highly unstable and on average falls apart back into the helium pair in only 10^{-16} seconds. However, there must be an equilibrium in this reaction, so that at any one time there will always be a tiny amount of ^{8}Be. If one of these ephemeral atoms can collide with another helium nucleus, the two can fuse to create stable carbon, ^{12}C, and a gamma ray. Such a reaction requires that three helium nuclei (*a* particles) come together practically simultaneously, and the event is known as the 3-*a* process. It can occur only if vigorous collisions happen very frequently and requires extremely high temperatures and pressures, which are unavailable in main sequence stars. The necessary heat is in fact so great that another a particle can occasionally penetrate a ^{12}C nucleus, producing ^{16}O (plus another gamma ray) and a mixture of the two elements.

With a new energy source, the state of degeneracy in the core is lifted and the core expands. In the act of stabilization, the luminosity actually drops, the envelope contracts, the surface heats, and the star descends to a point about halfway down the giant branch. Here, in a clump on the HR diagram, it will quietly burn its helium in a core that is nested within a shell of fusing hydrogen. The star is in a sort of short-lived helium-burning main sequence. Many well-known K giants, Arcturus and Aldebaran among them, are in this configuration, as are the four K giants of the Hyades HR diagram and many of the stars seen in the magnificent globular clusters.

The location of a dwarf depends on its chemical composition, and so does that of a helium-burning star. Those that are deficient in metals, the ones of the halo and of the globular clusters, will be displaced to the left of the clump. Small differences in mass will then cause them to spread into the distinctive horizontal branch. The ones that fall into the instability strip become the RR Lyrae variables, the low-amplitude oscillators with short-(under-one-day) periods.

The stars along the upper part of the intermediate main sequence, above 4 $M_{\odot}$, go through some of the same internal changes, but their behavior is somewhat different. Instead of a short cooling track on the HR diagram followed by a great rise in luminosity, they undergo a long period of cooling from class B or so to class K and then a smaller rise in luminosity. They populate much of the realm of the bright giants in the HR diagram, luminosity class II, and swell considerably, increasing their already large sizes by a factor of 50 or so. Following helium ignition, which is no longer explosive, they are not very stable and move back and forth on the diagram as they alternately heat and cool. The

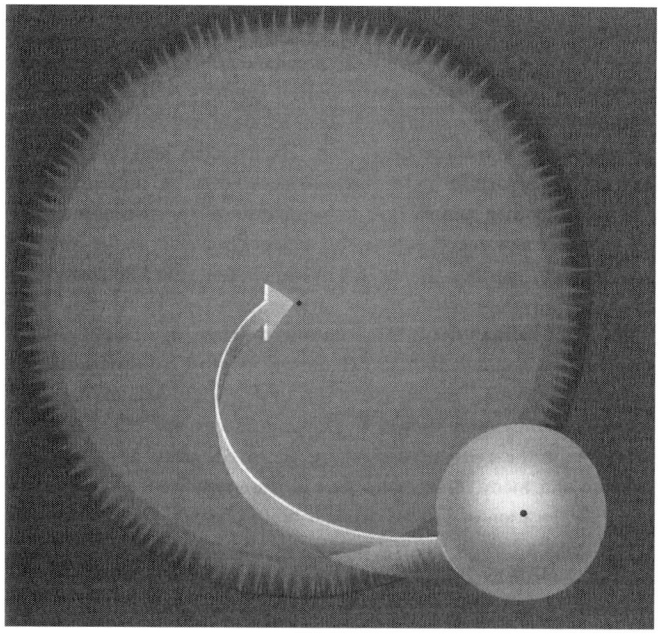

The degenerate core of a fully developed giant star—invisible on this scale without magnification—can just be seen on the right, where the dot in the center is blown up.

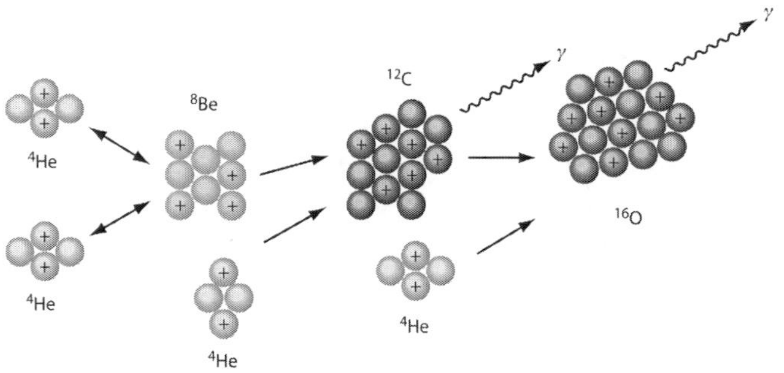

The 3-a reaction, described by Edwin Salpeter in the early 1950s, involves the near-simultaneous collision of three helium nuclei (a particles) to produce carbon, ^{12}C, and a gamma ray. The ^{8}Be nucleus formed by the collision of a pair of a particles decays almost instantly; the third nucleus must come along at just the right time. An occasional collision with yet another a particle will produce I6O and more gamma radiation.

transition from the lower part of the intermediate main sequence to the upper is not sudden but gradual, a continuum of behavior.

As the higher-mass stars move across the HR diagram, they also encounter the instability strip and are seen as Cepheids. This narrow zone is produced when layers of hydrogen and helium ionization deep within the stellar envelope trap and valve the outgoing radiation. When they hold it, the star shrinks. The shrinkage in turn changes the internal temperature structure, which alters the ionization balance, releasing the radiation—and the star expands. The result is a pulsation that makes the star alternately brighten and dim. The more massive stars are more luminous and larger and take longer to throb, explaining the period—luminosity diagram and the short periods of the less massive RR Lyrae stars.

The evolution to this point is impressive, to be sure, but it is only an overture. Grander things will come when even the helium runs out, and the stars begin to pace their last miles to expiration.

EVOLUTIONARY AGES AND THE GALAXY

Now the HR diagrams of star clusters like the Hyades, Pleiades, and the globular M 5 start to make some sense. A cluster is born with a full range of masses that occupy the entire main sequence. The Double Cluster in Perseus (h and χ Persei) is an example, a uniquely linked pair of systems born just yesterday whose main sequence runs all the way to the O stars. As a cluster ages, the massive stars die first, and the main sequence begins to disappear from the top down: The stars on the upper main sequence become supergiants, those on the intermediate section, giants.

With the aid of theory we can actually date a cluster by observation of where the dwarfs stop: h and χ are about 2 million years old. The Pleiades cluster is older: It has no O stars, but it does have B stars, showing it to have an age of about 10^8 years. Its half-sisters, the Hyades, with no main sequence stars brighter than class A, are older still—born some 10^9 years ago. M 67 in Cancer, at around 5 billion years, is one of the older open clusters known, with a main sequence that stops just brighter than the Sun. As a cluster ages, its giants keep pace with the main sequence and also descend in luminosity. Stellar evolution is fast enough so that the observed locus of giants on the HR-diagram approximately outlines the path of evolution.

Applying the theory of evolution to clusters provides a powerful way to study the Galaxy. We can date the age of the Earth and solar system from the oldest rocks, and we can tell the age of the Galaxy from the oldest clusters. The open clusters occupy the galactic disk exclusively, so we know that the plane of the Milky Way must be at least as old as the oldest of them, about 8 billion years. Unfortunately, open clusters are loose affairs. Even the richest contains only a few hundred or a thousand stars, and the internal gravitational forces that bind them together are not very strong. Over time, internal interactions fling stars away, and tides raised by the dense mass of the Galaxy cause further disruption. Most open clusters are rather young; it is rare for one to survive even as long as M 67 has. Consequently, these assemblies give only a lower limit to the disk's age.

Now we can understand the general HR diagram of the stars of the Galaxy as well. The giant branch is densely populated with stars, but bright giants and supergiants are scarce. The vast majority of stars that we see locally, those that appear on the HR diagram, are in the disk. They are being born all the time and have a variety of ages and masses. All the stars above G8 or so on the main sequence are able to produce giants or supergiants. The luminosity function, however, increases dramatically down the main sequence, so the space density of supergiants and giants must also strongly increase downward toward lower luminosity. The evolution of the lower-mass stars provides the large bulk of the giants, and we see a distinct, heavily populated branch streaming up and to the right from the middle of the main sequence. In principle, we could derive the age of the disk as we do the clusters, but the observations are too confused by a spread of masses, ages, chemical compositions, and stars mixing in from the thick outer disk and even from the inner halo. Nevertheless, from noting where the general giant branch intersects the main sequence, we can tell that the disk is at least as old as the oldest open clusters.

We fare better with the globular clusters. The majority are in the Population II bulge and halo. They are packed with stars and are so tightly bound together that they defy easy disruption. The best fit between theory and observation shows that the oldest of these magnificent groups were born about 14 to 16 billion years ago. They are older than the open clusters and clearly older than the stellar assembly within the disk. Since we find nothing of greater age, we identify that number with the age of the Galaxy itself. Because they are part of the constituency of the galactic halo, it is evident that the halo was the original component of the Galaxy and that the disk, with its current complement of O and B stars, came later. The set of noncluster halo subdwarfs, whose main sequence above class G long ago burned away, yields the same picture.

Since the globulars and subdwarfs have fewer metals than the disk, the number of heavier atoms clearly must somehow be increasing as the Galaxy ages. Stars are producing them—but how? We have already seen how helium and carbon are manufactured as by-products of stellar power generation. The general problem was solved in one of the landmark papers of twentieth-century astrophysics, published in 1957 by E. Margaret Burbidge, Geoffrey Burbidge, William Fowler, and Fred Hoyle (affectionately referred to as B^2FH), who described the various processes of nucleosynthesis, the creation of heavier atoms from lighter ones. As we turn to more advanced stages of stellar evolution and, in the next chapter, to the evolution of high-mass stars, we will begin to see how these processes work and how the new atoms arrive at the stellar surfaces.

GIANTHOOD REPRISED

Helium burning sustains only a short hiatus in the inevitable contraction. Once this fire gets started, the reaction rate is so fierce that the fuel supply does not last very long. Soon the last of the helium is gone from the central core, leaving a huge ball of carbon and oxygen, and the shrinkage starts all over again, with the same results. The internal temperature rises, and now the helium burning sweeps outward in a shell around the collapsing defunct core, fed from above by a shell that is still fusing hydrogen. Once again the outer layers begin to swell and cool. The star begins its second ascent of the giant branch, this time with a more powerful energy supply of not one but two nuclear-burning shells.

The second ascent rather parallels the first. The stars from the horizontal branch climb more or less asymptotically to the tracks taken on the first ascent, and this new phase has come to be known as the asymptotic giant branch, or AGB. (The first ascent is distinguished by calling it the red giant branch, or RGB.) Nuclear burning now proceeds in a strange fashion, the two shells switching on and off in sequence.

As the ascent begins and the star swells, the H-burning zone moves outward, cools, and shuts down, the luminosity coming from a thick shell of fusing helium. As brightening proceeds, the helium begins to bum away. The old H-burning shell contracts and re-ignites, feeding fresh helium into the zone around the still-contracting core. Then, with searing violence, the new helium shell explosively re-ignites, a nuclear bomb buried deep within the expanding star. The H shell retreats and shuts down to wait its turn, which will

The AGB Sun, brilliant and red, a huge wind screaming from its surface, rises over the Martian landscape—if anything remains of the little planet.

come when the new helium is done. These helium flashes or thermal pulses come closer and closer together as the star nears the top of its great-—and last—ascent. So much energy is supplied that the stars become quite luminous and attain immense sizes. The Sun will reach an absolute bolometric magnitude of -4 or so. It is expected to swallow Mercury and Venus both, and to reach to the orbit of the Earth itself. These stars are so swollen that the surface temperatures drop to 3000 or even 2500 K; their spectral classes are as extreme as M8.

To test the theory, its predictions must be compared to observation. By and large, the agreement is good, the theory predicting the correct general distribution of stars on the HR diagram and about the right luminosities. However, there are some "illuminating" differences, which let us know about processes that have not yet been correctly included. The theoretical RGB and AGB tracks show about the same low minimum temperatures; in reality, the first-ascent red giants never get this cool. Moreover, the observed AGB flattens out near the top. The theoretical RGB can be slid to the left to match reality by decreasing internal opacities or by increasing the sizes of the convection cells. The AGB can be brought into better agreement by also changing mass-loss rates. Unfortunately, none of these quantities is known very well, and we can do no more than make uncertain corrections that cannot actually be built directly into the theories.

As an AGB star brightens, its envelope becomes unstable. Somewhat like a Cepheid, it begins to valve the outflowing radiation and starts to pulsate. It is so huge that, unlike a Cepheid's, its oscillation period, instead of being a matter of days or weeks, is months or years. We then look into the sky to see a long-period, or Mira, variable. Mira, *o* Ceti, is the archetypal AGB star. This monster (appropriately found in the constellation Cetus) is the size of the orbit of Mars, and takes 330 days to expand and contract, changing its radius by a factor of 2. The Miras are everywhere—and are the most common type of intrinsic variable known. Periods range from a hundred days or so to over five years. The amplitudes, the differences between maxima and minima, can be as much as 10 visual magnitudes. Such large fluctuations are deceptive, however. The change in radius produces a notable variation in temperature. The star cools so much that a good part of its radiation disappears into the invisible infrared, rendering it quite faint. The bolometric amplitude, though still impressive, is much less.

The Sun is expected to swell to the size of the terrestrial orbit. If it overtakes our now-molten planet, the Earth will actually orbit inside the low-density envelope. Friction with the gases will cause it to lose orbital energy and spiral inward, eventually to be utterly destroyed along with Venus and Mercury. Mars will likely be spared, and the heat may be enough to render conditions springlike on the outer planets.

CHEMICAL ENRICHMENT

The outer layers of a giant star transport energy outward by convection in great circulating flows of gas that dip far down into its envelope. The thermal pulses set up internal convection cells that move the by-products of fusion upward. If the two convective layers can meet, freshly made elements can be brought to the top, and *the star will change the chemical composition of its surface.* Remember the stars of spectral class N, the deep red giants rich in carbon? Here is their origin. At the beginning of the AGB we will see an M star whose carbon-to-oxygen (C/O) ratio is about 0.4. There is so much oxygen that it combines readily with metals to produce the characteristic TiO bands. As the carbon, the dregs of helium burning, is swept up to the top, the C/O ratio climbs. Carbon has a strong affinity for oxygen, and when C/O climbs above 1, it ties up all the oxygen in CO and CO_2. The metallic oxides disappear, and the remaining carbon combines with itself to produce the powerful C_2 bands of class N. The carbon bands are very strong at short optical wavelengths and cut out most of the blue component of the emerging light, rendering the stars a rich, beautiful red.

The deep fusion process is no longer hidden from us. We are seeing chemical elements made before our very eyes. Absolute proof of such transmutation comes from observation of the element technetium, atomic number 43. It has no stable isotopes. Even the longest lived, ^{98}Tc, has a half-life of only 2 million years and therefore long ago disappeared from the Earth. Never buy stock in a technetium mine. But some stars, particularly those of the carbon-rich variety, show the stuff in their atmospheres! We believe it is ^{99}Tc, and it can exist only because it is being newly made and brought to the surface along with the carbon. The alchemists of old would have loved it.

The manufacture of technetium demonstrates a host of subsidiary nuclear reactions collectively called the s-process, which results from the slow capture of free neutrons onto heavy elements. Each neutron addition produces the next heavier isotope of the same element. If the isotope is stable, it eventually captures another neutron, and the mass number goes up another notch. If the nucleus is unstable, and if the capture rate is slow enough, it may beta-decay (the neutron changing into a proton with the ejection of an electron), increasing the atomic number to the next atom in the periodic table. However, before it decays, the nucleus may well catch another neutron, jumping the mass by another unit and producing the next higher isotope of the next element. In this way, an entire nuclear network of creation is established, the higher elements being made from the next lower ones. The networks can be included in stellar models and isotopic ratios predicted from neutron-capture rates (which depend on temperature and density) and the decay rates of the various isotopes.

The S stars represent a transition state between the oxygen and carbon-rich, where the abundances of the two are about equal. Titanium is much more abundant than zirconium, so that in M stars TiO predominates over its cousin ZrO. But zirconium has a much greater affinity for oxygen. Given only a few atoms of O and a mixture of zirconium and titanium,

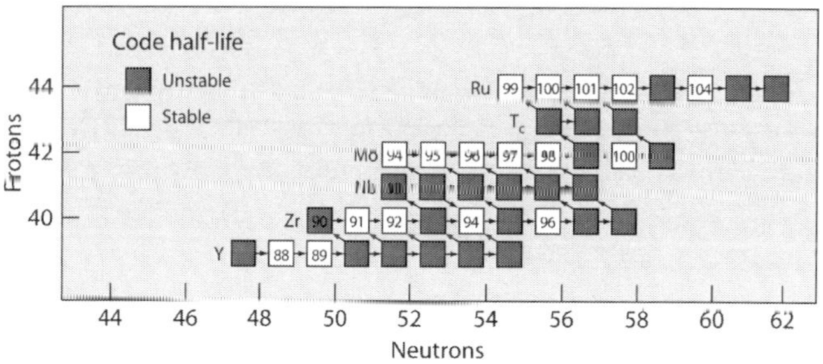

This s-process network shows how heavy elements can be made from lighter ones. The elements and isotopes are laid out in a chart of the nuclides (a nuclide is any single species of atomic isotope) in which atomic (proton) number is plotted against neutron number (atomic mass minus atomic number). Start with yttrium in its common form, ^{89}Y. It captures a neutron to become ^{90}Y and beta-decays into ^{90}Zr (the common metal zirconium). Although ^{90}Y is highly unstable, some of it will capture another neutron to become ^{91}Y, which then produces ^{91}Zr. The process works all the way to ^{94}Y. Beyond that, the isotopes of Y are violently unstable, and the isotopic upgrading stops. As the zirconium isotopes are produced, they immediately start capturing their own neutrons, so that ^{91}Zr is produced from both ^{91}Y and ^{90}Zr. But the Zr isotopes then decay into atoms of niobium, Nb. As a result, the isotopic ratios of Zr depend on neutron-capture rates relative to decay rates of the various isotopes. Molybdenum is then created from the niobium. Finally, the object of our attentions, ^{99}Tc, is manufactured by beta decay from ^{99}Mo. 100Tc can be produced only by neutron capture onto ^{99}Tc because ^{100}Mo is stable. Tc turns into ruthenium, and so on.

ZrO will appear first, even if the heavier metal is fairly rare. If C/O is close to 1, the C grabs all but a small part of the O, and what is left produces ZrO. Moreover, zirconium is an s-process element and is increasing in abundance along with the C. Consequently, the S stars are characterized by the ZrO bands. Not surprisingly, many of them display technetium lines as well. The M and carbon stars both also display enrichment in nitrogen, the result of burning by the carbon cycle. Although these stars are too cool for us to see helium lines, other evidence shows that this element, too, can be enriched.

Stellar models tell us the amount of fresh material that is brought to the stellar surface and, most importantly, where in the star it came from. This information enables prediction of the strengths of various spectrum lines. The absorptions produced by different isotopes of an atom fall at nearly identical wavelengths and cannot ordinarily be separated—for example, we cannot tell one isotope of Tc from another. But *molecular* isotopic wavelength differences are relatively large. The absorptions produced by ^{12}CO and ^{13}CO or by ^{90}ZrO and ^{91}ZrO can be distinguished, allowing very detailed observational checks on the models. Although exact agreement can hardly be claimed, the results are enough in accord with theory so that we have a good deal of confidence that we know what is going on deep inside the stars as they evolve. The main problem is the convection that brings the new stuff to the surface. From observational evidence, the carbon stars seem to come mostly from bodies that started in the neighborhood of 2 or 3 $M_{\odot}$. However, our lack of knowledge about convection keeps us from being able to predict the exact mass ranges responsible for the carbon stars as well as accurate elemental and isotopic yields. If stellar astrophysicists had a last wish, it might be to have a theory of convection.

MASS LOSS

All stars, by their very natures, must lose mass—even the Sun, albeit at a paltry 10^{-14} $M_{\odot}$ per year. The rates among the giants are perhaps a million or more times greater, and on the AGB mass loss is driven at a still higher rate. The spectra of Mira stars show hydrogen emission lines that are interpreted as shock waves generated by the violence of pulsation (a shock wave is an acoustic overpressure caused by something that moves through a gas faster than the speed of sound—a supersonic aircraft provides a familiar example). The shocks, together with high stellar luminosity, loft gas off the surface into the relative cold of space. Some of the gas condenses into small solid grains, loosely called dust. Radiation pressure acting on the dust pushes it away, and the dust drags the gas along with it at a speed of a few tens of kilometers per second. The result is an outflowing dirty wind that near its maximum may reach an astonishing rate of 10^{-5} $M_{\odot}$ per year. The stars are in this state for more than 10^5 years and will lose substantial parts of themselves back into space. Between its future RGB and AGB states, the Sun will come close to cutting itself in half, and a more massive star, one that is near the upper bound of the intermediate main sequence, may lose an astounding 80 percent of its mass.

A Mira itself loses so much mass that it is surrounded by a huge expanding cloud that is visible in the infrared and radio parts of the spectrum. In the IR, the spectral signatures of dust grains around Miras reflect the composition of the underlying star. If it is carbon-rich, we will see silicon carbide grains, and if it is of the oxygen-rich M variety, silicates (silicon-oxygen compounds) appear. Radio observations are even more fascinating. The circumstellar shells of the M stars display powerful emission lines of hydroxyl, OH, and the systems are consequently called OH/IR stars. The OH is a natural maser, the microwave version of the now-familiar laser that works by pumping electrons into high-energy states. When they cascade downward, a powerful beam of light can be produced. In the lab, the laser is energized by an outside energy source such as electricity; in space, the radiation from the Mira is responsible. We also see masering lines of water and of silicon monoxide.

Part of the cloud is going away from us and part toward us. The lines are then split by the Doppler shift, and we can measure the expansion velocity. Since the OH line strengths depend on the Mira's luminosity, they will also vary. And because it takes considerable time for the starlight to reach the shell, the near side of the cloud will reflect the variations first. The time difference in variation between the near and far sides depends on the cloud's physical diameter and the speed of light. From the physical and angular sizes, we find the precious distance and can derive the stellar luminosity. These stars are treasuries of information.

The carbon stars show different but equally bizarre characteristics. Carbon forms complex organic molecules, for which conditions in space are highly favorable. In the most extreme example, a carbon star complex called IRC +10 216 (infrared carbon star, 10° declination, 21.6 hours right ascension), some 20 organic molecules are found, including such exotic ones as HC_7N and CH_3CN.

The flow of gas may be so strong that, like IRC +10 216, the star may bury itself in its own effluvia and in the optical part of the spectrum disappear from view, leaving us with only infrared and radio radiation to mark its position. As the circumstellar shell departs, it carries with it the fresh nitrogen, carbon, helium, and s-process elements that were created earlier in the star's evolution. There are thousands of these OH/IR and carbon stars. Not only are they recycling much of their matter back into space, but together they also cause a continuing large-scale enrichment of the interstellar gases, out of which new stars are always being formed. We now begin to see a reason for the relation between stellar metal

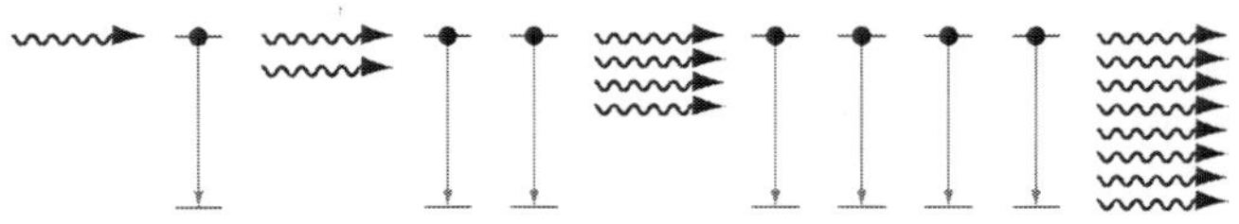

Not only can a photon cause an upward jump of an electron, it can stimulate an electron in an upper orbit to jump downward (left). If it does, two photons go flying away in the same direction in phase with one another. In a normal gas, absorptions and both spontaneous and stimulated emissions are in balance. If we can somehow overpopulate or pump the upper level, each photon generated produces another, resulting in a cascading beam of laser light.

content and age!

Mass loss may be the Earth's salvation. As solar gravity is diminished, our battered planet's orbit will slowly become larger, and the expanding Sun may not be able to reach us. Mercury and Venus are still doomed, however. Their mass may actually contribute significantly to the dust within the wind. Moreover, the great luminosity of the evolved Sun will probably melt most of the comets in the inner comet belt (just beyond Neptune), augmenting the wind's water content. It is sobering to think that the dust and the water masers we see around AGB stars may, at least in part, be a signature of the destruction of a planetary system.

The internal temperature is still going up, and we may think that at some point the carbon-oxygen mixture in the degenerate core will fire and start to burn to something heavier. But because the star has lost so much mass, the compression stops short of the required conditions. Mass loss will eventually strip away the envelope to reveal the dense dregs, the remnants that we long ago recognized as 40 Eri B and Sirius B.

PLANETARY NEBULAE

While scanning the skies 200 years ago, William Herschel came across an odd-looking body in the constellation Aquarius, a somewhat circular fuzzy patch of light now known as NGC (New General Catalogue) 7009. He called this object—and several others that he later discerned—planetary nebulae, as they reminded him of the extended disks of planets. He soon also discovered that at their centers they commonly had single stars that he realized were intimately connected with the nebulosity.

The clue that led to the explanation of the planetaries was served up in 1864 by the English spectroscopist William Huggins, who discovered a trio of emission lines that proved them to be made of a thin gas. One, at a wavelength of 4861 Å, was thereafter identified with common hydrogen, but the others, at 4959 and 5007 Å, defied classification. For a time, they were thought to be produced by an as-yet-undiscovered element, which was named nebulium. After the periodic table of the elements was filled in, it was obvious that there was no place for "nebulium" and that some unknown mechanism was causing the production of the lines. As the spectrum was explored further, several more of these unidentifiable emissions were found, particularly bright ones appearing at 3868 Å, 3727 Å, and near Hα at 6584 and 6548 Å.

The solution to the mystery, and the explanation of the mechanism that lights the planetaries, was offered by Hermann Zanstra and Ira S. Bowen around 1929. Zanstra showed that the hydrogen lines were produced as a result of photoionization by ultraviolet light emanating from the central star. When an electron in the ground state of hydrogen absorbs a photon with a wavelength of 912Å (the Lyman limit) or less, it will gain enough energy to escape the proton. A nebula is a sea of these free protons and electrons. When the two kinds of particles collide, there is a good chance that the electrons will be captured into upper orbits.

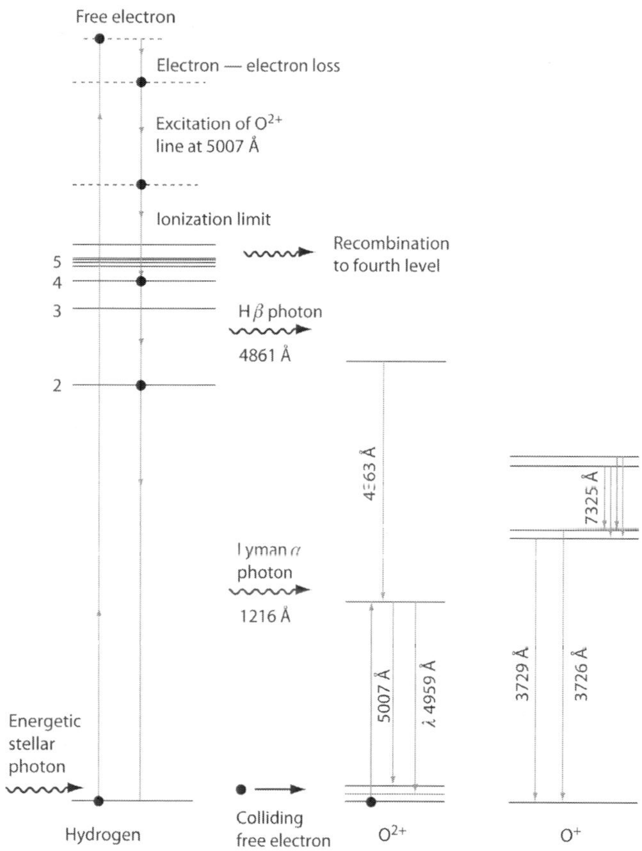

Recombination lines of hydrogen are formed when an electron is ejected (left) from an atom after the absorption of a stellar ultraviolet photon. Its energy is then greater than the ionization limit, and it flies free in the gas. It loses some energy to other electrons, in the process establishing the kinetic temperature, and then hits an 0+2 ion, knocking its electron into an upper orbit. The free electron subsequently recombines with a proton and skips down the energy ladder to produce hydrogen emission lines. The excited oxygen electron eventually drops back down to produce one of the so-called nebulium lines at 4959 or 5007 [?]. Other such lines, initially thought to be theoretically forbidden, are shown for both 0+2 and 0+. Chemical compositions can be derived by applying basic theory and appropriate data on collision, recapture, and cascade probabilities (scales of 0+2 and 0+ are doubled for clarity).

Jumps to the ground state will then produce the observed emission lines. The re-formed atoms subsequently absorb more stellar photons and repeat the cycle. Helium and even oxygen, nitrogen, and neon produce these recombination lines.

Zanstra showed that in a dense nebula, every stellar ultraviolet photon will produce one ionization, which will be followed by one recapture and one observable Balmer photon (produced by an electron that lands in hydrogen's second orbit). The number of these is proportional to the ultraviolet luminosity of the star, whereas the visual magnitude is proportional to the visual luminosity. If the star is assumed to be a simple blackbody, which it generally appears to be, the ratio of the brightness of the observed nebular hydrogen emissions to the visual stellar brightness yields the effective, or Zanstra, temperature. Zanstra—who referred to this procedure of detecting otherwise invisible ultraviolet radiation as "space research at low cost"—found that the central stars of the planetaries were enormously hot. Modern work shows that they are the hottest stars known, with Zanstra temperatures that start at 30,000 K or so and extend to over 200,000 K. These are also among the most luminous stars in the Galaxy, many between 1000 and 10,000 times brighter than the Sun, but because most of their energy is radiated in the UV, we do not much notice them.

Bowen looked inward, to the laboratory. He used ultraviolet experiments to chart out atomic orbital structures and found that the "nebulium" lines were actually produced by energy levels of doubly ionized oxygen that lie just a bit above the ground state. By the simplest approximations of the rules of quantum mechanics, transitions between these states are not allowed, and the resulting lines are "forbidden." Only when higher-order approximations of the equations are taken into account do we see that the jumps are not truly forbidden, just unlikely. The electrons are excited into the upper levels by collisions with those electrons that were freed from hydrogen. The forbidden lines are not seen in the laboratory because the density of the gas is too high and they are weak relative to other emissions. But in the nebulae, conditions are just right, and they overpower even the recombination lines. The forbidden lines at 3727 Å, 3868 Å, and the pair in the red were subsequently found to be created by similar transitions of ionized oxygen (O^+), neon (Ne^{+2}), and nitrogen (N^+).

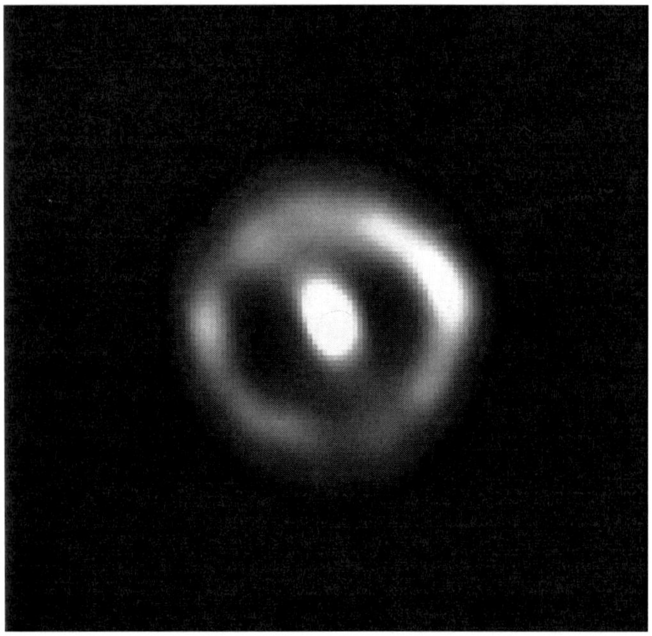

Just over a year after the explosion, Hubble observed a cloud expanding around Nova Cygni 1992. The star in the center is the ordinary dwarf that is still passing matter onto the white dwarf too faint to be detected.

The theory of the production of the recombination and forbidden lines is so straightforward that we can easily use them to derive not just the central star temperatures, but also nebular densities, temperatures, and chemical compositions. We find typical densities of 100 to 10,000 atoms per cubic centimeter, comparable to the best vacua produced on Earth, and kinetic temperatures of 10,000 to 20,000 K. Compositional analyses show that these objects tend to have elevated levels of helium, nitrogen, and carbon. In a few cases, the enrichment is quite high, double the basic solar helium content, and 10 times the nitrogen.

The planetaries exhibit an almost bewildering variety of structures and are among the loveliest objects to be found in the Galaxy. Some are nearly round, others display a distinctly bipolar structure, and yet others are so complicated as to defy simple description. Many have double or even triple shells. In size they range from small and compact, nearly stellar, to dimensions so huge that they span the spaces between stars.

It was not until the second half of this century that astronomers were finally able to fit these graceful objects into the flow of stellar evolution. On both its ascents of the giant branch, an evolving star loses mass. By the time it is near the top of the AGB what is left of the star is surrounded by an enormous expanding cloud. As luminosity climbs, the mass-loss rate increases as well, so the inner part of the cloud is denser than the outer. Finally comes the big moment when nearly all the envelope is removed, not quite revealing the nuclear-burning shells. The chances are that this event happens when the helium shell is turned off, so there is now a degenerate C-0 core that is successively nested in an inert helium shell, a burning hydrogen shell, and a thin though quite large hydrogen envelope about the size of the present Sun. Now, instead of coming from a distended giant, the wind must blow from a smaller star with a higher surface gravity, and it thins and increases its speed. It will eventually blow as fast as 4500 km/s. This fast wind, acting much like a snowplow, shovels the mass lost earlier into a relatively dense shell with a mass a few tenths that of the

A Hubble Space Telescope image reveals the graceful sweeps of ionized gas that characterize the Orion Nebula, 420 pc away and 7 pc across. At the center is the Trapezium, the brightest of whose four stars θ^1 Ori C (O6 V), is largely responsible for the ionization of the object. A zoom-in on a portion of it shows complex filamentary structures.

Sun. While this inner ring floats away at about 20 km/s, the hydrogen envelope of the central star thins, dissipated from above by the wind and eaten into from below by nuclear burning. The effective temperature rises at constant luminosity. When it hits about 30,000 K it produces enough ultraviolet light to ionize the surrounding mass, and suddenly a new planetary nebula blossoms into the sky.

Deep imaging will frequently show the resulting planetary buried within a vast mass of previous ejecta. The structure of a nebula apparently depends on asymmetries present in the original mass-loss flow that are magnified by the compressing fast hot wind. The composition of the object, nitrogen- or carbon-rich, tells us the kind of AGB star that produced it, giving us a direct look into the old stellar envelope and providing an excellent check on the theory of stellar structure and evolution.

The star heats to a maximum that depends on mass, 100,000 K or greater. After 10,000 years or so (depending on mass) the H envelope is nearly gone, the nuclear burning is banked if not shut down, and the star begins to cool and dim. By this time, the nebula is a tenth or two of a parsec across. Finally, after some 50,000 years, it has become so large and thin that it disappears from view. The giant has now returned a share of its matter back into interstellar space, adding gifts of helium, nitrogen, and carbon, and perhaps a load of heavier s-process elements. The next generation of stars will benefit from this largesse. What remains is a hot, naked, white dwarf. There is evidence that not all of the white dwarfs go through this stage, but it is clear that the majority do. Those that do not somehow manage to pass directly from the horizontal branch; we do not know how.

WHITE DWARFS AT LAST

As the end products of the intermediate main sequence, white dwarfs are everywhere. Huge numbers roam free in the Galaxy, as will the degenerate descendant of the Sun. They do not obviously populate the sky only because they are so faint; even the brightest is only eighth magnitude. It took billions of years to make the AGB star—then in a blink it is gone, replaced by its defunct, degenerate core. Now nothing remains for it but to cool forever, to become a clinker in the celestial ash heap.

Although the huge densities of a million grams per cubic centimeter make these stars strange enough, there are more surprises. When the spectrum of Sirius B was first observed, the star was classified as type A because of the great strengths of its hydrogen lines. Sirius B is hot, however, and actually has a temperature that would place it in class B, which ordinarily exhibits helium lines, too—-yet none at all are found. Other white dwarfs that do display helium absorptions were called class B. All real B stars have hydrogen lines, but these did not. The white dwarfs are chemically differentiated into two broad groups. The hydrogen- and helium-rich degenerates are respectively called DA and DB after their initial classifications, but there is little relation to temperature: Each

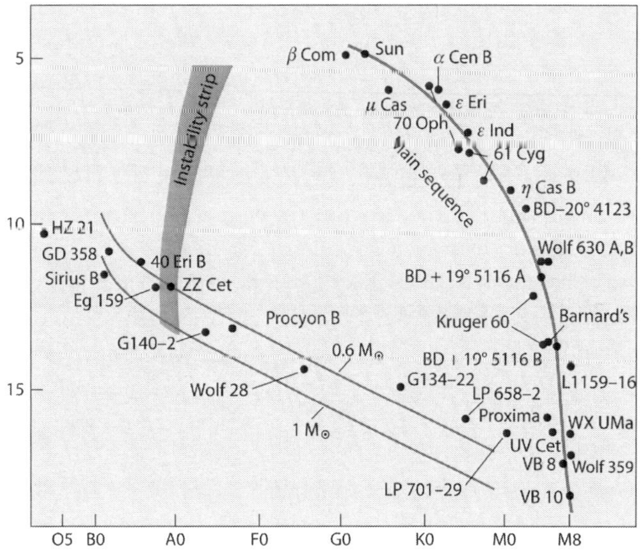

The white dwarfs are seen cooling along their final evolutionary tracks in this detail from the HR diagram shown on page 89. Theoretical tracks for [???] have been superimposed. White dwarfs are not classifiable by ordinary rules, so they have been placed here according to their temperatures and the temperature scale of the main sequence. The more massive white dwarfs are less luminous, because higher gravity gives them smaller surface areas; at the lower end the stars will crystallize, but the lifetime of the Galaxy is not yet long enough for any of them to have cooled to invisibility.

The colros of summer's lovely Trifid Nebula, M 20 (above), differentiate and H II region (red) from a reflection nebula (blue). The Rosette Nebula in Monoceros (left) is a fine example of a Stromgren sphere. Superimposed upon the apparent surface of the nebula are large numbers of black spots (globules) and long dark strings ("elephant trunks")—actually foreground clouds of opaque dust. (photo courtesy of James McGaha, Grasslands Observatory)

kind spreads out across the whole temperature sequence. Even though the cooler white dwarfs can exhibit no lines of either element because of their low temperatures, we are confident that the difference extends to them as well.

If you could stand on the Sun (an interesting thought), you would weigh triple what you do on Earth, the result of a 300,000-fold increase in mass and a 100-fold increase in radius. On a white dwarf, however, the radius is so tiny that a typical 70-kg human would weigh an astonishing 600,000 kg! The force of gravity is so high that the heavier atoms, including those of helium, are dragged to lower depths in the atmosphere and the hydrogen floats to the top. The result is a pure hydrogen gas and a DA star, no matter what the temperature. Given this diffusion process, we know that a helium-rich DB star can contain no hydrogen at all, or it would be seen. It must have lost its entire hydrogen envelope during evolution out of the AGB state. We can actually see this differentiation start with the planetary central stars, but we do not know why one star loses its hydrogen skin and another does not. Stranger still, there is a gap between 45,000 K and 30,000 K in which there are no DB stars. Apparently, the degenerates can change from one type to another as they cool! Helium may be lofted back to the surface of a DA, the hydrogen of a DA may be lost through a weak wind, or a DB may accumulate hydrogen from the interstellar medium. But we really do not understand.

The magnetic fields possessed by some of the stars are more remarkable yet. The Zeeman effect causes the absorptions of the magnetic A dwarfs, which have fields up to a few thousand times that of the Earth, to be split by a few tenths of an Ångstrom. In the class of magnetic white dwarfs, the splitting can be so huge that the lines become unrecognizable, Hα and Hβ spread out over much of the visible spectrum. To produce these prodigious separations, fields of the order of *hundreds of millions* of times that of Earth are required. Somehow, nature squeezes the stellar field along with the star to produce such intense magnetic concentrations. Again, we do not know why or how, nor why some white dwarfs are magnetic and others show no fields at all.

Had enough weirdness? Allow one more. As the stars cool to effective temperatures of roughly 4000 K they start to *crystallize*! We cannot see this development actually take place because the atmosphere is still a gas, but the theory seems to make it necessary. The cooling time of a white dwarf is enormously long, and it takes billions of years to descend to this point on the HR diagram. The time required for a star to move *off* the diagram is longer than the age of the Galaxy. No white dwarf ever born has ever disappeared. They are all still there and visible to us. In principle, we can use the end of the white dwarf sequence, near 4000 K, to date the Galaxy. Current work gives a lower limit of about 10 billion years, consistent with the ages derived by other means.

The most massive white dwarf known is only somewhat heavier than the Sun, reflecting the profound distinction between the upper and intermediate main sequences mentioned earlier. Now it is time to look at the reason. In 1930, a future Nobel-prize-winning physicist named Subrahmanyan Chandrasekhar traveled from his native India to Cambridge University to begin graduate studies. During the voyage he explored the theory of the internal structures of the white dwarfs and degenerate matter. He realized that under extreme conditions, the velocities of the most energetic electrons could approach the speed of light. It was necessary then to take into account the general theory of relativity. When he did, he discovered that the laws that govern degeneracy change. If the internal pressure could be elevated enough, degenerate electrons could no longer keep a star buoyed up. It would collapse if the degenerate mass could exceed what is now known to be 1.4 $M_\odot$. Above this Chandrasekhar limit, no white dwarf can exist.

As initial mass increases upward along the main sequence, so must the mass of the core. The dividing line between the upper and intermediate main sequence occurs at an initial mass of about 8 $M_\odot$, where the core reaches the Chandrasekhar limit. Above 8 solar masses, some quite wonderful things happen.

BINARIES

The white dwarfs hold yet another surprise that renders them wonderfully visible. The majority of stars reside in binary systems, and because the masses of the two partners are usually different, some interesting things can happen over the course of evolution. Look first at our old friend Sirius. The white dwarf B component has less than half the mass of the brilliant primary. Since it had to evolve first, it must once have been the more massive, and clearly has lost over half its initial matter, providing direct proof of mass loss. Sirius A is something of an Am star, mildly enriched in metals.

We believe that when Sirius B was a giant, it passed some of its enriched surface material onto Sirius A through its outflowing wind. A whole class of giant oddballs called barium stars seem to have been made in this way. But this kind of mild interaction is nothing compared to what else can happen. In 1975, a brilliant "new star," or nova (from the Latin for "new"), flared up in the constellation Cygnus to rival first-magnitude Deneb. Dozens of these celestial explosions are seen every year, though most are so far away as to be invisible to the naked eye. The process begins with main sequence companions. The more massive component starts to evolve. Unlike the Sirius pair, the two are so close that the envelope of the giant actually encompasses its neighbor. Orbital energy will be lost, and the pair will spiral closer together. By the time the first star finishes its evolution, the two are so intimate that the white dwarf raises tides in the remaining main sequence star, tides so great that the ordinary dwarf fills its gravitational equilibrium surface and starts to send fresh hydrogen cascading down onto the degenerate. The hydrogen does not flow directly to the companion star, but first accumulates into an orbiting accretion disk from which it is transferred to the surface. The layer of new hydrogen builds up until the temperatures become so high that fusion is initiated, the surface explodes, and the nova (a terrible misnomer, as the event is caused by an *old* star) lights the sky. It disappears after a few weeks or months, and after some years we can photograph the expanding cloud. The process repeats itself, and our descendants a hundred thousand years hence may see another.

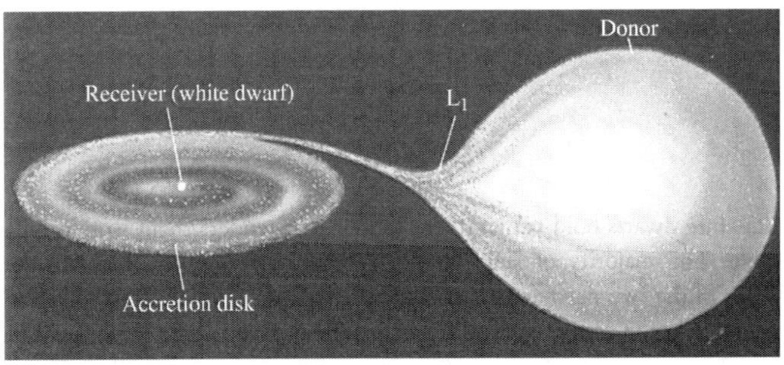

One member of a binary fills an equilibrium surface over which the gravitational strengths from the two are equal. The large one can then pass matter to the small one via an accretion disk through point L1.

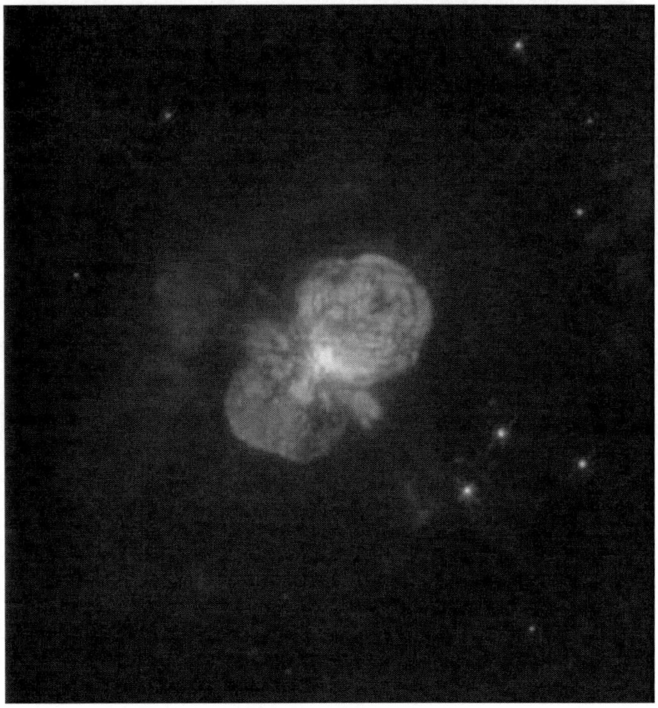
Some 150 years ago, Eta Carinae pumped out so much matter that it enveloped itself in a massive cloud.

Yet another scene: The giant companion of a white dwarf expands and throws copious matter into an accretion disk. The disk, or a spot on the star where the matter actually falls, is so hot that it produces considerable ultraviolet radiation, enough to ionize the surrounding matter. A spectrogram of the star then shows a weird mixture of TiO bands on a red continuum, emission lines of H, He, and even some forbidden features. This symbiotic star may even occasionally produce novalike detonations and brighten considerably, burning for years. A few generate wonderful surrounding nebulae reminiscent of the planetaries, but much more complex.

So concludes the picture of the evolution of these stars. We begin here to see how the Galaxy and the stars are inextricably locked together, each one feeding the other. More wonders await in the next chapter, where we will examine the amazing high-mass stars.

CASE 2

CATASTROPHE: THE SPECTACULAR LIVES AND DEATHS OF THE HIGH-MASS STARS

Some January evening, step outside to admire Orion, his belt riding the celestial equator. From it hangs his mighty sword. Focus on the middle star, θ^1. Raise a pair of binoculars and discover that it is quite fuzzy; a small telescope easily shows an intricate cloud of gas, the Orion Nebula, the archetype of the gaseous or diffuse nebula. In the center is a quartet of stars, a gravitationally bound group called the Trapezium. One, θ^1 Ori C, is notably brighter than the others. Its spectrum shows it is a hot main sequence 06 star with a temperature near 40,000 K. Its luminosity is a quarter-million times that of the Sun, and it produces so much ultraviolet radiation that it is responsible for ionizing a cloud with a mass of several hundred Suns out to a distance of nearly 4 parsecs.

Orion and its great nebula exemplify the properties of the rare stars of the upper main sequence. They are enormously luminous. Their effective temperatures must be greater than 25,000 K—spectral classes hotter than B2—in order to produce enough ultraviolet photons to ionize a surrounding nebula. They cling not only to diffuse nebulae, but also to the OB associations that dot the thin galactic plane, a pattern that testifies to their youth. The associations are observed to be expanding and disintegrating, and the nebulae are the remnant clouds of interstellar gas out of which the stars were born. Even moving at several tens of kilometers per second, the O and B stars never get very far away from their origins before they begin to die. Spectral class B2 corresponds to a 10-solar-mass star, close to the nominal lower limit for the upper main sequence. The main sequence stars that illuminate the gaseous nebulae are not expected to produce white dwarfs. What they *will* produce is the main subject of this chapter.

THE DIFFUSE NEBULAE

Since the most obvious manifestations of the great O stars are the diffuse nebulae, it is appropriate to look at these wonders in more detail and to make a small survey of the brighter ones. Many are easily observable with only a small telescope or even binoculars. Half a dozen even made it into the famous list of bright nonstellar objects compiled by the eighteenth-century comet-hunter Charles Messier; the Orion Nebula is M 42.

Stars are descended from the gas and dust clouds of interstellar space. The less massive ones have little effect on their parent clouds and live so long that they escape to roam free in space. The birth of an O star, however, is marked by a vast zone of ionized hydrogen gas, the H II region or diffuse nebula. The Orion Nebula is actually a small blister on the edge of a giant cold, neutral cloud that has a mass 100,000 times solar and occupies most of the southern part of the constellation. These bubbles of ionization are called Stromgren spheres after the Swedish-born American astrophysicist Bengt Stromgren (1908–1987). The star maintains high ionization to the point at which the ultraviolet photons are exhausted, whereupon the cloud suddenly becomes neutral. One of the best examples of such a structure is in Orion's eastern neighbor Monoceros, a beautifully symmetrical billow called the Rosette Nebula. Within it is a dense cluster of stars, and both nebula and cluster are centered in the huge Monoceros OB 2 association (the second OB association in Monoceros). Both it and the Orion Nebula show superimposed dark clouds that contain dust grains mixed within the gas, the same kind of dust that gives the Milky Way its structure.

The northern summer sky, which contains the thicker part of the Milky Way, is richer ground. In Sagittarius is the most visible of all diffuse nebulae, the Lagoon, M 8. Just above it is the delightful Trifid Nebula, M 20. Better than any other, it illustrates the effect of stellar temperature and the division between the upper and intermediate zones of the main sequence. The top part glows bright red, the color produced by the Hα recombination line, showing that the gas is ionized and that the illuminating star must be hot and massive. The lower part contrasts smartly in blue. This part of the nebula is illuminated by a cooler (but still blue) B star. It throws out insufficient ionizing photons, and therefore the gas remains neutral. The blue starlight, however, is scattered by dust grains that are always intermixed with the gas to produce a reflection nebula. Proceeding northward into Cygnus we find H II regions that are more spread out. Near Deneb is the North America Nebula, NGC 7000, which can actually be outlined on a dark night with the naked eye.

The southern hemisphere has even richer treasures, most notably the Eta Carinae Nebula. This huge complex is lit by enormously massive stars and contains both the weird supergiant after which it is named and the great luminary of the Galaxy itself, HD 93129A. For the greatest of all H II regions, however, we have to leave our own Galaxy and go to the Large Magellanic Cloud. There, off to one side, is the Tarantula Nebula, so immensely bright that it actually received a Flamsteed number, 30 Doradus. This object, called a giant H II region, is so large that if placed at Orion Nebula's location 420 pc away, it *would fill the entire constellation.*

For reasons we do not really understand, giant H II regions seem to inhabit the smaller galaxies and the spirals that have wide-open arms. Our own Milky Way has none. They are so large that they can be illuminated only by huge crowds—clusters—of O and B stars. Some of the most massive stars known are in 30 Dor, brilliant O stars that will one day evolve with devastating effect.

SUPERGIANTS

Before it reaches its finale, however, an O star evolves first into a supergiant. The name implies that these stars are just extreme examples of the giants; they are not. True, there is a continuum in empirical luminosity classification from the subgiants through the red giants to the red supergiants—just look at any of the HR diagrams. In an evolutionary context, though, the red supergiants are profoundly different. At their peaks, the more massive of the AGB stars may reach absolute bolometric luminosities that are actually greater than those of the lesser supergiants, and even by visual magnitude they can be rivals. Yet they remain giants because they will eventually produce white dwarfs. The supergiants, which have evolved from high-mass stars, above 8 or 10 solar masses, will not.

The relation between the upper main sequence and the supergiants is evident from stellar distributions. The giants, progeny of the intermediate main sequence, are found everywhere, even far out into the galactic halo. Supergiants, on the other hand, are spread thinly along the plane of the Milky Way just like the O and B stars. Furthermore, they inhabit the same OB associations; red Antares is part of Scorpius OB2 and Mu Cephei, a member of Cepheus OB2. Since the red supergiants are still close to their place of origin, the time scale of evolution to the red supergiant state must be very short, measured only in millions of years, again confirming evolutionary theory. Finally, we can find red supergiants coupled to class B companions. Since to have evolved first they must initially have been the more massive of the pair, they have to be descendants of B or even O dwarfs.

The internal evolution that develops the supergiants is at first similar to that leading to the giants. As hydrogen is consumed in the deep cores of these stars, the points that represent them on the plane of the HR diagram move up and to the right, giving considerable breadth to the main sequence. By the time the initial fuel has disappeared, most of them have changed into Bl or B2 dwarfs, and their effective temperatures have cooled to the point where they can no longer ionize any surrounding gas. Diffuse nebulae, if any are present, will shortly turn into reflection nebulae. As the now-quiet helium cores contract under gravity's squeeze, the stars depart the main sequence and cool at nearly constant luminosity. They first become blue supergiants like Orion's Rigel and then pass into type A, to appear like Deneb.

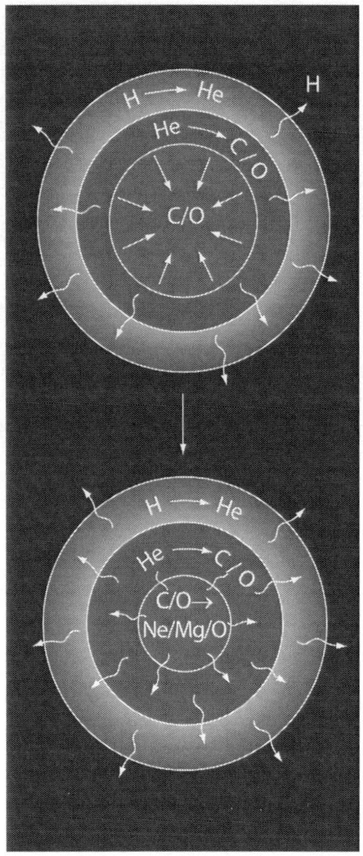

The evolution of a supergiant differs from that of a giant. The carbon—oxygen core contracts and the carbon eventually beams to burn to a mixture of neon and magnesium and oxygen. The entire nuclear-burning region, which includes shells of fusing helium and hydrogen, is well past the Chandrasekhar limit, and is not degenerate. No white dwarf can form.

From about 40 solar masses down, the stars continue cooling all the way through class K. As they enter M, the core temperatures reach ISO million degrees, and the helium will again fire up and burn via the 3-α process into carbon and then oxygen. The stars have now arrived in the realm of the classic red supergiants and are monsters like Antares, Betelgeuse, and Mu Cephei. Yet the red supergiants never get as cool as the effective temperature advanced AGB stars, another difference between the two kinds. Because of rapid evolution through the middle part of the diagram, F and G supergiants are rare. As a result, these massive evolved stars divide into two broad classes, cool red and hot blue. The ratio of the two groups is an important test of astrophysical theory, which passes. [missing word here] Below about 12 M□ the evolutionary tracks on the HR diagram cross the instability strip and the stars become the most luminous Cepheids, first during the transit from blue to red and then as their tracks loop back and forth.

Above 40 M□ the evolutionary pattern changes, largely as a result of extreme mass loss. The stars do not become red, but instead fire their helium within class A or the hot end of F. There are few to start with and their evolution is rapid, so there are not many of them to admire. Finally, whatever the mass might be, the helium runs out, resulting in a star with a carbon-oxygen core surrounded by burning shells of helium and hydrogen. The speed of the process is astonishing. A 20-M□ star will complete hydrogen burning in only 8 million years, and helium burning in a mere 1 million. A million years may seem like forever to us, but remember that the Sun takes 10 *billion* years just to complete hydrogen burning.

After this point, the internal evolution of upper and inter-

Supernova 1987A was a gleaming jewel set into the Large Magellanic Cloud near the Tarantua Nebula. The star that produced it, a B1 supergiant of 18 solar masses, developed an iron core that suddenly became a neutron star, collapsing with energy sufficient to blast the outer layers into space. A much-magnified view two years later showed the star faded to 13th magnatude. Light from the blast had hit gas ejected earlier from the evolving B star, however, creating a great, glowing ring.

mediate main sequence stars is totally different. The AGB stars quit burning at their centers when they develop degenerate carbon—oxygen cores. Their masses are too low to allow the interiors to get hot enough to fuse the carbon into other elements. Core contraction then stops as a result of degeneracy pressure. In contrast, the supergiants are so massive that the burning interiors are beyond the Chandrasekhar limit. However, their higher temperatures and lower densities keep them from becoming degenerate and collapsing. When their contracting carbon cores reach an incredible billion degrees, the carbon begins to fuse, creating another stabilizing energy source and a growing mixture of neon, magnesium, and oxygen. From around 20 solar masses down, C burning takes place among the red supergiants. It is impossible to identify specific stellar candidates. The phase takes such a short period of time that their number must be very small. In the upper reaches of mass, above about 40 $M_{\odot}$, the stellar evolutionary tracks loop all the way back across the HR diagram, and the stars are reborn as blue supergiants before carbon is ignited.

The end of evolution is not yet at hand. The Ne-Mg-O mixture will eventually burn into something else, the fusion machine creating heavier elements until it is stopped by sudden death. But there are yet more acts before the curtain is lowered. The supergiants are so luminous that they lose mass at a prodigious rate, their extreme winds producing some of the stranger celestial performers.

TRANSFIGURATION

Mira, the prototype of a giant long-period variable, displays emission lines of hydrogen in its spectrum and is therefore known as an M7 IIIe star. The pioneers of spectral classification gave the O stars last place in the stellar alphabet because many of them were also seen to have emission lines. It was not until well after tens of thousands of stars had been typed that Antonia Maury realized on the basis of continuity of the absorption lines that the O stars should be placed at the beginning of the sequence, not at the end. The O stars display such a variety of emissions that the simple letter "e" does not do justice to them. Some have emissions of hydrogen, others display nitrogen and helium; the latter have f rather than e appended to their spectral classes. The e and f suffixes signal powerful mass loss, a unifying characteristic of supergiants.

For the most massive stars, the Oe and Of states are the first observational stages following the main sequence. As they brighten, these stars begin losing mass and become surrounded by circumstellar shells that produce the emissions. Matter is being stripped away. At the same time, some of the by-products of thermonuclear burning can make their way to the surface to enhance the nitrogen lines that give rise to the Of class. The most intrinsically luminous star known, HD 93129A, which is involved in the Eta Carinae Nebula, is an O3 If supergiant.

In his *Uranometha* (1603), Johannes Bayer named a relatively obscure star in Cygnus "P"; it is one of the few that still retain their old roman-letter designations. The spectrum of this star is outlandish, showing weird emission lines that are flanked to the violet by absorptions. These features are the special signatures of dense winds. The emissions again show that a hot, low-density gas surrounds the star. But along the line of sight the wind is superimposed on the stellar surface, and it is dense enough to create deep absorptions that are Doppler-shifted to shorter wavelengths. All emission-absorption combinations of this kind have become known as P Cygni lines. They are a fount of information. The point of minimum wavelength at which we see absorption corresponds to the matter that is coming directly at us, allowing the determination of the wind's velocity. From the strengths of the absorptions and emissions, the amount of matter in the wind can be found, which when combined with the velocity yields the mass-loss rate. The Of stars, those near the beginning of their evolution, display P Cygni features in the deep ultraviolet parts of their spectra. Here we find spectrum lines that arise from the lowest orbits, the ground states, in which most of the atoms reside. At these wavelengths the gas is sufficiently opaque to form the absorption

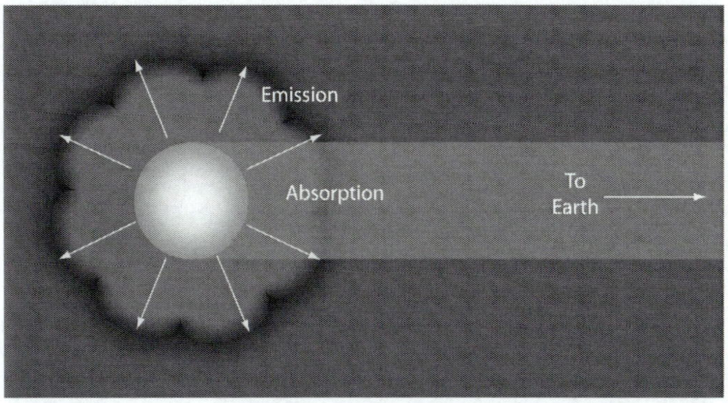

The weird spectrum of P Cygni has emission lines coupled to absorptions on their violet edges. Such lines are formed by a rapidly expanding cloud of gas that results from mass loss. Most of the low-density cloud will produce emissions; however, some of the matter is flowing directly at the Earth with the star in the background, and will instead create absorptions. The high velocity of the gas will Doppler-shift the absorptions to shorter wavelengths, producing emission lines flanked on the blue side by absorption lines.

components. The extreme Doppler shifts of the absorptions indicate winds that can blow in excess of 4000 km/s. In extreme cases, the stars create visible bubbles and have powerful impacts on their surroundings.

P Cygni itself, a Bl lae supergiant ("la" for the brightest of the super-giants, "e" for hydrogen emission), is one of the brightest stars in the Galaxy, with an absolute bolometric magnitude approaching -11. It is an extreme example of a windy star. There is so much matter surrounding it that even the optical lines, those that arise from the less populated higher orbits, can form absorption components. It is losing matter at the astonishing rate of 10^{-4} solar masses per year. Since Homeric bards composed the *Iliad* the star has pumped a third of a solar mass into space! And it does not do so quietly. In 1600, before P Cygni acquired its name, it had blossomed to first magnitude and was recorded as a nova, which it definitely is not.

It appears to be related to an even odder star, Eta Carinae, after which the vast southern nebula was named. Eta itself is now visually obscure, of the sixth magnitude and only barely visible to the naked eye. But in 1844 it was brilliant, second only to Sirius. Then it began to fade, and only a dozen years later had dropped out of sight to seventh magnitude. It has only brightened a bit since, and even more remarkable, its spectrum has changed too. In the 1890s it appeared as an F supergiant. Now in the optical part of the spectrum it displays neither absorptions *nor* emissions. Instead it is hidden in a dusty nebula called the homunculus that radiates its own emission lines.

But a star of class F cannot possibly ionize a nebula. We believe that η Car actually contains a masquerading B supergiant that has lost so much mass that it has simply buried itself in its own effluvia. The class F spectrum was not that of the star, but of the thickening surrounding shell. Filaments of gas around the bright nebula show directly that this kind of activity has been going on for a thousand years or more. One hundred fifty years ago, Eta Carinae suddenly erupted, producing an extraordinary burst of matter in a wind outflow as high as a tenth of a solar mass per year in which as much as a solar mass may have been lost. Today we see the result of the event as an immense bipolar flow set perpendicular to a thick central waistband. The culprit itself is among the most massive stars in the Galaxy, weighing in at 100 Suns.

Eta Carinae seems to be an example of one of the rarest breeds of stars, the luminous blue variables (LBVs). P Cygni is another, and so may the star Rho Cassiopeiae, which in 194S dropped from magnitude 4.S to 6 as its spectrum wandered from F to M; it even developed TiO bands! Again, the phenomenon responsible was probably a dusty shell caused by a mass outflow rate of 10^{-4} solar masses a year. These stars are so bright that they can easily be seen over intergalactic distances. Four of them, originally called Hubble-Sandage variables after their discoverers, are seen in the nearby spiral galaxy M 33. Like Rho Cassiopeiae, one of them recently dropped 3.5 magnitudes as its spectrum changed from F to M. The small number in a whole galaxy reflects the rarity of the stars at the top of the main sequence. We have no idea what causes such erratic changes in mass-loss rates.

No less strange are the Wolf-Rayet stars. These show only broad emission lines and no trace of the most common of all elements, hydrogen, although helium abounds. Mass loss seems to have been so vigorous that the stars *have lost their entire hydrogen envelopes*. They come in two very distinct varieties, carbon-rich (WC) and nitrogen-rich (WN). The WN stars have a normal complement of carbon and oxygen, but the WCs have no nitrogen at all. The ionization level of the Wolf—Rayet stars is very high; lines of He^{+}, C^{+3}, and N^{+3} dominate. The temperatures required to produce such ions relate these stars strongly to class O, although the emissions are so powerful that we are really unaware of the stellar surface, if one actually exists. The strong broad lines indicate current mass-loss rates of 10^{-5} to 10^{-4} $M_{\odot}$ per year.

Several Wolf-Rayet stars are members of binary systems, and one, γ Velorum, is even visible to the naked eye. From orbital analysis, we find that their masses are high, averaging about 20 $M_{\odot}$, but nowhere near the mass limit. Their companions, however, are invariably O stars that are closer to the top, from which we deduce that the W-R stars have lost at least 40 percent of their original matter, consistent with the current mass-outflow rate. The stars have lost so much of themselves and are so hot and luminous that many are surrounded by extraordinary gaseous nebulae of their own making, massive analogues to planetary nebulae.

We think that the two varieties may be related to different modes of nuclear fusion. The high nitrogen in the WN stars is a by-product of the carbon cycle, which explains why carbon is also present. On the other hand, the carbon-rich variety may be the result of the next stage, helium burning. If that is the case, then the WC should succeed the WN; but the masses of the two average about the same, so we cannot prove on that basis that evolution has taken place from one kind to the other.

The blue main sequence and supergiant stars have higher maximum luminosities than do those on the red side. There seems to be a distinct upper envelope on the HR diagram, called the Humphreys-Davidson limit, that slopes downward toward lower temperatures among the O and B stars and then levels off at about B3. The flat portion corresponds to stars of about 40 solar masses. Below that limit, stars can evolve horizontally all the way across the diagram, changing from blue supergiant to red. Much above 40 $M_{\odot}$, however, the enormous luminosities produce such high mass-loss rates that the stars' redward evolutionary progressions on the HR diagram stall within class B, and they are prevented from becoming red supergiants. The LBVs and other stars with extensive surrounding shells lie on the Humphreys-Davidson limit, P Cygni and probably η Carinae right on the high-mass descending slope. They are now stripping away their hydrogen envelopes, and before our eyes may be turning themselves into Wolf-Rayet stars. In the distant future astronomers may see the most massive component of Eta Carinae as such a star. It will be accompanied by its less massive class O companions, which will not yet have had time to evolve, much as we see so many WR stars today.

SUPERNOVAE

Wolfgang Schuler, a resident of Wittenberg, must have been amazed the night of November 6, 1572 (by the Julian calendar) when he looked up to find a new resident shining among the stars of Cassiopeia. Five days later that eminent surveyor of the skies, Tycho Brahe, saw the same object for the first time, his view of it almost certainly delayed by obscuring Baltic Sea clouds. The star overwhelmed its stellar neighbors, at its peak reaching apparent magnitude -4, rivaling Venus, and visible in full daylight. Though many people saw it, only one thought to record its progress, and it is known in commemoration as Tycho's Star. After 100 days it was still as bright as Vega, magnitude zero, and it did not completely fade from sight until March 1574. Ordinary novae are not all that uncommon; in this century alone eight have attained first or second magnitude. However, the brilliance of Tycho's set it apart. Another, which appeared in October 1604 in Ophiuchus, was similarly striking—this one described by the great Kepler, Tycho's one-time assistant. Kepler's Star outshone Jupiter and reached apparent magnitude -2.5, finally disappearing in the winter of 1605. More than sheer brilliance made these two special, though their true natures were not to be recognized for another 330 years.

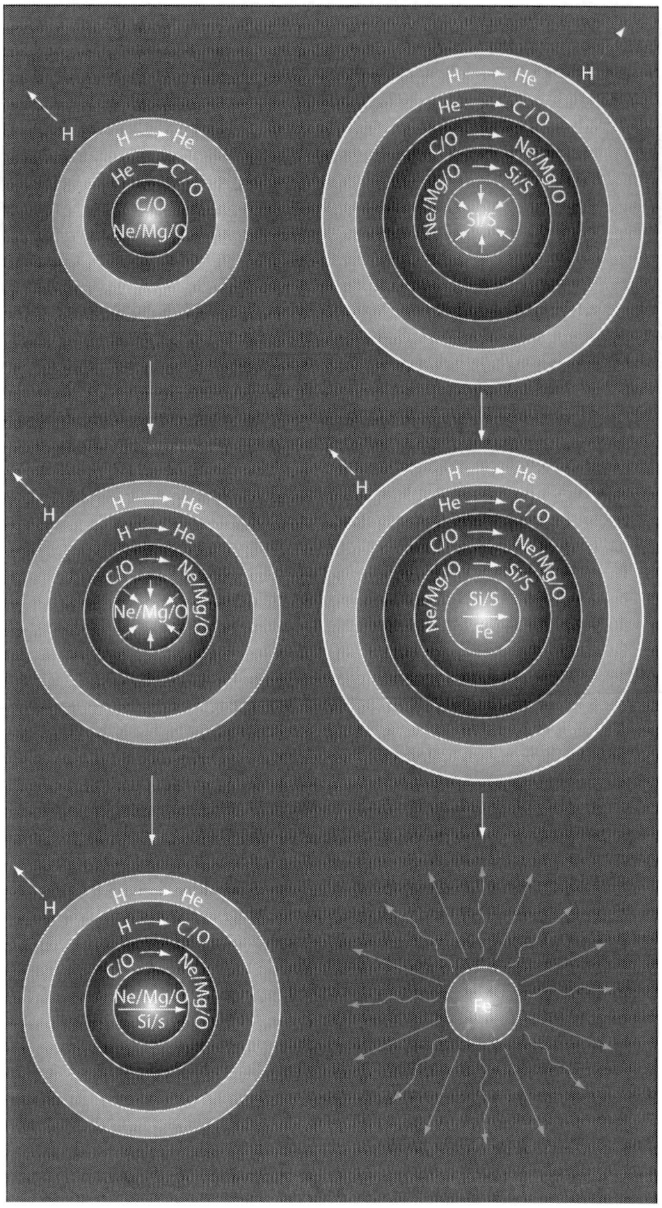

A simplified cross section of a developing supernova shows a layered structure. Carbon first burns to neon and magnesium as above. The dead Ne-Mg-0 core contracts until it is hot enough to fuse into a mixture of silicon and sulfur, the silicon eventually burning to iron. The whole affair is wrapped in a huge blanket of hydrogen (the shells are not drawn to relative scale). Most of the burning takes place at the bases of the individual shells. In reality there are many intermediate isotopic products. Iron cannot fuse into anything; once formed, in a fraction of a second it undergoes a catastrophic collapse that sends a blast wave ripping through the outer stellar layers. A supernova is born.

Another brilliant so-called nova, S Andromedae, erupted in the Andromeda Galaxy, M 31, in 1885. Others were seen in similar spiral nebulae. When astronomers finally realized in the 1920s that these fuzzy objects were actually distant external galaxies, it became clear that the erupting stars were not ordinary novae and that there was a whole new class of stellar violence out there, one with unimaginable power. The M 31 event was only 2.5 magnitudes—a factor of 10—

fainter than the entire stellar system. These supernovae do not involve a simple thermonuclear reaction on a stellar surface, but the near or even total destruction of the star itself.

Supernovae are rare beasts. The Chinese and Japanese recorded a brilliant one in Taurus in 1054 that may have reached apparent magnitude -5. Chinese records show two other likely possibilities that occurred in 1006 and 1181. None has been seen in our Milky Way since Kepler's Star, making five in the millennium, randomly distributed. From the theory of stellar evolution, we actually expect anywhere from one to four per century in the Galaxy, but most have probably been hidden behind thick clouds of galactic dust. The local scarcity of supernovae has forced astronomers to examine them in other galaxies, but their distances are so great that we can see little detail. The only saving grace is that there are so many other galaxies that a few supernovae are discovered each year. Every astronomer alive has dreamed of seeing at least one within our own system. Thus it was with great relish that we watched a supernova blossom in the Large Magellanic Cloud in February 1987. Even though 52,000 pc away—nearby on a cosmic scale—it had such power that it reached third magnitude. Within a few days it was evident that the source was an ordinary 18-solar-mass Bl supergiant called Sk -69°202, confirming—but for a time confounding—theories of how massive stars are supposed to die.

These wonders of the sky have been watched carefully since the 1930s, when Fritz Zwicky of Mount Wilson began to classify them, dividing them into two quite different groups. The more common, Type I, can reach an astounding absolute magnitude of -18 or -19 and decay relatively quickly after outburst. If one of them were to go off at the distance of Vega, it would equal the brightness of a hundred full Moons. Type I events also have strange spectra with no hydrogen lines, their definitive characteristic. Expansion velocities from P Cygni lines reach a remarkable 10,000 km/s, 3 percent the speed of light. Type II supernovae, a couple of magnitudes fainter, usually show a plateau in their light curves, and do have hydrogen lines; their expansion velocities are about half those observed in Type I.

The places of origin of the two types are also different. Type I events occur in galactic disks, where most of the stars reside, but also in elliptical galaxies and in the bulges and halos of spirals, showing them to be related to lower-mass stars. Type II explosions, on the other hand, confine themselves to galactic disks and to spiral arms where only the massive stars are found. So with some consternation we find Types I and II respectively associated with Populations II and I. When we look more closely, we see the types and subtypes proliferate. The most important of these is a division of Type I into the la, the original Zwicky class, and 1b, which although they occur in galactic arms still do not have hydrogen lines in their spectra.

Work by a generation of astronomers and physicists has allowed an understanding of how these stellar bombs work. A Type II explosion is the natural end of a massive star. As nuclear burning proceeds, the particles in the nuclei become progressively more tightly bound. Each burning stage can provide less total energy and consequently lasts a shorter period of time. It took but a million years to fuse the helium of a 20-solar-mass red supergiant into carbon. Carbon burning (which produces neon and magnesium) then takes less than 100,000 years. When the carbon is gone, the core resumes its relentless contraction and heats until the oxygen residue fires up and begins to burn into silicon and sulphur: *That takes less than 20 years.* (The process is in fact much more complex and has several intermediate steps.) Then, *in a week*, the silicon turns to iron. The temperature is over 3 billion degrees, and the reactions generate more *energy* in neutrinos than they do in photons. The supergiant is now layered like an onion as each stage of nuclear burning moves outward in a shell around an iron core of nearly 1.4 solar masses.

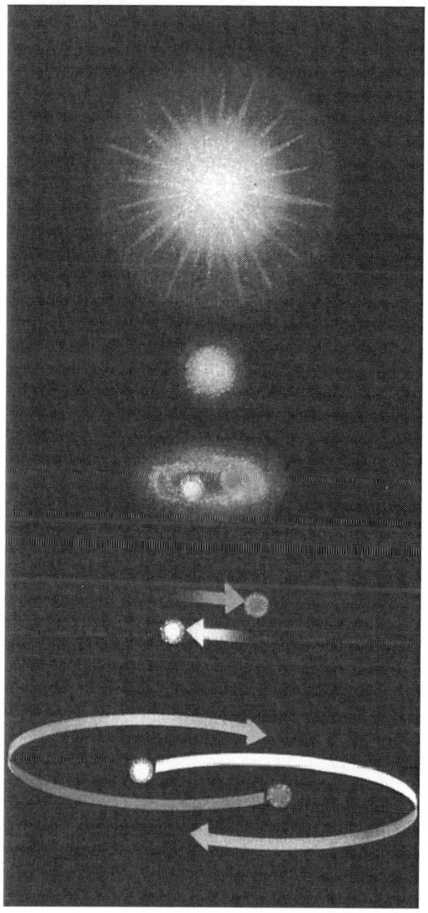

Two white dwarfs find themselves in proximity as a result of prior evolution, in which each red giant phase encompassed the other star and brought it nearer. Radiation of gravity waves (a [?] effect) slowly draws them closer, after which a common tidal envelope drags them together. Because the combined mass exceeds the Chandrasekhar limit, the now-single star collapses and explodes, probably annihilating itself.

Iron is the most tightly bound of all atomic nuclei. No energy can be gained by its fusion: It is the end of the line. As the silicon-burning phase comes to an end, the core, about the size of Earth, is near the Chandrasekhar limit and is briefly supported by degenerate electrons. The iron nuclei now become subject to attack. The density is so high that the electrons start to combine with them to form manganese, and the heat is so fierce that extremely energetic gamma

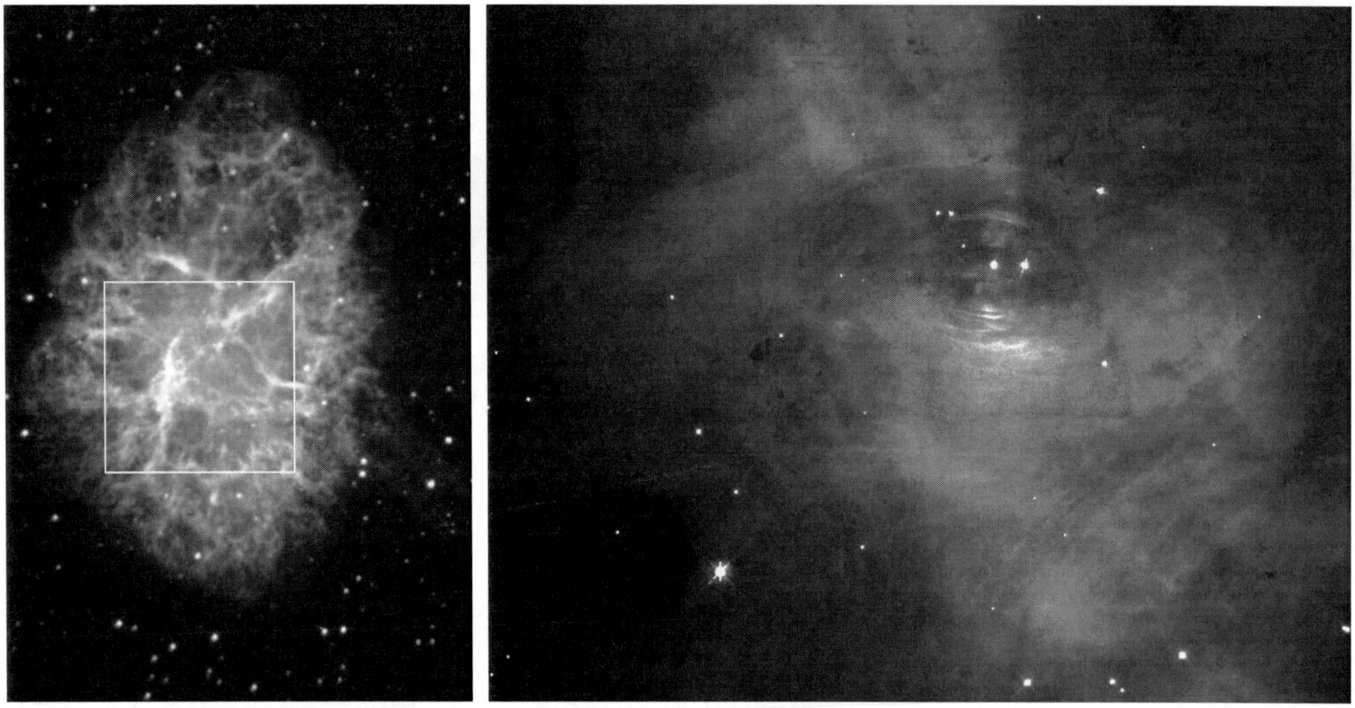

Light from the Crab pulsar, visible over the entire electromagnetic spectrum. Hubble shows us the pulsar's powerful effect on the surrounding gas, a scene that changes over a period of months.

rays penetrate them and begin to break them back down into helium nuclei. With electron degeneracy support and gamma-ray energy both being removed from the interior, the core contracts faster and faster, then goes into a catastrophic collapse. The star has lived for 10 million years. In less than a tenth of a *second* the iron core flies inward at a quarter the speed of light to a sphere only 100 kilometers across. The gravitational energy released is beyond imagining. In that blink of an eye the star expends over 10^{46} joules—more than 99 percent of it in neutrinos. This exceeds the energy output of all the other stars in the Universe; it is 100 times more than the Sun has radiated in its lifetime.

The density near the center now becomes so great that the protons and electrons of the inner part begin to coalesce into neutrons, which condense into a ball ultimately no more than 10 to 20 km across. At the center, the temperature may initially be 200 billion degrees. The sudden implosion creates a shock wave that propagates outward. The envelope is so dense that even the neutrinos, which can normally penetrate a light-year of lead, have to fight to get out, contributing critical pressure to the blast wave that is blowing the remainder of the star apart. Only the central core is left, a shrunken star supported by the pressure of degenerate neutrons.

In the envelope's nuclear fury vast numbers of neutrons are created, which rapidly attach themselves to highly radioactive isotopes. This r-process can build heavy isotopes that the s-process of the giants cannot reach. Several solar masses of stellar material, enriched in the elements created in both the supergiant and the supernova phases, are then suddenly returned to interstellar space at several thousand kilometers per second.

Evidence for this immense nuclear activity lies in the light curve. The nuclear reactions burn the exploding envelope all the way to ^{56}Ni, a radioactive isotope with a half-life of only six days. The nickel decays into ^{56}Co, which after 57 days produces an excited nuclear state of ^{56}Fe. This isotope then relaxes into normal iron with the release of a gamma ray, which heats the gas. The star then dims in accord with the nuclear decay rates, demonstrating a huge mass of newly made metal blasting back into space.

What of the-even brighter Type I supernovae? Because they appear in the old galactic halos, they cannot be caused by massive stars. Instead, we believe these events to be the final acts of white dwarfs in binary systems. There are two related possibilities. An ordinary nova is produced when a main sequence companion is tidally stretched and feeds matter onto the surface of a white dwarf, which then explodes and returns to normal. If the degenerate is close to the Chandrasekhar limit, it is possible that before a surface explosion can occur the incoming mass may push the star over the edge. Then degeneracy pressure can no longer hold it up, and the white dwarf collapses—the heat causing the whole star to ignite in a vast nuclear bomb. In a Type II supernova, the energy is generated by gravitational collapse that drives nuclear reactions; in the brighter Type I variety it is generated by the nuclear reactions themselves, evidence for which is again seen in the decay of ^{56}Co into iron. No iron or neutron core can survive; the star annihilates itself.

The other possibility involves double white dwarfs. The stars are first brought close together during red giant evolution when the expanding envelope of one embraces the other. As the white dwarfs orbit, they emit gravity waves, outward-propagating disturbances in their gravitational fields (they have been predicted by relativity theory but have never been directly observed). This emission causes a loss of energy that makes the stars slowly spiral together. Eventually, they merge as a result of tidal interaction, exceed the Chandrasekhar limit, and—-as in the first model—explode. Tycho's and Kepler's stars are both thought to be Type I.

Type Ib events seem to be core-collapse supernovae like Type Us that somehow lost their hydrogen envelopes during evolution. Possibly they were gravitationally stripped as supergiants by a white dwarf or even a main sequence companion. Such action is going on even now in the massive supergiant VV Cephei, which is passing matter onto an O dwarf. Or Type Ib supernovae may once have been Wolf-Rayet stars.

Collectively, the supernovae are partly responsible for the synthesis of the elements beyond helium, and are the sole creators of the heaviest. Supernovae may be rare, but they are potent. In the 13-billion-year lifetime of the Galaxy, maybe a quarter billion have occurred. If each releases an average of 10 solar masses (Type II more, Type I much less), the events have returned the equivalent of 2 billion Suns into space, a notable fraction of the Galaxy's total mass.

Supernova 1987A provided dramatic proof of the theories of Type II supernovae. It was first seen on February 24 by the Canadian astronomer Ian Shelton, who had been taking survey photographs of the Large Magellanic Cloud. The object was already brilliant and visible to the naked eye; it was actually photographically recorded on the rise by Robert McNaught, an Australian amateur, nearly a day earlier. But that is not the most remarkable part of the story.

Lying deep in the Kamioka zinc mine in Japan and in the Morton salt mine under Lake Erie are two massive pools of dark water designed to detect flashes of light from decaying protons. The Kamioka detector has been used to observe neutrinos from the Sun. At $7^h\ 35^m\ 35^s$ on February 23, two hours before McNaught caught the supernova, the photocells that line the walls of the Kamioka detector registered that 11 neutrinos had smashed into the detector's water pool and that they had penetrated the Earth from the direction of the Large Magellanic Cloud; six seconds later, within the error of measurement for the instrument in Japan, 8 more hit the Morton mine. Given the incredibly small chance of an interaction between a neutrino and an atom, some 10 billion neutrinos must have passed through every square centimeter of the Earth (and of *you*). This number was (within the standard margins of error) that expected for a core-collapse supernova of a 20-solar-mass star 52,000 pc away. The detectors had caught the exact moment of the collapse of the core even before we saw the surface of the star respond to the event! There have been few triumphs of science to equal this one.

However, a surprise was in store. Everyone expected that the exploding star would be a red supergiant, a really massive star that had returned to the blue state, or perhaps a luminous blue variable—Eta Carinae had long been considered a prime candidate. But no one anticipated that the first nearby supernova event since the invention of the telescope would be an everyday Bl supergiant with a relatively modest mass. For a time astronomers thought that Sk -69°202 might be just a foreground star, and that a red supergiant lurked behind it. But the two-hour delay between the neutrino arrival and the beginning of the optical outburst was consistent with the relatively small radius appropriate to a B star. The light curve, moreover, was strange, not at all typical of a Type II event. Instead of coming quickly to a peak and then decaying, the star first dropped in brightness and then leisurely took nearly *two months* to come to maximum. Were the theories of internal structure and evolution wrong?

The explanation lies in the low metal content of the Magellanic Clouds. In our Galaxy this star *would* have been a swollen red supergiant, but in the Large Magellanic Cloud it had shrunk, a result of low atmospheric opacity. The small size also resulted in very high temperatures at the optical surface of the expanding debris and the production of the radiation in the invisible X-ray and ultraviolet. As the stellar surface cooled, more radiation entered the visible and the star continued to brighten to the eye in an extreme example of a decreasing bolometric correction. Four months after the explosion, a tenth of a solar mass of radioactive cobalt became visible in the debris. The light curve then dropped as the cobalt decayed into the same amount of iron. We even saw the resultant gamma rays.

The aftermath of a supernova is as extraordinary as the event itself, both in the material that is blown outward and the mass that remains behind. We look at the exploded debris first.

SUPERNOVA REMNANTS

Number one on Charles Messier's list is an easily visible fuzzy ball near Zeta Tauri, which because of its filamentary appearance is called the Crab Nebula. As early as 1921 it was identified with the great Chinese "guest star" of 1054, the brightest supernova on record, which appeared in about the same location. By 1941 the association was secure. Over the years astronomers have watched the filaments moving outward from the center at a rate of about 0.2 seconds of arc per year. To reach its current angular diameter of 3 minutes of arc, it should have begun its expansion near the year 1100. The Crab produces hosts of emission lines from which a radial expansion velocity of 1300 km/s is derived. Combination of the angular and radial expansion rates yields a distance of about 2000 pc. Allowing for some dimming of the light by interstellar dust, the absolute visual magnitude must have been around -17, about right for a Type II supernova. Over nine hundred years after the great blast, we are seeing the enriched churning debris, a supernova remnant (SNR) returning to space whence it came.

Visual remains like the Crab are scarce. To the south of ϵ Cygni is the bright Cygnus Loop, which consists principally of two arcs of filamentary gas 2° apart—the Filamentary and the Veil nebulae—and is the remnant of a star that exploded some 10^5 years ago. In the southern hemisphere, we find the huge Gum Nebula (named after its discoverer, Colin Gum) sprawled among the stars of Argo. The debris of Tycho's and Kepler's stars are far less visible, consisting of only a few expanding wisps.

Most supernovae, whether Type I or Type II, occur in the galactic plane where optical observations are seriously impeded by the absorbing effects of thick clouds of interstellar dust. To study SNRs seriously, we must use radio waves, which penetrate the dust without attenuation. In the early days of radio astronomy sources were named in order of discovery with roman letters following the constellation name; the Crab, for instance, is also called Taurus A. In contrast to their feeble optical appearances, the SNRs as a class are exceedingly luminous in the radio. Cassiopeia A, the remnant of an unobserved supernova in the seventeenth century, is one of the brightest radio sources in the sky. Some ISO of these objects are now "seen," including Tycho's, Kepler's, and the supernova of the 1006 event.

The Crab, the first clearly defined SNR, is the more uncommon of two basic kinds. It is filled with brightly radiating gas; the other kind consists of hollow shells. In the youngest ones, we see mostly the gas that was blown out of the supernova. As they age, the shock wave created by the detonation begins to sweep up ambient interstellar gas, incorporating it into the stellar material. By the time they are old, all we see—as in the Cygnus Loop—is the outline of the shock, now diminished in speed to a few hundred kilometers per second. SNR structures are, thus, controlled primarily by the prior distribution of interstellar gas. Where it is dense, the outwardly moving gas of the SNR will be retarded; where it is thin, the gas will break through. The enriched material of the explosion is then effectively merged with the interstellar medium.

SNRs were quickly found to have distinguishing radio spectra. The distribution of energy in the continuous radio spectrum of a diffuse nebula like Orion or the Rosette is relatively constant with frequency, but that from an SNR falls precipitously as frequency increases. This radiation is caused by electrons moving in a magnetic field at nearly the speed of light, reminiscent of the movements of particles in the kind of atomic accelerator called a synchrotron. As they spiral around the field lines they emit energy in the direction of their motions. Synchrotron radiation must then be polarized (that is, the waves must oscillate in the same direction), as observed. The higher the speeds and energies of the electrons, the fewer there are of them, thus explaining the drop in intensity with frequency. For most SNRs, the expanding shock wave is ultimately responsible for the radiation. It both compresses the weak ambient magnetic field of the Galaxy to a sufficient strength and provides the energy needed to accelerate the electrons. The Crab Nebula is so energetic that the synchrotron radiation extends right into the optical spectrum, where it makes a visible-amorphous background to the filaments.

Optical and ultraviolet spectra generally show that the filaments radiate recombination lines of hydrogen and/or helium and also forbidden lines, mostly [O III] and [S II], typically indicating temperatures and densities of 10,000 K and 100 atoms/cm^3. In addition, we see some very highly ionized species associated with the shocks, in which temperatures extend to 1 million degrees. Even higher values are found within the hollowed-out, expanding shells. SNRs are one of the dominant lights of the X-ray sky, just as they are in the radio domain, a remarkable testimony to their huge energies. The X-ray emission shows heated gas with temperatures over *10 million degrees*. The Crab, however, is different. Its X-ray spectrum is dominated by the synchrotron mechanism, showing electrons moving at near 99 percent the speed of light.

This incredibly hot gas is expanding away into interstellar space. Long after the SNR has dissipated, a huge, hot bubble remains. Eventually it connects with expanding bubbles from other SNRs, and they weave a tapestry of hot loops and tunnels, its temperature measured in the hundreds of thousands of degrees, that pervades the entire Galaxy. The gas has the power to compress the interstellar medium and thus seems to be one of the triggers that can begin the formation of stars, the old creating the new.

PULSARS AND NEUTRON STARS

In Type II events, the iron core collapses inward, a phenomenon predicted as long ago as 1934 by Baade and Zwicky. Baade even tentatively identified the stellar remains in the Crab with a 13th-magnitude star that has no absorption lines, but he had no way of proving that the two were actually connected; that had to wait until 1967.

Anthony Hewish, a British radio astronomer, had designed a telescope to look for rapid changes in the signal strengths of pointlike radio sources, "twinkling" caused by the passage of radio waves through the solar wind. A graduate student working on the project, Jocelyn Bell, found instead a strange source in the constellation Vulpecula that turned on with a sharp pulse that recurred every 1.337011… seconds. Sometimes the pulses faded away and were completely invisible for months. She noted, however, that when they returned, they occurred at just the same intervals as before. That the source was celestial in origin was quickly obvious from its passage across the sky in a sidereal day. Even a distance could be measured. Light and radio waves travel at c only in a vacuum. In a medium they go more slowly, which is the cause of refraction. Thin as it is, the ionized interstellar gas slows radio waves and also disperses them, lower frequencies having slower speeds. Sure enough, the lower the frequency, the later the pulses were seen to arrive. An estimate of the density of interstellar space put the source—designated a pulsar—300 pc away.

Bell and Hewish knew *where* the pulsar was, but not *what* it was. Was the mechanism producing such regular signals natural, or was it evidence of another distant civilization, perhaps some kind of beacon? The only physical system that could account for such regularity was rotation, but what could be rotating fast enough to produce the observed period? The answer lay in the prescient prediction of neutron stars by the Russian physicist Lev Landau in 1932 and by Baade and Zwicky in 1934. To rotate that fast, a body would have to be very small, only 10 or so kilometers in diameter—just the size expected for the collapsed stellar remnant of a supernova. The pulsar must be a degenerate neutron star that had well over a solar mass condensed within the size of a small town spinning somewhat faster than once per second.

The clincher came when a pulsar was discovered in the Crab Nebula, one with a record-setting short period of 0.03106 seconds, clearly relating the pulsars to supernovae. The search was finally narrowed in the optical when astronomers examined the candidate stars in the Crab's center with a rapidly recording photoelectric photometer. One star was seen to glow brightly for a fraction of a second, then *disappear*. This was the one predicted by Baade 50 years earlier.

The scientific journals quickly filled with more discoveries about these odd characters, as their number grew to the 400 or so known today. The periods have a wide distribution, from that of the Crab up to about 4 seconds. Furthermore, they are not entirely regular; individual pulses can fall slightly off the average and flicker about the mean. Most pulsars are slowing down, albeit at a very slow rate. The Crab increases its period by about a ten-thousandth of a percent a day. There are also glitches, wherein the pulsar will suddenly speed up a bit only to resume its steady decay. The masses (derived from orbital effects in binary pulsars) are around 1.4 $M_{\odot}$, close to the value expected if their progenitors are supernovae.

We now have a good general explanation for all these phenomena. A normal atom is mostly empty space. Only one part in 10^{15} is actually filled with protons or neutrons. A neutron star of 1.4 $M_{\odot}$ only 10 to 20 km across can exist because the space is nearly squeezed out, resulting in a density near 10^{12}gm/cm^3, a million times denser than a white dwarf. You would attain the same density by shoving the mass of the Earth into a typical sports stadium. The rapid spin of the pulsar owes itself simply to the conservation of angular momentum. As the radius of the collapsing star decreases after a supernova, its spin velocity increases. Moreover, the magnetic field of the one-time star is collapsed along with the matter, attaining a strength some 10^{13} times that of the Earth.

The magnetic field axis of a pulsar, like the Earth's, is oblique to the rotation axis, so that it wobbles around. The moving magnetic field is thought to produce a powerful electric field that accelerates electrons along the magnetic axis at a speed nearly that of light. The result is a tightly concentrated beam of electromagnetic radiation that revolves as the star spins, appearing something like a skewed lighthouse beacon. If the Earth happens to be in the way, we receive a pulse of radiation: if not, no star is perceived at all. If the angle between the magnetic axis and rotation axis is high, we may catch a glimpse of the other pole as well as it goes past, providing us with an interpulse, such as we receive from the Crab.

The radiation derives its energy from the stellar rotation. Consequently, the pulsar must slow down with time; the youngest, like the Crab, rotate the fastest and have the highest energies, explaining why the Crab pulsar is seen not only at optical wavelengths but in the X-ray as well Indeed, the pulsar is thought to be the source of the magnetic field and high-speed electrons that give the Crab its extraordinary power. Only the youngest pulsars will be associated with SNRs because the expanding clouds dissipate long before the spinning star runs out of energy. Most SNRs have no visible pulsars, either because the star was entirely destroyed (in a Type 1 blast, for example) or because the rotation axis is not oriented to bring the beam past the Earth. It is also possible that some pulsars are produced in a stellar collapse that does *not* generate a supernova. When all the possibilities are considered, the number of observed pulsars is consistent with the galactic supernova production rate. Together, the supernovae and their stellar and gaseous remnants may be the origin of the mysterious cosmic rays, high-speed atomic nuclei constantly crashing into the Earth with energies that dwarf those created in our greatest atomic accelerators.

The observed glitches are a result of the strange internal construction of a pulsar. The little star is no longer gaseous. Its surface is thought to be a crystalline solid. Inside, the matter is in a superfluid state: that is, it has no viscosity, no friction within itself. The interior can be spinning at a rate different from that of the crust. Vortices within then interact chaotically with the crust to force the outside temporarily to spin faster. Still, our ignorance runs deep; the various theories are only partial ones, and no one yet knows the actual mechanism for the production of the beamed radiation from the magnetic field.

If this picture is not strange enough, consider another breed, the millisecond pulsar. The fastest known rotates an astonishing 885 times per second, about 30 times faster than the Crab. The explanation—as so often for bizarre behavior seen in normal stars—is stellar duplicity. The more massive member of a binary system explodes in a supernova and becomes a pulsar, which slowly decays away. The companion survives and eventually begins its own evolution. As it expands, it sends matter raining down onto the old pulsar, providing a kick that makes the pulsar spin faster, to extraordinary speeds. The energies are now so high that we see a rare millisecond X-ray pulsar, which becomes a millisecond radio pulsar after the companion has evolved into a white dwarf. We can see the effect of such a companion as orbital motion produces a varying Doppler shift in the pulse arrival times.

There are, however, two millisecond pulsars clearly *without* binary companions, apparently negating the theory. A third pulsar solves the mystery. This one is an *eclipsing* system, and from the observations of its orbit we can see that its companion has both a very low mass and a large radius. Apparently, this pulsar—dubbed the Black Widow—is in the process of consuming its mate, as the two lone millisecond pulsars have done already.

CHAPTER 11
WHY ARE BLACK HOLES SO STRANGE?

Henceforth space by itself, and time by itself, are doomed to fade away into mere shadows, and only a kind of union of the two will preserve an independent reality.
—Hermann Minkowski, 1908

The universe consists of matter and energy moving through *space* with the passage of *time*. Up to this point in the book, we have discussed the concepts of space and time as though they are absolute and distinct—just as they appear in everyday life. But what if this appearance is deceiving?

About a century ago, Albert Einstein discovered that space and time are not what they appear to be. Instead, space and time are intertwined in a remarkable manner described in Einstein's *Theory of Relativity*. Because space and time are such fundamental concepts, understanding relativity is important to understanding the universe.

The theory of relativity is *not* difficult to understand, despite popular myths to the contrary. It does, however, require us to think in new ways. That is our task in this chapter and the next. By the time we are finished, you will see that Einstein brought about a revolution in human thinking with many important ramifications for understanding our place in the universe.

2.1 EINSTEIN'S REVOLUTION

Albert Einstein was born on March 14, 1879, in Ulm, Germany. As a toddler, he was so slow in learning to speak that relatives feared he might have brain damage. Even when he began to show promise, it was in unconventional ways. Young Albert dropped out of high school at the suggestion of a teacher who told him that' he would "never amount to anything." Nevertheless, he was admitted to college in Switzerland, in large part because of his proficiency in mathematics.

Einstein hoped to teach after his college graduation, but his Jewish heritage and lack of Swiss citizenship deterred prospective employers. He therefore took a job with the Swiss patent office, starting in 1901. He devoted his free time to the study of unsolved questions in physics.

From the standpoint of history, Einstein burst onto the scene in 1905, when he earned his Ph.D. and published five papers in the *German Yearbook of Physics*. Three of the papers solved three of the greatest mysteries in physics at the time. The first paper dealt with something called the "photoelectric effect," and essentially presented the first concrete evidence of the wave-particle duality of light. The second paper explained why suspended particles in water jiggle even after the water has been still for a long time (an effect known as "Brownian motion"). Einstein's analysis led to the first direct measurements of molecular motion and molecular and atomic sizes. The third paper introduced the special theory of relativity, which forever changed our understanding of space and time. (To mark the one-hundredth anniversary of the publication of these papers, the United Nations designated the year 2005 as "World Year of Physics.")

Einstein returned to Germany in 1913, where he soon published his general theory of relativity. After some of its central predictions were confirmed by observation in 1919 [Section S3.4], Einstein's fame began to extend beyond the world of science. He received the Nobel Prize for Physics in 1921 and was by this time a household name.

Einstein was a visiting professor at the California Institute of Technology when Hitler came to power in Germany. Recognizing Hitler's evil, Einstein decided to remain in the United States. He played an important role in convincing President Franklin Roosevelt to start the Manhattan Project to build an atomic bomb. Nevertheless, Einstein was a committed pacifist who supported the project only because he feared that Hitler would otherwise develop an atomic bomb first. He expressed great dismay when the bomb was used against Japan, and he spent much of the rest of his life arguing for a worldwide agreement to ban the further manufacture and use of nuclear weapons. He also worked to promote human rights and equality for men and women of all races and nationalities. Einstein died in Princeton, New Jersey, on April 18, 1955. More than 50 years later, he remains the most famous scientist of modern times.

What are the major ideas of special relativity?

We often talk about the theory of relativity as though it were a single theory, and in some sense it is. However, Einstein actually developed his theory in two parts that he published separately. The **special theory of relativity**, published in 1905, shows that space and time are intertwined but does not deal with the effects of gravity. The **general theory of relativity**, published in 1915, offers a surprising new view of gravity—a view that we will use to help us understand topics such as the expansion and fate of the universe and the strange objects known as *black holes*. In essence, the special theory is "special" because it deals only with the special case in which we ignore the role of gravity, while the general theory is "general" because it applies with or without gravity. We will focus on the special theory of relativity in this chapter and on the general theory in Chapter S3.

Einstein developed his theories in large part with the aid of "thought experiments"—experiments that could be carried out in principle but that would be very difficult in practice. Thought experiments allow us to think through the consequences of simple ideas or statements. Let's start with a thought experiment designed to show you why the special theory of relativity is so surprising.

Imagine that, with the aid of a long tape measure, you carefully measure the distance you walk from home to work and find it to be 5.0 kilometers. You wouldn't expect any argument about this distance. For example, if a friend drives her car along the same route and measures the distance with her car's odometer, she ought to get the same measurement of 5.0 kilometers—as long as her odometer is working properly. Likewise, you would expect agreement about how much time it takes you to walk this distance. Suppose your friend continues driving and you call her on a cellular phone just as you leave your house at 8:00 A.M. and again just as you arrive at work at 8:45 A.M. You'd certainly be surprised if she argued that your walk took an amount of time other than 45 minutes.

Distances and times appear absolute and distinct in our daily lives. We expect everyone to agree on the distance between two points, such as the locations of home and work. We also expect agreement about the time between two events, such as leaving home and arriving at work. Thus, it came as a huge surprise to everyone when Einstein showed that these expectations are not strictly correct: With extremely precise measurements, the distance you measure between home and work will be *different* from the distance measured by a friend in a car, and you and your friend will also disagree about the time it takes you to walk to work. At ordinary speeds, the differences will be so small as to be unnoticeable. But if your friend could drive at a speed close to the speed of light, the differences would be substantial.

Disagreements about distances and times are only the beginning of the astonishing ideas contained in the special theory of relativity. In the rest of this chapter, we will see how this theory leads to each of the following ideas:

- No information can travel faster than the speed of light (in a vacuum), and no material object can even reach the speed of light.
- If you observe anyone or anything moving by you at a speed close to the speed of light, you will conclude that time runs more slowly for that person or moving object. That is, a person moving by you ages more slowly than you, a clock moving by you ticks more slowly than your clock, a computer moving by you runs more slowly than your similar computer, and so on.
- If you observe two events to occur simultaneously, such as flashes of light in two different places at the same time, a person moving by you at a speed close to the speed of light may not agree that the two events were simultaneous.
- If you carefully measure the size of something moving by you at a speed close to the speed of light, you will find that its length (in the direction of its motion) is shorter than it would be if the object were not moving.
- If you could measure the mass of something moving by you at a speed close to the speed of light, you would find its mass to be greater than the mass it would have if it were stationary. As we will see, Einstein's famous equation, $E = mc^2$, follows from this fact.

Although these ideas of special relativity may sound like science fiction or fantasy, a vast body of observational and experimental evidence supports their reality. They also follow logically from a few simple premises. If you keep an open mind and think deeply as you read this chapter, you'll soon understand the basic ideas of relativity.

What is "relative" about relativity?

You've probably heard people claim that Einstein taught that "everything is relative," but this claim is simply not true. The theory of relativity takes its name from the idea of *relativity of motion*. You can see why motion must be relative with another simple thought experiment.

Imagine a supersonic airplane that flies at a speed of 1,670 km/hr from Nairobi, Kenya, to Quito, Ecuador. How fast is the plane going? At first, this question sounds trivial—we have just said that the plane is going 1,670 km/hr.

But wait. Nairobi and Quito are both nearly on Earth's equator, and the equatorial speed of Earth's rotation is the same 1,670 km/hr speed at which the plane is flying (see Figure 1.12). Moreover, the east-to-west flight from Nairobi to Quito is opposite the direction of Earth's rotation (Figure 52.1). Viewed from the Moon, the plane would therefore appear to stay put *while Earth rotated beneath it*. When the flight began, you would see the plane lift off the ground in Nairobi. The plane would then remain stationary while Earth's rotation carried Nairobi away from it and Quito toward it. When Quito finally reached the plane's position, the plane would drop back down to the ground.

We have two alternative viewpoints about the plane's flight. People on Earth say that the plane is traveling westward across the surface of the Earth. Observers in space say that the plane is stationary while Earth rotates eastward beneath it. Both viewpoints are equally valid.

In fact, there are many other equally valid viewpoints about the plane's flight. Observers looking at the solar sys-

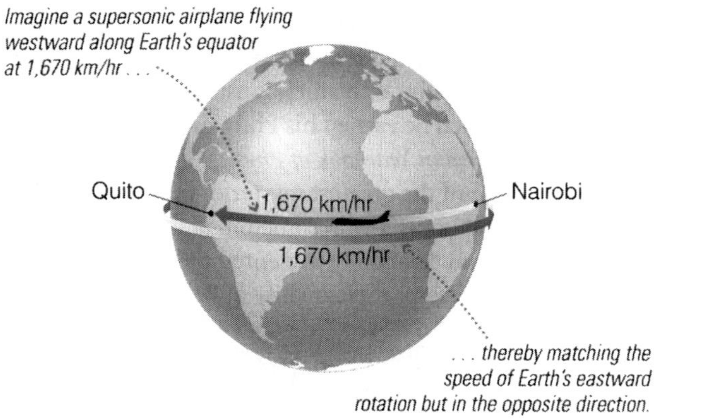

Figure 2.1 A plane flying at 1,670 km/hr from Nairobi to Quito (westward) travels precisely opposite Earth's eastward rotation. Viewed from space, the plane appears to remain stationary while Earth rotates underneath it.

tem as a whole would see the plane moving at a speed \ of more than 100,000 km/hr—Earth's speed in its orbit around the Sun. Observers living in another galaxy would see the plane moving at about 800,000 km/hr with the rotation of the Milky Way Galaxy. The only thing all these observers would agree on is that the plane is traveling at 1,670 km/hr *relative* to the surface of the Earth.

The airplane example shows that questions like, "Who is really moving?" and "How fast are you going?" have no absolute answers. Einstein's *theory of relativity* tells us that measurements of motion, as well as measurements of time and space, make sense only when we describe whom or what they are being measured relative to.

THINK ABOUT IT

Suppose you are running on a treadmill and the readout says you are going 8 miles per hour. What is the 8 miles per hour measured relative to? How fast are you going relative to the ground? How fast would an observer on the Moon see you going? Describe a few other possible viewpoints on your speed.

The Absolutes of Relativity

The theory of relativity tells us that motion is always relative, but it does *not* say that *everything* is relative. In fact, the theory claims that two things in the universe are absolute:

1. The laws of nature are the same for everyone.
2. The speed of light is the same for everyone.

As we'll see shortly, all the astounding ideas of relativity follow directly from these two seemingly innocuous statements.

Making Sense of Relativity

One reason relativity has a reputation for being difficult to grasp, despite its underlying simplicity, is that most of its ideas and consequences are not obvious in everyday life. They become obvious only when we deal with speeds close to the speed of light or, in the case of general relativity, with gravitational fields far stronger than that of Earth. Because we don't commonly experience such extreme conditions, we have no *common* sense about them. Thus, to say that relativity violates common sense is not really accurate. The theory of relativity is perfectly consistent with everything we have come to expect in daily life.

Making sense of relativity requires only that you learn to view your everyday experiences from a new, broader perspective. Fortunately, you learned to broaden your perspective in a similar way before. At a very young age, you learned "common sense" meanings for *up* and *down*: *Up* is above your head, *down* is toward your feet, and things tend to fall down. One day, however, you learned that Earth is round. When you looked at a globe with the Northern Hemisphere on the top, you were immediately confronted with a **paradox** a situation that *seems* to violate common sense or to contradict itself. Your common sense told you that Australians should fall off the Earth (Figure 2.2a), but they don't.

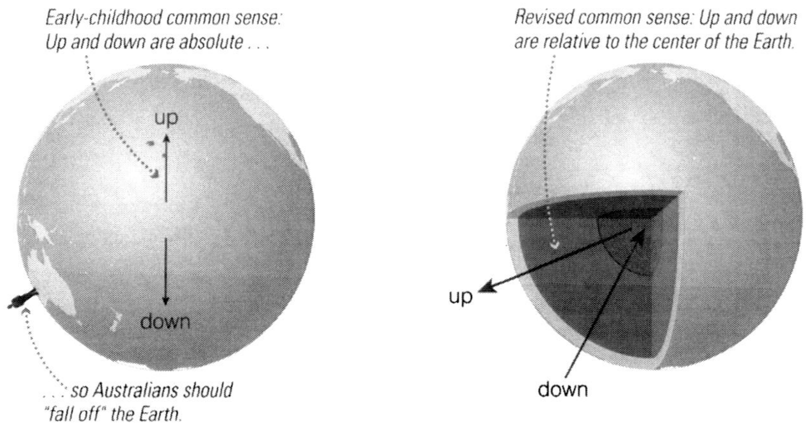

Figure 2.2 Learning that Earth is round helps children revise their "common sense "understanding of *up* and *down*.

To resolve this paradox, you were forced to accept that your "common sense" about *up* and *down* was incorrect. You therefore revised your common sense to accept that *up* and *down* are determined relative to the center of the Earth (Figure 2.2b).

THINK ABOUT IT

Why do you suppose that most maps show the Northern Hemisphere on the top and the Southern Hemisphere on the bottom? If you hung your map upside down and rewrote the words so they read right side up, would the map be equally valid?

As a teenager, Einstein wondered what the world would look like if he could travel at or beyond the speed of light. He inevitably encountered paradoxes when he thought about this question. Ultimately, he resolved the paradoxes only when he recognized that our common sense ideas about space and time must change if we are to extend them to the realm of very high speeds or very strong gravitational fields. Just as we all once learned a new common sense about up and down, we now must learn a new common sense about space and time.

2.2 RELATIVE MOTION

We have already discussed the basic ideas of relativity, but we have not yet explained how Einstein arrived at these ideas. In this section, we will begin to explore the theory of relativity by following a series of thought experiments through to their logical conclusions.

How did Einstein think about motion?

We will use slightly different thought experiments than the ones Einstein himself used, though the ideas will be essentially the same. Just as Einstein did, we will base all our thought experiments on the assumption that the two absolutes of relativity are true.

The first of the two absolutes, that the laws of nature are the same for everyone, is probably not surprising. If you're on an airplane with the shades drawn during a very smooth flight, you won't feel any sensation of motion. Thus, you should expect to get the same results from any experiments you perform on the airplane that someone else would get performing the same experiments on the ground. These equivalent results demonstrate that the laws of nature are the same in the moving airplane and on the ground. In the language of relativity, the ground and the airplane represent different **reference frames** (or *frames of reference*); we say that two people share the same reference frame if they are *not* moving relative to each other. The idea that the laws of nature are the same for everyone means that they do not depend on your reference frame.

The second absolute of relativity, that the speed of light is the same for everyone, is far more surprising. In general, we expect people in different reference frames to give different answers for the speed of the same moving object. For example, suppose you roll a ball down the aisle of an airplane. The ball rolls slowly in your airplane reference frame. However, the ball moves much faster in the reference frame of a person on the ground, because the ball has the airplane's speed in addition to the speed at which it rolls down the aisle (Figure 2.3a). Now, suppose instead that you turn on a flashlight and measure the speed of the emitted light. The idea that the speed of light is the same for everyone means that a person on the ground will measure exactly the same speed for the light beam, even though the light was emitted inside the moving airplane (Figure 2.3b). In other words, people in different reference frames can disagree about the speeds of material objects, but everyone always agrees about the speed of light, regardless of where the light comes from.

How do we know that everyone always measures the same speed of light? Einstein reached this conclusion because it was the only way he could resolve the paradoxes he encountered when he thought about traveling at speeds close to the speed of light. But observations and experiments are the ultimate judge of any scientific theory. Many observa-

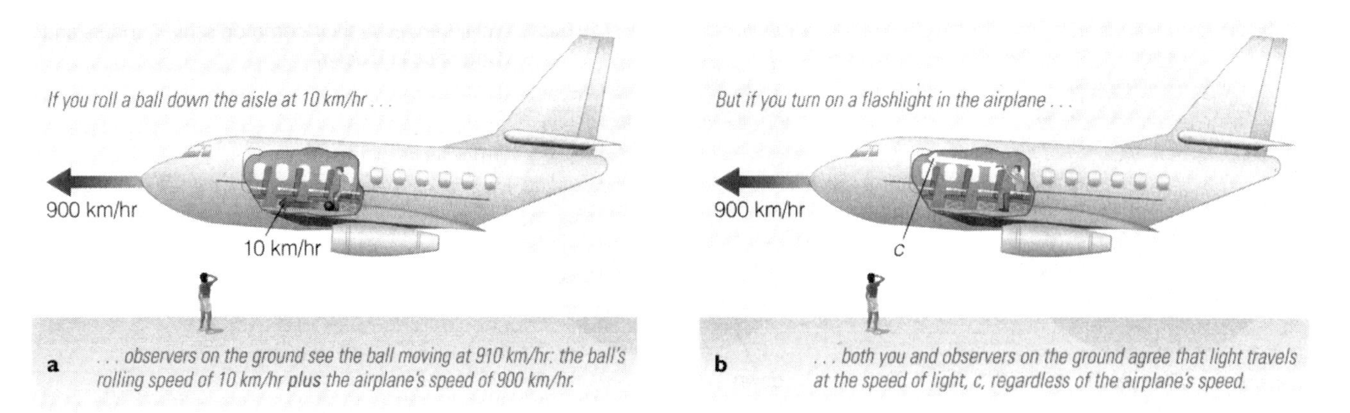

Figure 2.3 Unlike the speed of material objects, the speed of light is the same in all reference frames.

tions and experiments, some of which we'll study later in the chapter, have verified that the speed of light really is an absolute in nature. Thus, *the absoluteness of the speed of light is an experimentally verified fact.*

Experimental evidence also supports all the other ideas of relativity that we listed in the beginning of this chapter, which is why the theory of relativity is so well accepted. We'll discuss some of this experimental evidence later in the chapter. First, however, we'll see why these ideas follow logically from the two absolutes of relativity.

What's surprising about the absoluteness of the speed of light?

The idea that everyone always measures the same speed of light may not sound earth-shattering, but Einstein soon realized that it leads to far-ranging consequences—consequences that force us to let go of many of our intuitive beliefs about how the universe works. We will investigate the implications of the absoluteness of the speed of light by constructing a series of thought experiments about relative motion viewed from different reference frames. Because special relativity does not deal with the effects of gravity, the thought experiments are easiest to visualize by imagining that we are in spaceships in deep space, far from any planets or stars and drifting along without engine power. Everything in and around these spaceships is weightless and floats freely, so we call the reference frames of these spaceships **free-float frames** (or *inertial reference frames*).

Thought Experiments at Ordinary Speeds

As we will see shortly, the consequences of relativity become clear when we deal with spaceships or other objects moving at very high speeds; that is, speeds that are a significant fraction of the speed of light. We generally do not notice anything strange at the more ordinary speeds we experience in everyday life. Nevertheless, before we turn to high-speed thought experiments, it's useful to start with some thought experiments involving more familiar speeds, so that you can become familiar with the basic method we will use at both low and high speeds.

Thought Experiment 1 (Figure 2.4)

Imagine that you are floating freely in a spaceship. Because you feel no sensation of motion, you perceive yourself to be at rest, or traveling at zero speed. As you look out your window, you see your friend Jackie in her own spaceship, moving away at a constant speed of 90 km/hr. How does the situation appear to Jackie?

We can answer the question by logically analyzing the experimental situation. Jackie has no reason to think she is moving. Like you, she is in a free-float frame, floating freely in a spaceship with the engines off. Jackie will therefore say that *she* is at rest and that you are moving away from her at 90 km/hr.

Both points of view—yours and Jackie's—are equally valid. You both would find the same results for any experiments performed in your own spaceship, and your research would lead you to exactly the same laws of nature. You could argue endlessly about who is really moving, but your argument would be pointless because all motion is relative.

Thought Experiment 2 (Figure 2.5)

We begin with the same situation as in Thought Experiment 1, but this time you put on your space suit and strap yourself to the outside of your spaceship. You happen to have a baseball, which you throw in Jackie's direction at a

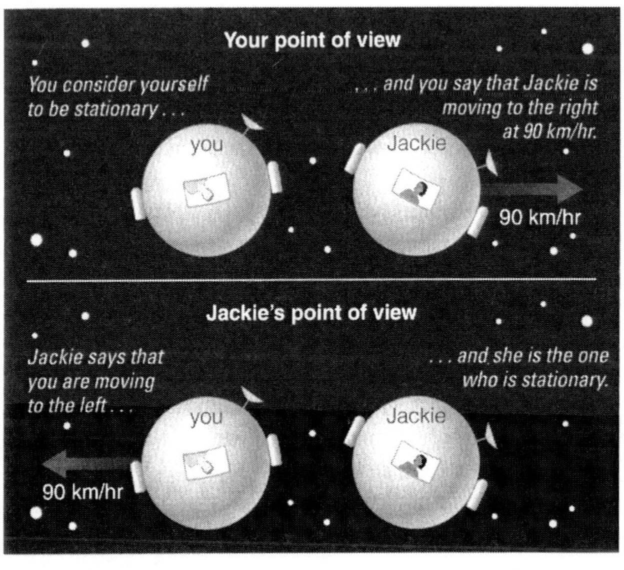

Figure 2.4 Thought Experiment 1.

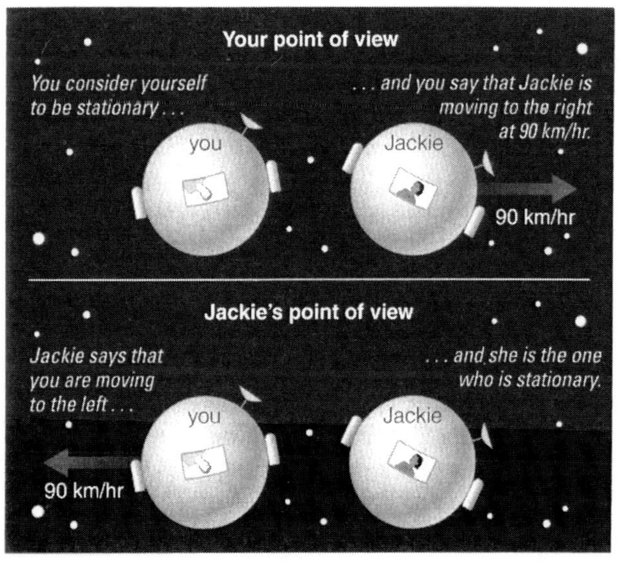

Figure 2.5 Thought Experiment 2.

speed of 100 km/hr. How fast is the ball moving relative to Jackie?

From your point of view, Jackie and the ball are both going in the same direction. Jackie is going 90 km/hr and the ball is going 100 km/hr, so the ball is going 10 km/hr faster than she is. The ball will therefore overtake and pass her.

From Jackie's point of view, *she* is stationary and *you* are moving away from her at 90 km/hr. Thus, she would say the ball is moving even before you throw it, since it would be going with you at your speed of 90 km/hr. Once you release the ball, she therefore sees it moving toward her at 10 km/hr—the ball's speed of 100 km/hr relative to you *minus* the 90 km/hr at which you are moving relative to her. In other words, although you and Jackie disagree about the actual speed of the ball, you both agree that the ball will pass her at a *relative* speed of 10 km/hr.

Thought Experiment 3 (Figure 2.6)

This time you throw a baseball in Jackie's direction at 90 km/hr. From your point of view, the ball is traveling at exactly the same speed as Jackie. Therefore, you'll see the ball forever chasing her through space, neither catching up with nor falling behind her.

From Jackie's point of view, the 90 km/hr at which you threw the ball exactly matches your 90 km/hr speed away from her. Therefore, the ball is *stationary* in her reference frame. Think about this for a moment: Before you throw the baseball, Jackie sees it moving away from her at 90 km/hr because it is in your hand. At the moment you release the baseball, it suddenly becomes stationary in Jackie's reference frame, floating in space at a fixed distance from her spaceship. Many hours later, after you have traveled far away, Jackie will still see the ball floating in the same place. If she wishes, she can put on her space suit and go out to retrieve it, or she can just leave it there. From her point of view, it's not going anywhere, and neither is she.

THINK ABOUT IT

In Thought Experiment 3, suppose instead that you throw the ball in Jackie's direction at 80 km/hr. In that case, what will Jackie see the ball doing? Next, suppose that Jackie is moving *toward* you rather than *away* from you. What would she see the ball doing in that case?

Thought Experiments at High Speeds

The absoluteness of the speed of light did not come into play in our first three thought experiments because the speeds of the objects were small compared to the speed of light. For example, 100 km/hr is less than *1 ten-millionth* of the speed of light. Now let's raise the speeds much higher and explore the strange consequences of the absoluteness of the speed on light.

Thought Experiment 4 (Figure 2.7*)

Imagine that Jackie is moving away from you at 90% of

*For simplicity, this and most other figures in this chapter ignore effects of length contraction, which would make the moving ships shorter in their direction of motion; only Figure S2.13 shows the length contraction effect.

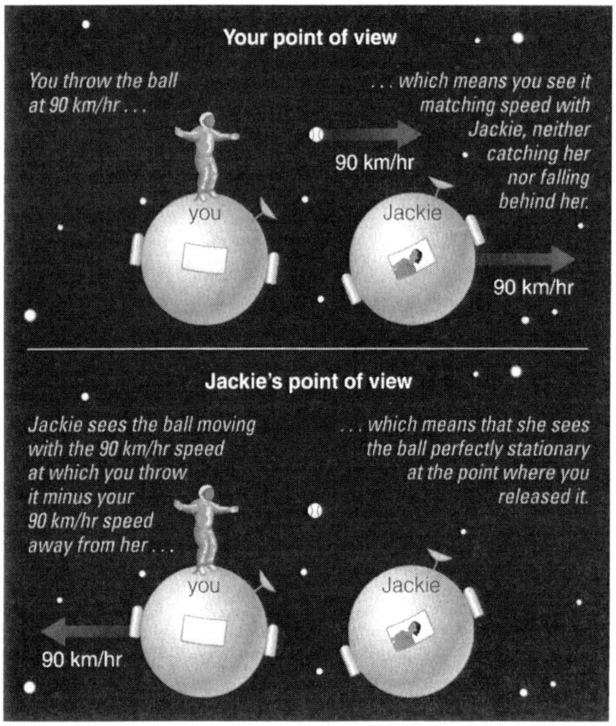

Figure 2.6 Thought Experiment 3.

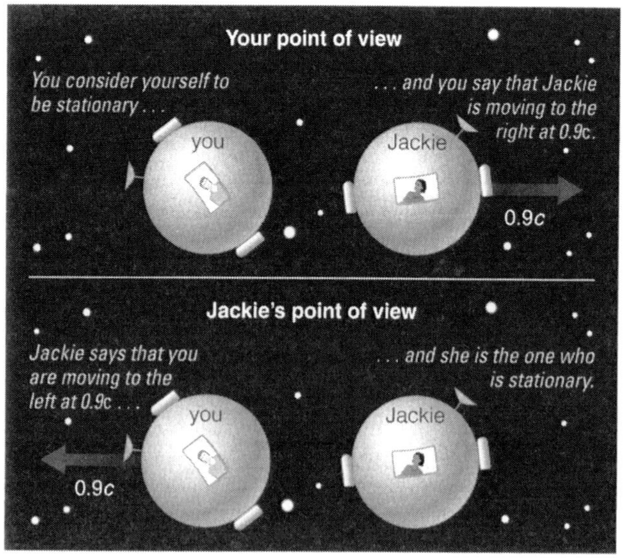

Figure 2.7 Thought Experiment 4.

the speed of light, or $0.9c$. (Recall that c is the symbol for the speed of light, which is about 300,000 km/s.) How does the situation appear to Jackie?

Other than the much higher speed, this situation is just like that in Thought Experiment 1. Jackie perceives *herself* to be at rest and sees *you* moving away from her at $0.9c$.

Thought Experiment 5 (Figure 2.8)

Now, instead of throwing a baseball, you climb out of your spaceship and point a flashlight in Jackie's direction. How fast is the beam of light moving relative to Jackie?

From your point of view, Jackie and the flashlight beam are both going in the same direction. Jackie is going at 90% of the speed of light, or $0.9c$, and the light beam is going at the full speed of light, or c. Thus, you see the light beam going $0.1c$ faster than Jackie.† Nothing should be surprising so far.

From Jackie's point of view, *she* is stationary and *you* are moving away from her at $0.9c$. Following the "old common sense" used in our earlier thought experiments, we would expect Jackie to see the light moving toward her at $0.1c$— the light's speed of c minus your speed of $0.9c$. *But this answer is wrong!* Relativity tells us that the speed of light is always the same for everyone. Therefore, Jackie must see the beam of light coming toward her at c, *not* at $0.1c$.

In this case, you and Jackie no longer agree about her speed *relative* to the speed of the light beam: She'll see the light beam pass by her at the speed of light, c, but you'll see it going only $0.1c$ faster than she is going. By our old common sense, this result sounds preposterous. However, we found it by using simple logic, starting with the assumption that the speed of light is the same for everyone. As long as this assumption is true—and, remember, the absoluteness of the speed of light is an experimentally verified fact—our conclusions follow logically.

†You can't actually see a light beam moving forward. When we say that you "see" the beam moving at the speed of light, we really mean that you would find it to be moving at this speed if you made a careful measurement with instruments at rest in your reference frame.

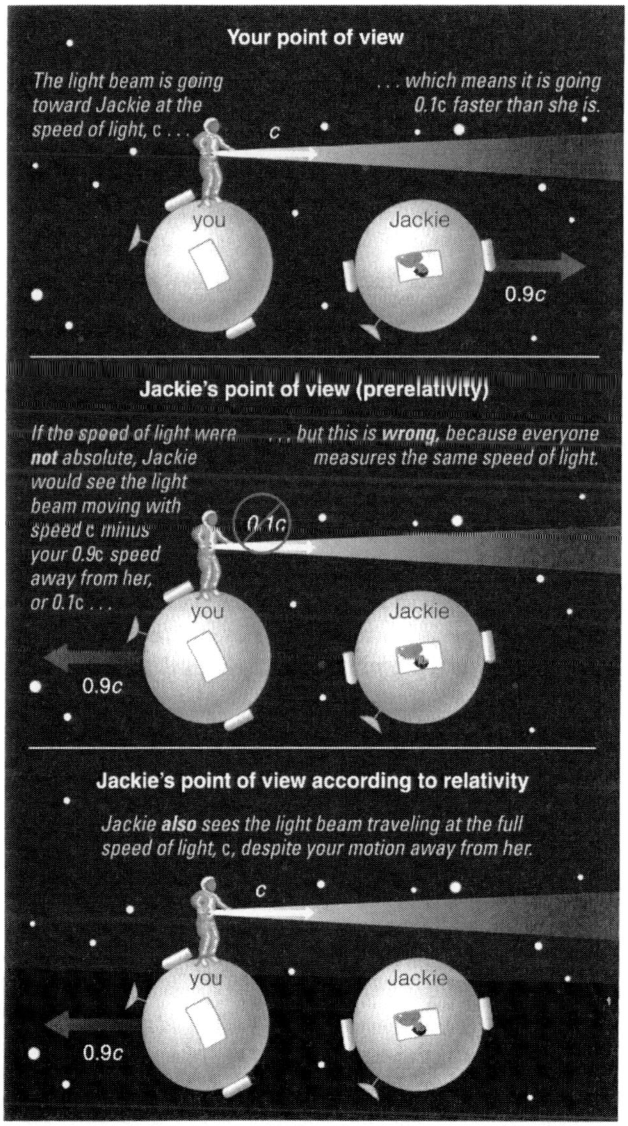

Figure 2.8 Thought Experiment 5.

THINK ABOUT IT

Suppose Jackie is moving away from you at a speed of $0.99999c$, which is short of the speed of light by only $0.00001\,c$, or 3 km/s. At what relative rate would you see a light beam catch up with her? How fast will Jackie see the light going as it passes?

WHY CAN'T WE REACH THE SPEED OF LIGHT?

You might be wondering what Jackie would see if she were moving away from you at the speed of light or faster—which is essentially the same question Einstein asked when he wondered how the world would look if he could travel at or beyond the speed of light. However, once we accept the absoluteness of the speed of light, it follows that neither Jackie nor you nor any other material object can ever reach the speed of light, let alone exceed it.

Thought Experiment 6

You have just built the most incredible rocket imaginable, and you are taking it on a test ride. Soon you are going faster than anyone had ever imagined possible—and then you put the rocket into second gear! You keep going faster and faster and faster. Here is the key question: Are you ever traveling faster than the speed of light?

Before we answer this question, the fact that all motion is relative forces us to answer another question: In what reference frame is your speed being measured? Let's begin with *your* reference frame. Imagine that you turn on your rocket's headlights. Because the speed of light is the same for everyone, you must see the headlight beams traveling at the speed of light—which means they are racing away from your rocket at a speed you'll measure to be 300,000 km/s. The fact that you'll see your headlight beams racing away is true no matter how long you have been firing your rocket engines. In other words, you cannot possibly keep up with your own headlight beams.

The fact that you cannot outrace your headlight beams, combined with the fact that the speed of light is the same for everyone, means that no observer can ever see you reach or exceed the speed of light. Observers in different reference frames will measure your speed differently, but *all* observers will agree on two key points: (1) Your headlight beams are moving out ahead of you, and (2) these light beams are traveling at c = 300,000 km/s. Clearly, if you are being outraced by your headlights and if the headlights are traveling at the speed of light, you must be traveling *slower* than the speed of light (Figure 2.9). It does not matter who is measuring your speed. It can be you, someone on Earth, or anyone else in any other reference frame. No one can ever observe you to be traveling as fast as a light beam.

In case you are still not convinced, let's turn the situation around. Imagine that, as you race by some planet, a person on the planet turns on a light beam. Because the speed of light is absolute, you will see the light beam race past you at c = 300,000 km/s. The person on the planet will also see the light traveling at c = 300,000 km/s and will see the light outrace you. Again, everyone will agree that you are traveling slower than the speed of light.

The same argument applies to any moving object, and it is true with or without headlights. All light travels at the speed of light, including the light that reflects off an object (allowing us to see it) and the infrared light that even cool objects emit as thermal radiation [Section 5.41]. As long as the speed of light is absolute, no material object can ever keep up with the light it emits or reflects, which means no material object can reach or exceed the speed of light. Building a spaceship to travel at the speed of light is not a mere technological challenge—it simply cannot be done.

2.3 THE REALITY OF SPACE AND TIME

In the first section of this chapter, we listed five key ideas of special relativity. Each of these ideas is actually a prediction that arises from the two basic assumptions of the theory—that the laws of nature are the same for everyone and that everyone always measures the same speed of light. So far, we have shown only how relativity predicts that no ma-

SPECIAL TOPIC: WHAT IF LIGHT CAN'T CATCH YOU?

If you're like most students learning about relativity for the first time, you're probably already looking for loopholes in the logic of our thought experiments. For example, confronted with Thought Experiment 6, you might be tempted to ask, "What happens if you're traveling away from some planet faster than light, so the light from the planet can't catch you?" While it's surely true that light couldn't catch you if you were going faster than the speed of light, it also makes the question moot: If you can't see light from the planet, there is no way for you to know that the planet is there, so you couldn't actually make any measurement that would show your speed to be greater than the speed of light.

In fact, what relativity really tells us about the speed of light is that it is a limit on the speed at which *information* can be transmitted. There are numerous circumstances in the universe in which an object may *seem* to be exceeding the speed of light. However, these circumstances do not provide any means of sending information or objects at speeds faster than the speed of light.

As an example, imagine that you held a laser light and swept it across the sky between two stars that are separated by an angular distance of 90° and are each 10 light-years away from Earth. About 10 years from now, you'd see your laser light first make a dot (an extremely dim one!) on the surface of one star and a few seconds later make a dot on the surface of the second star. Thus, your laser dot would seem to have traveled the many light-years from the first star to the second star in just a few seconds—which means at a speed far in excess of the speed of light. However, while the laser dot can carry information from you to each of the individual stars, it clearly cannot be transmitting any information from one star to the other. (After all, you held the laser, not someone at either star.) As a result, there's no violation of relativity.

Other examples of things that *seem* to move faster than the speed of light arise frequently in the world of quantum mechanics. According to quantum principles, measuring a particle in one place can (in certain specific circumstances) affect a particle in another place *instantaneously* —even if the particle is many light-years away. In fact, this process has been observed in laboratories over short distances. This instantaneous effect of one particle on another may at first seem to violate relativity, but it does not. The built-in randomness of quantum mechanics prevents this technique from being used to transmit useful information. Moreover, if we wish to confirm that the second particle really was affected, we will have to receive a signal carrying information about the particle—and that signal can travel no faster than the speed of light.

terial object can reach or exceed the speed of light. Let's continue down the path we've started, so that we can see why our old conceptions of time and space must be revised.

How does relativity affect our view of time and space?

The special theory of relativity tells us that measurements of time, length, and mass can all be different in different reference frames. It also tells us that different observers can disagree about whether two events are simultaneous and about the speeds of different objects. Strange as these ideas may sound, we can continue to use simple thought experiments to show why they follow directly from the premises of relativity. We'll begin by considering how relativity affects our view of time, then move on to other effects.

Time Differs in Different Reference Frames

In preparation for our next thought experiment, imagine that you are on a moving train tossing a ball straight up so that it bounces straight back down from the ceiling. How do the path and speed of the ball appear to an observer along the tracks outside the train?

As shown in Figure 2.10, the outside observer sees the ball going forward with the train at the same time that it is going up and down, so the ball's path always slants forward.

Because the outside observer sees the ball moving with this forward speed in *addition* to its up-and-down speed, while you see the ball going only up and down, he would measure a *faster* overall speed for the ball than you would.

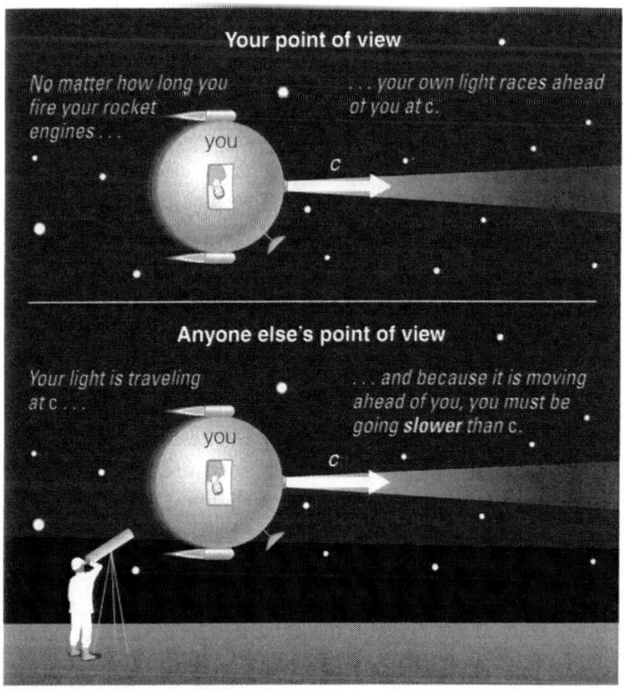

Figure 2.9 Thought Experiment 6

If the train is moving slowly, the outside observer sees the ball's path slant forward only slightly and would say that the ball's overall speed is only slightly faster than you would report. If the train is moving rapidly, the ball's path leans much farther forward, and the ball appears to be going considerably faster to the outside observer than to you.

Thought Experiment 7 (Figure 2.11)

Inside her spaceship, Jackie has a laser on her floor that is pointed up to a mirror on her ceiling. She momentarily flashes the laser light and uses a very accurate clock to time how long it takes the light to travel from the floor to the ceiling and back. As Jackie zips by you at a speed close to the speed of light, you observe her experiment and, using your own very accurate clock, you also time the laser light's trip from Jackie's floor to her ceiling and back.

Because Jackie, the laser, and the mirror are all moving from your point of view, the light's path looks slanted as it goes from floor to ceiling and back, just as the ball tossed up and down in the train followed a slanted path to an outside observer. Thus, from your point of view, the light travels a *longer* path in going from the floor to the ceiling than it does from Jackie's point of view, just as the ball in the train took a

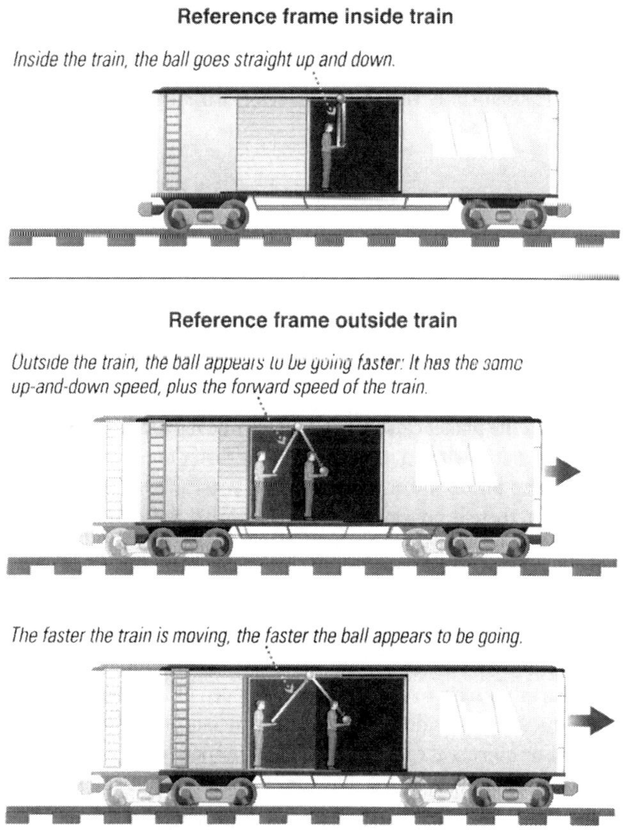

Figure 2.10 A ball tossed straight up and down inside a moving train appears to an outside observer to follow a slanted path. Thus, the outside observer sees the ball traveling at a faster total speed than does the person inside the train.

longer path from the outside observer's point of view.

By our old common sense, this would be no big deal. You and Jackie would agree on how long it takes the light to go from the floor to the ceiling and back, just as you and an outside observer would agree on how long it takes you to toss a ball up and down on a train. You would explain the light's longer path by saying that the light is moving faster relative to you than it is moving relative to Jackie, just as the outside observer sees the tossed ball moving faster because of the forward motion of the train. However, because we are dealing with *light*, you and Jackie must both measure its speed to be the *same*—the speed of light, or 300,000 km/s—even though you see its path slanted forward with the movement of the spaceship. And this fact has an astonishing implication.

Remember that, at any given speed, including the speed of light, traveling a longer distance must take a longer time. Because both you and Jackie see the light traveling at the same speed, but you see it traveling a longer distance, *your clock must record more time* than Jackie's as the light travels from floor to ceiling and back. And because you can see Jackie's clock as well as your own, you must see her clock running at a slower rate than yours in order for it to end up reading less elapsed time after the light completes its round trip. In other words, you will see Jackie's clock running more slowly than your own. Moreover, it doesn't matter what kind of "clock" you and Jackie use to measure the time for the light trip. *Anything* that can measure time will be going slower in Jackie's reference frame than in yours, including mechanical clocks, electrical clocks, heartbeats, and biochemical reactions. Our astonishing conclusion: From your point of view, *time itself* is running slower for Jackie.

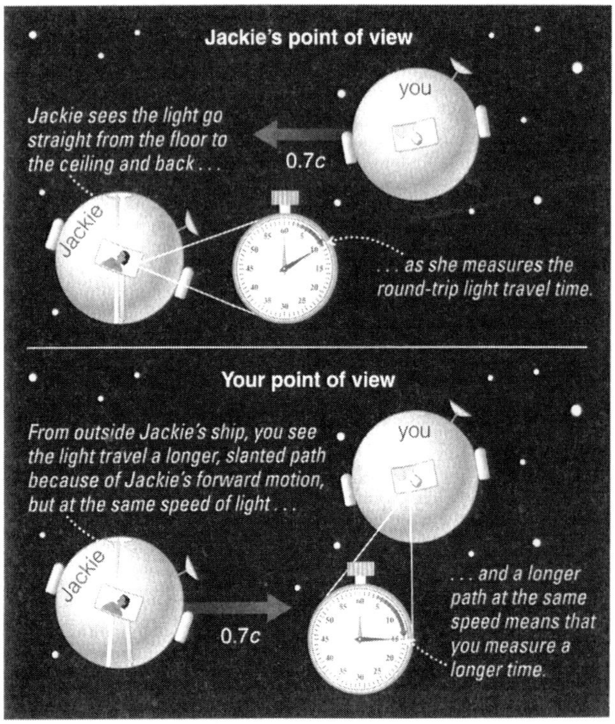

Figure 2.11 Thought Experiment 7

How much slower is time running for Jackie? It depends on her speed relative to you. If she is moving slowly compared to the speed of light, you will scarcely be able to detect the slant of the light path, and your clock and Jackie's clock will tick at nearly the same rate. The faster she is moving, the more slanted the light path will appear to you and the greater the difference between the rate of her clock and that of yours. Generalizing, we reach the following conclusion:

From your point of view, time runs more slowly in the reference frame of anyone moving relative to you. The faster the other reference frame is moving, the more slowly time passes within it.

This effect is called **time dilation** because it tells us that time is *dilated*, or expanded, in a moving reference frame.

The Relativity of Simultaneity

Our old common sense also tells us that everyone must agree on whether two events happen at the same time or one event happens before another. For example, if you see two apples—one red and one green—fall from two different trees and hit the ground at the same time, you expect everyone else to agree that they landed at the same time (assuming you've accounted for any difference in the light travel times from the two trees). If you saw the green apple land before the red apple, you would be very surprised if someone else said that the red apple hit the ground first. Well, prepare yourself to be surprised, because our next thought experiment will show that observers in different reference frames will not necessarily agree about the order or simultaneity of events that occur in different places.

Before we go on, note that observers in different reference frames *must* agree about the order of events that occur in the *same* place. For example, suppose you grab a cookie and eat it. In your reference frame, both events (picking up the cookie and eating it) occur in the same place. Thus, it could not possibly be the case that someone else would see you eat the cookie before you pick it up.

Thought Experiment 8

Jackie has a brand-new, extra long spaceship, and she is coming toward you at a speed of 90% of the speed of light, or $0.9c$ (Figure 2.12a). She is in the center of her spaceship, which is totally dark except for a flashing green light at its front end and a flashing red light at its back end. Suppose that you see the green and red lights flash at *exactly the same time*, with the flashes occurring at the instant that Jackie happens to pass you. During the very short time that the light flashes are traveling toward you, Jackie's forward motion carries her toward the point where you saw the green

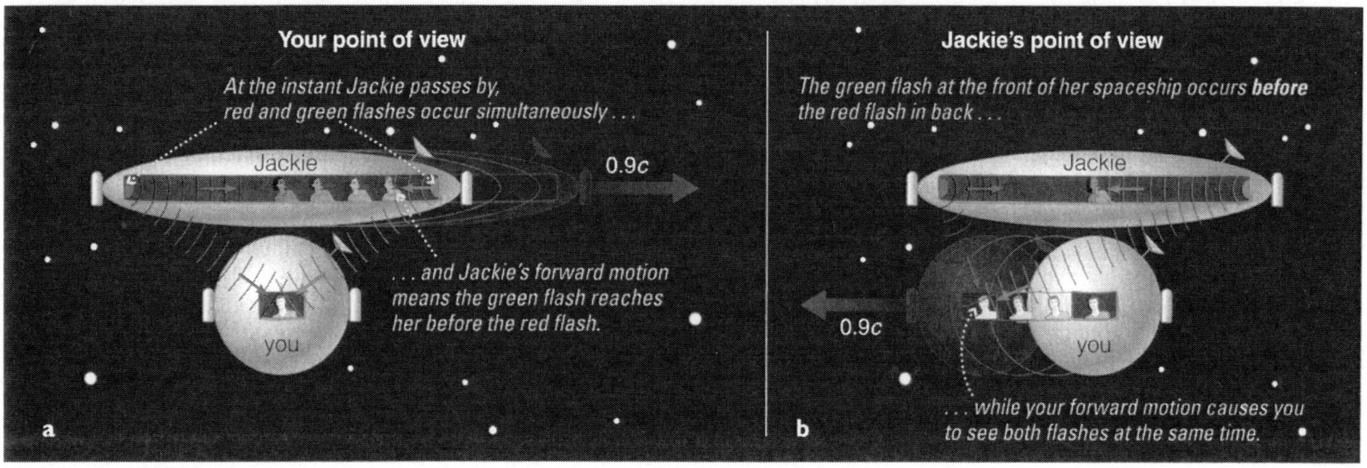

Figure 2.12 Thought Experiment 8

flash occur. Thus, the green flash will reach her before the red flash, which means you'll see her illuminated first by green light and then by red light. So far, nothing should be surprising: You see the two flashes simultaneously because you are stationary, while Jackie sees the green flash before the red flash because of her forward motion. But let's think about what happens from Jackie's point of view, in which she is the one who is stationary and you are the one who is moving.

Remember that motion cannot affect the reality of events that occur in the same place at the same time. Thus, no one can argue with the fact that you saw the green and red flashes at the same time, and no one can disagree with the fact that the green light reached Jackie before the red light. However, this agreement about events has an important implication: Because Jackie considers herself to be stationary in the center of her spaceship, the only way that the green light can reach her before the red light is if the green flash occurred first. In other words, she must see the green light flash first, and the red light flash a tiny bit later. Note that this also allows her to explain why you see the two flashes simultaneously: According to Jackie, your forward motion means the green light has to travel a little farther to reach you than the red light, so you end up seeing the two lights at the same time even though the green flash occurred first (Figure 2.12b).

Let's summarize the situation: You both agree on the indisputable reality of what each of you saw—that is, you saw the two flashes at the same time, while Jackie saw the green flash before the red flash. However, you say this reality occurs because the flashes really did occur at the same time and Jackie was moving toward the green flash, while Jackie says it occurs because the green flash really occurred first while you were moving away from it. Who is right? All motion is relative, so both of you are equally correct. In other words, the flashes that are simultaneous in your reference frame occur at different times in Jackie's reference frame. That is what we mean when we say that the order or simultaneity of events depends on your frame of reference.

Effects of Motion on Length and Mass

The fact that time is different in different reference frames also implies that lengths (or distances) and masses are also affected by motion, although the reasons are a bit subtler. The following two thought experiments use the idea of time dilation to help us understand the effects on length and mass.

Thought Experiment 9

Jackie is back in her original spaceship, coming toward you at high speed. As usual, both you and she agree on your *relative* speed. You disagree only about who is stationary and who is moving. Now imagine that Jackie tries to measure the length of your spaceship as she passes by you. She can do this by measuring the time it takes her to pass from one end of your spaceship to the other; recall that length = *speed* × *time*.

In your reference frame, you say that Jackie is moving by you and therefore you see Jackie's time running slow-

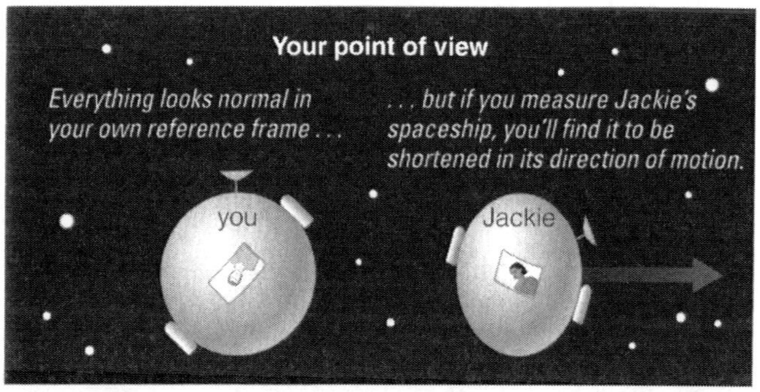

Figure 2.13 People and objects moving relative to you are contracted in their direction of motion. The diagram shows only your point of view, but Jackie's view would be equivalent: She would see herself at "normal" size and you contracted in your direction of motion.

ly. Because you agree on her relative speed but her clocks record less time than yours as she passes from one end of your spaceship to the other, she must measure the length of your spaceship to be *shorter* than you measure it to be. Now, remember that your viewpoint and Jackie's viewpoint are equally valid. The fact that the laws of nature are the same for everyone means that if she measures your spaceship to be shorter than it is at rest in your reference frame, you must also measure hers to be shorter than it would be at rest, an effect called **length contraction**. Figure 2.13 shows that lengths are affected only in the direction of motion. Her spaceship is shorter from your point of view, but its height and depth are unaffected. Generalizing, we reach the following conclusion:

From your point of view, the lengths of objects moving by you (or the distances between objects moving by you) are shorter in their direction of motion than they would be if the objects were at rest. The faster the objects are moving, the shorter the lengths.

Thought Experiment 10

To see how motion affects mass, imagine that Jackie has an identical twin sister with an identical spaceship, and suppose that her sister is at rest in your reference frame while Jackie is moving by you at high speed. At the instant Jackie passes by, you give both Jackie and her sister identical pushes (Figure 2.14). That is, you push them both with the same force for the same amount of time. By our old common sense, we'd therefore expect both of them to accelerate by an identical amount, such as gaining 1 km/s of speed relative to you. But we must instead apply our new common sense, which tells us something different.

Remember that, because Jackie is moving relative to you and her sister, you'll see Jackie's time running more slowly than yours and her sister's. Thus, Jackie's clock shows that she feels the force of your push for a *shorter* time than does her sister. For example, if you and Jackie's sister measure the duration of the push to be 1 microsecond, Jackie's clock will show the duration to be less than 1 microsecond. Moreover, because Jackie feels the force of your push for a shorter time, the push must have a *smaller* effect on Jackie's velocity than it has on her sister's velocity. In other words, you'll find that your push has less effect on Jackie than on her sister, despite the fact that you gave them identical pushes. According to Newton's laws of motion [Section 4.2], the same push can have a smaller effect on Jackie's velocity only if her mass is *greater* than her sister's mass. This effect is sometimes called **mass increase:***

From your point of view, objects moving by you have greater mass than they have at rest. The faster an object is moving, the greater the increase in its mass.

Mass increase provides another way of understanding why no material object can reach the speed of light. The faster an object is moving relative to you, the greater the mass you'll find it to have. Thus, at higher and higher speeds, the same force will have less and less effect on an object's velocity. As the object's speed approaches the speed of light, you will find its mass to be heading toward infinity. No force can accelerate an infinite mass, so the object can never gain that last little bit of speed needed to push it to the speed of light.

Velocity Addition

We have just one more important effect to discuss: As the next thought experiment shows, you and Jackie will disagree about the speed of a material object moving relative to both of you. The only speed you will agree on is the

*Newton's second law can be stated either as "force = mass x acceleration" or as "force = rate of change in momentum." (Recall that momentum = mass x velocity.) For more advanced work, most physicists prefer the second form, and they therefore prefer to think of the effect in relativity as momentum increase rather than mass increase. The two viewpoints are equivalent for the level of discussion in this book.

Figure 2.14 Thought Experiment 10

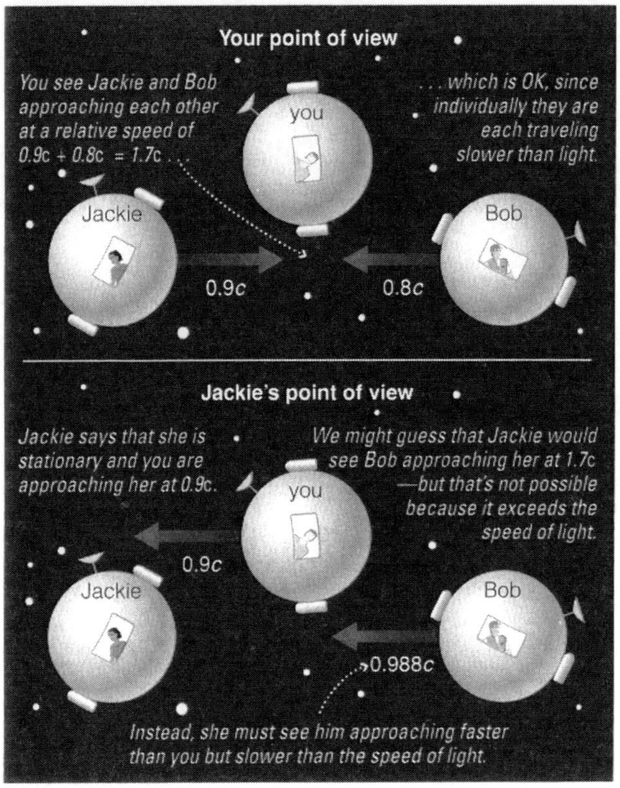

Figure 2.15 Thought Experiment 11.

speed of light.

Thought Experiment 11 (Figure 2.15)

Jackie is moving toward you at $0.9c$. Your friend Bob jumps into his spaceship and starts heading in Jackie's direction at $0.8c$ (from your point of view). How fast will Jackie see him approaching her?

By our old common sense, Jackie should see Bob coming toward her at $0.9c + 0.8c = 1.7c$. We know this answer is wrong, however, because $1.7c$ is faster than the speed of light. Jackie must see Bob coming toward her at a speed less than c, but he will be coming faster than the speed of $0.9c$ at which she sees you coming. Thus, she'll conclude that Bob's speed is somewhere between $0.9c$ and c. (In this case, it turns out that she'll see him coming at $0.988c$; see Mathematical Insight 2.2.)

Do the effects predicted by relativity really occur?

The clear logic of our thought experiments has shown that all the major predictions of special relativity follow directly from the absoluteness of the speed of light and from the fact that the laws of nature are the same for everyone. However, remember that logic alone is not good enough in science; our conclusions remain tentative until they pass observational or experimental tests. Does relativity meet the test?

The Absoluteness of the Speed of light

The first thing we might wish to test is the surprising premise of relativity: the absoluteness of the speed of light. In principle, we can test this premise by measuring the speed of light coming from many different objects and going in many different directions and verifying that the speed is always the same.

The speed of light was first measured in the late 1600s, but experimental evidence for the *absoluteness* of the speed of light did not come until 1887, when A. A. Michelson and E. W. Morley performed their now-famous *Michelson-Morley experiment.* This experiment showed that the speed of light is not affected by Earth's motion around the Sun. Countless subsequent experiments have verified and extended the results of the Michelson-Morley experiment. The speed of light coming from opposite sides of the rotating Sun and from the orbiting stars in binary systems has also been measured, with the same result: The speed of light is always the same.

Experimental Tests of Special Relativity

Although *we* cannot yet travel at speeds at which the effects of relativity should be obvious, tiny subatomic particles can reach such speeds, thereby allowing us to test the precise predictions of the formulas of special relativity.

Let's first consider one way of testing time dilation. In machines called *particle accelerators,* physicists accelerate subatomic particles to speeds near the speed of light and study what happens when the particles collide. The colliding particles have a great deal of kinetic energy, and the collisions convert some of this kinetic energy into mass-energy that emerges as a shower of newly produced particles. Many of these particles have very short lifetimes, at the end of which they decay (change) into other particles. For example, a particle called the $\pi+$ ("pi plus") meson has a lifetime of about 18 nanoseconds (billionths of a second) when produced at rest. But $\pi+$ mesons produced at speeds close to the speed of light in particle accelerators last much longer than

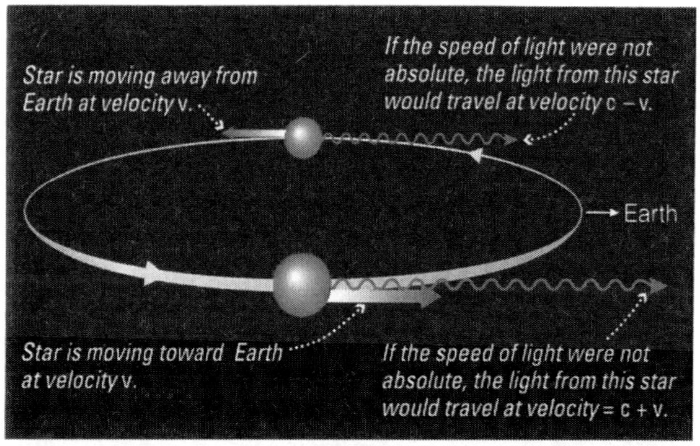

Figure 2.16 If the speed of light were not absolute, the speed at which light from each star in a binary system comes toward Earth would depend on its velocity toward us in the binary orbit.

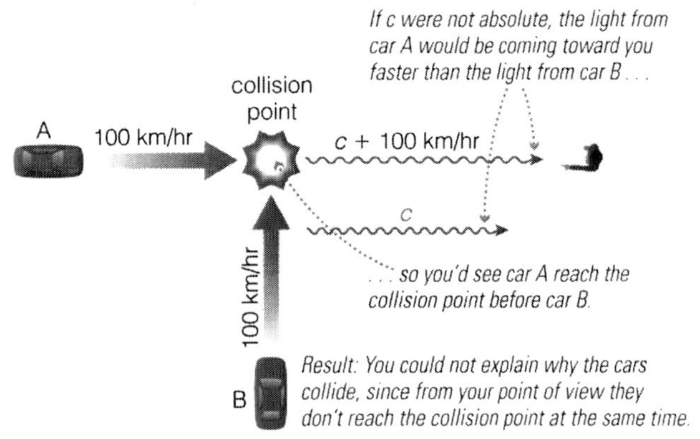

Figure 2.17 If the speed of light were not absolute, the light from a car coming toward you would approach you faster than the light from a car going across your line of sight.

18 nanoseconds—and by precisely the amount predicted by the time dilation formula.

The same experiments allow the mass increase formula to be checked as well. The amount of energy released when high-speed particles collide depends on the particle masses and speeds. Just as relativity predicts, these masses are greater at high speed than they are at low speed—again by just the amount predicted by Einstein's formulas.

Particle accelerators even offer experimental evidence that nothing can reach the speed of light. It is relatively easy to get particles traveling at 99% of the speed of light in particle accelerators. However, no matter how much more energy is put into the accelerators, the particle speeds get only fractionally closer to the speed of light. Some particles have been accelerated to speeds within 0.00001% of the speed of light, but none have ever reached the speed of light.

Although the effects of relativity are obvious only at very high speeds, modern techniques of measuring time are so precise that effects can be measured even at ordinary speeds. For example, a 1975 experiment compared the amount of time that passed on an airplane flying in circles to the time that passed on the ground. Over 15 hours, the airborne clocks lost a bit under 6 nanoseconds to the ground clocks, matching the result expected from relativity. More recent experiments using the Space Shuttle have confirmed time dilation as well. All in all, special relativity is one of the best-tested theories in physics, and it has passed every experimental test to date with flying colors.

A Great Conspiracy?

Perhaps you're thinking, "I still don't believe it." After all, how can you know that the scientists who report the experimental evidence are telling the truth? Perhaps physicists are making up the whole thing as part of a great conspiracy designed to confuse everyone else so they can take over the world!

What you need is evidence that you can see for yourself. So . . . How about nuclear energy? Einstein's famous formula, $E = mc^2$, which explains the energy released in nuclear reactions, is a direct consequence of the special theory of relativity—and one that you can derive for yourself with a bit of algebra (see Mathematical Insight 2.3). Every time you see film of an atomic bomb, or use electrical power from a nuclear power plant, or feel the energy of sunlight that the Sun generated through nuclear fusion, you are really experiencing direct experimental evidence of relativity.

Another test you can do yourself is to look through a telescope at a binary star system. If the speed of light were not absolute, the speed at which light from each star comes toward Earth would depend on its velocity toward us in the binary orbit (Figure 2.16). Imagine, for example, that one star is currently moving directly away from us in its orbit. If we added and subtracted speeds according to our old common sense, light from the star at this point would approach us at speed $c - v$. Some time later, when the same star is moving toward us in its orbit, its light would approach us at speed $c + v$. This light would therefore tend to catch up with the light that the star emitted from the other side of its orbit. If the orbital speed and distance were just right, we might see the same star on both sides of the orbit at once! More generally, because the light from each star would come toward us at a different speed from each point in its orbit, we would see each star in multiple positions in its orbit simultaneously, making each star appear as a short line of light rather than as a point—*if* the speed of light were not absolute. Thus, the fact that we always see distinct stars in binary systems demonstrates that the speed of light *is* absolute.*

Finally, you can explore the paradoxes that would occur if the speed of light were not absolute. For example, imagine that two cars, both traveling at about 100 km/hr, collide at an intersection (Figure 2.17). You witness the collision from far down one street. If the speed of light were not absolute, the light from the car that was coming toward you would have a speed of c + 100 km/hr, while the light from the other car would approach you only at a speed of c. You therefore would see the car coming toward you reach the intersection slightly *before* the other car and thus would see events unfold differently than the passengers in the car or eyewitnesses in other locations. On Earth, this difference would be scarcely noticeable, because 100 km/hr is only about *one-millionth* the speed of light. However, the difference would increase if you could watch from very far away. For example, suppose you had a super telescope and watched such a collision on a planet located 1 million light years from Earth: Because the light from the first car (the one coming toward you) is com-

*This conclusion would not follow if light waves were carried by a medium in the same way that sound waves are carried by air. Scientists in the 1800s believed that such a medium, which they called the ether, permeated all of space. The Michelson-Morley experiment ruled out the existence of such a medium, so we are left with the conclusion that c is absolute.

ing at a speed one-millionth of the speed of light faster than the light from the second car, over a distance of 1 million light-years it would end up reaching you a full *year* ahead of the light from the second car. In other words, you would see the first car reach the intersection a *year* before the second car. This poses a paradox: From the viewpoint of the passengers in the cars, they have collided, yet you saw one car reach the collision point long before the other car had even started its journey! If you wish to believe that images show what really does happen, then the only way to avoid the paradox is to assume that we should not have added the car's speed to the speed of light; that is, we must assume that the speed of light is absolute.

THINK ABOUT IT

The preceding paradox presents us with a choice. If the speed of light is not absolute, different people can witness the same events in very different ways. If the speed of light is absolute, measurements of time and space are relative. Einstein preferred the latter choice. Do you? Explain.

Of course, like any scientific theory, the theory of relativity can never be *proved* beyond all doubt. However, it is supported by a tremendous body of evidence, some of which you can see for yourself. This evidence is real and cannot be made to disappear. If anyone ever comes up with an alternative theory, the new theory will still have to explain the many experimental results that seem to support relativity so well.

2.4 TOWARD A NEW COMMON SENSE

We've used thought experiments to show that our old common sense doesn't work, and we've discussed how actual experiments verify the ideas of our thought experiments. But we haven't yet figured out what new common sense should replace the old. Fortunately, while it may take you many years to become completely comfortable with the ideas of relativity, it's relatively easy to see what you need to do to make yourself comfortable.

How can we make sense of relativity?

Perhaps surprisingly, another thought experiment that may at first make everything seem even more bizarre will help us understand what is really going on, and thereby lead us toward a new common sense.

Thought Experiment 12

Suppose Jackie is moving by you at a speed close to the speed of light. From our earlier thought experiments, we know that you'll measure her time as running slow, her length as having contracted, and her mass as having increased. But what would *she* say?

From Jackie's point of view, she's not going anywhere—you are moving by *her* at high speed. Because the laws of nature are the same for everyone, she must reach exactly the same conclusions from her point of view that you reach from your point of view. That is, she'll say that *your* time is running slowly, *your* length is contracted, and *your* mass has increased!

Now we have what seems to be a serious argument on our hands. Imagine that you are looking into Jackie's spaceship with a super telescope. You can clearly see that her time is running slowly because everything, she does is in slow motion. You send her a radio message saying, "Hi, Jackie! Why are you doing everything in slow motion?" Because the radio message travels at the absolute speed of light, Jackie has no trouble receiving your message,* and she responds with her own radio message back to you.

As you listen to her response, you'll hear it in slow motion—"Hheeeellllloooo tthhheeerrr"—thus verifying that her time is running slowly. However, if you record the entire message and use your computer to speed up her voice so that it sounds normal, you'll hear Jackie say, "I'm not moving in slow motion, *you are!*"

You can argue back and forth all you want, but it will get you nowhere. Then you come up with a brilliant idea. You hook up a video camera to your telescope and record a movie showing that Jackie's clock is moving more slowly than yours. You put your movie into a very fast rocket and shoot it off toward her. You figure that when the movie arrives and she watches it, she'll have visual proof that you are right and that she really is moving in slow motion.

Unfortunately, just before you can declare victory in the argument, you learn that Jackie had the same brilliant idea: A rocket arrives with a movie that she made. You put it into your movie player and watch what appears to be clear proof that Jackie is right—her movie shows that *you* are in slow motion! Apparently, Jackie really is seeing your time run slowly, just as you are seeing her time run slowly.

How can this be? Think back to our earlier discussion of up and down, and imagine that an American child and an Australian child are talking on the telephone. The Australian

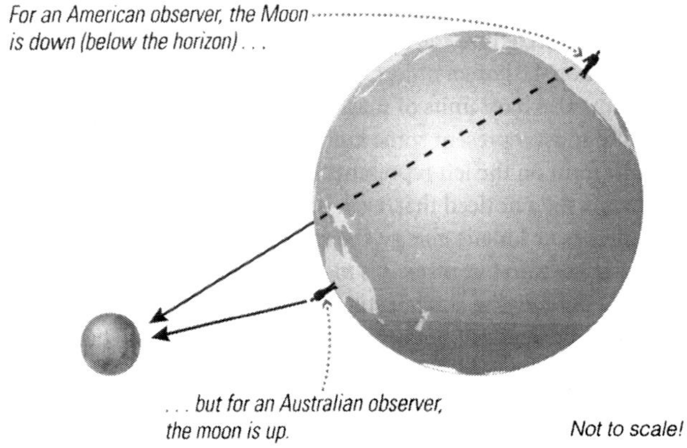

Figure 2.18 The Moon is up for the Australian observer but down for the American observer.

says, "Isn't the Moon beautiful up in the sky right now?" The American replies, "What are you talking about? The Moon isn't up right now!" According to childhood common sense about up and down, the two children appear to be contradicting each other and could argue endlessly. However, once they realize that up and down are measured *relative* to the center of the Earth, they realize that the argument stems only from incorrect definitions of *up* and *down* (Figure 2.18). They are both talking about the same Moon in the same place, and they argued only because they were using childhood common sense about up and down rather than the correct definitions of *up* and *down*.

In much the same way, the argument between you and Jackie arises because you are using the old common sense in which we think of space and time as absolutes and expect the speed of light to be relative. The theory of relativity tells us that we have it backward. The speed of light is the absolute, and time and space are relative.

By the new common sense, the fact that you and Jackie disagree about whose time is running slowly is no more surprising than the fact that the two children disagree about whether the Moon is up or down. The disagreement is meaningless because it relies on inadequate definitions of time and space. What counts are *results*, and every experiment that you perform will agree with every experiment that Jackie performs. You are both experiencing the same laws of nature, albeit in ways different from those our old common sense would have suggested.

*Of course, because of the Doppler effect [Section 5.5], she'll find that the radio waves are blueshifted from the frequency at which you send them as you come toward her, and redshifted as you move away from her. You'll find the same for her radio broadcasts: But as long as you each tune your radios to the correct Doppler-shifted frequencies, you'll have no problem hearing the broadcast message.

How does special relativity offer us a ticket to the stars?

One consequence of accepting the special theory of relativity is the fact that we'll never be able to travel to distant

*Given that you see the Earth-Vega distance contracted, shouldn't you also claim that it is *Earth's* time running slowly, rather than yours? This question underlies the so-called *twin paradox*, in which you make the trip while your twin sister stays home on Earth. The resolution comes from the fact that, because you must turn around at Vega, you effectively change reference frames relative to Earth at least once during your trip. A careful analysis of the changing reference frames is beyond the scope of this book, but it turns out that you do indeed measure less total time than your stay-at-home twin.

SPECIAL TOPIC: MEASURING THE SPEED OF LIGHT

It is difficult to measure the speed of light, because light travels so fast. If you stand a short distance from a mirror and turn on a light, the reflection seems to appear instantaneously. Such observations led Aristotle (384–322 B.C.) to conclude that light travels at infinite speed, a view that many scientists still held as recently as the late 1600s.

One way to make the measurement easier is to place a mirror at increasingly greater distances. If the speed of light truly were infinite, the reflection would always appear instantaneously. However, if it takes time for the light to travel to and from the mirror, you should eventually find a delay between the time you turn on the light and the time you see the reflection. Galileo tried a version of this experiment using the distance between two tall hills, but he was unable to detect any delay (instead of using a mirror, he stationed an assistant on the distant hill to signal back when he saw the light). He concluded that the speed of light, if not infinite, was too fast to be measured between hills on Earth with the technology of his day.

Galileo's idea was extended to space in 1675 by the Danish astronomer Olaus Roemer. Using the four largest moons of Jupiter as his "mirrors," Roemer successfully measured a delay in the time it took for light from the Sun to be reflected and thereby made the first measurement of the speed of light. His technique worked because, by that time, the orbital periods of Jupiter's large moons were well known. Thus, it was possible to predict the precise moments at which Jupiter would eclipse each of its moons. Roemer discovered that the eclipses occurred progressively earlier than expected during those times of the year when Earth was moving toward Jupiter in its orbit and progressively later when Earth was moving away from Jupiter. He realized that the eclipses actually were occurring at the predicted times, but that the light was taking longer to reach us when Earth was farther from Jupiter. His observations proved that the speed of light is finite and allowed him to estimate its precise value. Using estimates of the Sun-Earth and Sun-Jupiter distances that were available in 1675, Roemer calculated the speed of light to be 227,000 km/s. Redoing his calculations using the presently known values of these distances yields 300,000 km/s.

As technology advanced, it became possible to measure light travel time between mirrors at much closer distances. In 1849 and 1850, the French physicists Fizeau and Foucault (also famous for the Foucault pendulum) performed a series of experiments using rotating mirrors to measure the speed of light much more precisely. Modern devices for measuring the speed of light take advantage of the wave properties of light, particularly the fact that light waves can interfere with one another. Such devices, called *interferometers*, were refined by A. A. Michelson in the early 1880s and used in the Michelson-Morley experiment. Details of how these experiments work can be found in many physics texts.

stars at a speed faster than the speed of light. However, while this fact might at first make distant stars seem forever out of reach, time dilation and length contraction actually offer a "ticket to the stars"—if we can ever build spaceships capable of traveling at speeds close to the speed of light.

Suppose you take a trip to the star Vega, about 25 light-years away, in a spaceship that travels at a speed very close to the speed of light—say, at 0.999*c*. From the point of view of people on Earth, the trip to Vega must take you just over 25 years, since you are traveling at nearly the speed of light for a distance of 25 light-years. The return trip will take another 25 years. Thus, if you leave in the year 2025, you will arrive at Vega in the year 2050 and return to Earth in 2075.

However, from your point of view on the spaceship, you can legitimately claim that you and your spaceship never go anywhere; it is Earth and Vega that make the trip. That is, from your point of view, you remain stationary while Earth rushes away from you and Vega rushes toward you at 0.999*c*. You'll therefore find the distance from Earth to Vega contracted from its rest length of 25 light-years; with the length contraction formula, the contracted distance turns out to be just over 1 light-year. Thus, because Vega is coming toward you at 0.999*c* and has only 1 light-year to travel to reach you from your point of view, you'll be at Vega in only about 1 year. Your return trip to Earth will also take about 1 year, so the round-trip time is only about 2 years from your point of view (Figure 2.19). If you leave at age 40, you'll return as a 42-year-old.

Although it sounds like a contradiction by our old common sense, both points of view are correct. If you leave in 2025 at age 40, you'll return to Earth at age 42—but in the year 2075. That is, while you will have aged only 2 years, all your surviving friends and family will be 50 years older than when you left.

You could make even longer trips within your lifetime with a sufficiently fast spaceship. For example, the Andromeda Galaxy is about 2.5 million light-years away, so the round trip to any star in the Andromeda Galaxy would take at least 5 million years from the point of view of observers on Earth. However, if you could travel at a speed within 50 parts in 1 trillion of the speed of light (that is, $c - 5 \times 10^{-11} c$), the trip would take only about 50 years from your point of view. You could leave Earth at age 30 and return at age 80—but you would return to an Earth on which your friends, your family, and everything you knew had been gone for 5 million years.

Thus, in terms of time, relativity offers only a one-way ticket to the stars. You can go a long distance and return to the *place* that you left, but you cannot return to the epoch from which you left.

THE BIG PICTURE

Putting Chapter S2 into Context

In this chapter, we have studied Einstein's special theory of relativity, learning that space and time are intertwined in remarkable ways that are quite different from what we might expect from everyday experience. As you look back on our new viewpoint about space and time, keep in mind the following "big picture" ideas:

- The ideas of relativity derive from two simple facts: The laws of nature are the same for everyone, and the speed of light is absolute. The thought experiments in this chapter simply derived the consequences of these facts.
- Thought experiments are useful, but the ultimate judge of any theory is observation and experiment. The theory of relativity has been extensively tested and verified.
- Although the ideas of relativity may sound strange at first, you can understand them simply and logically if you allow yourself to develop a "new common sense" that incorporates them.
- Space and time are properties of the universe itself, and the new understanding of them gained through the theory of relativity will enable you to better appreciate how the universe works.

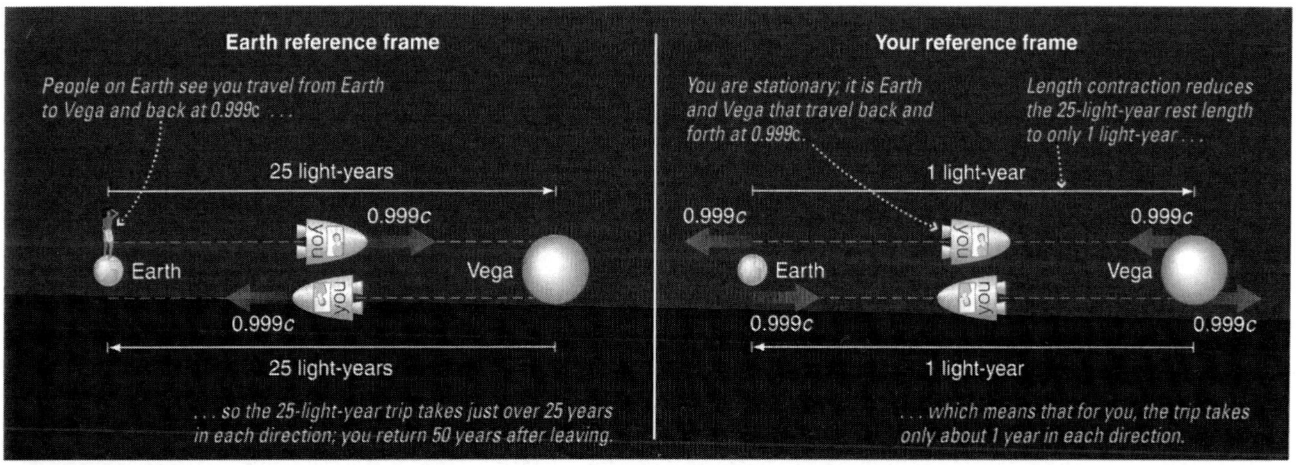

Figure 2.19 At high speed, a traveler to a distant star will age less than people back home on Earth.

SUMMARY OF KEY CONCEPTS

2.1 EINSTEIN'S REVOLUTION

What are the major ideas of special relativity?

Special relativity tells us that different observers can measure time, distance, and mass differently, even though everyone always agrees on the speed of light. It also tells us that no material object can reach or exceed the speed of light (in a vacuum).

What is "relative" about relativity?

The theory of relativity is based on the idea that all *motion* is relative. That is, there is no correct answer to the question of who or what is really moving, in the universe, so motion can be described only for one object relative to another.

2.2 RELATIVE MOTION

How did Einstein think about motion?

All the predictions of the special theory of relativity are based on two assumptions: (1) The laws of nature are the same for everyone; and (2) the speed of light is the same for everyone.

What's surprising about the absoluteness of the speed of light?

In our everyday lives, we expect velocities to add simply; for example, an observer on the ground should see a ball in an airplane traveling at the airplane's speed plus the ball's speed. However, this is not true for light, which everyone always measures to travel at the same speed.

Why can't we reach the speed of light?

Light always travels at the same speed, so your own light (the light that you emit or reflect) is always moving ahead of you at the speed of light. All other observers will also see your light moving at the speed of light—and because it is moving ahead of you, the observers will always conclude that you are moving slower than the speed of light.

2.3 THE REALITY OF SPACE AND TIME

How does relativity affect our view of time and space?

If you observe an object moving by you at high speed, you'll find that its time is running more slowly than yours, its length is shorter than its length when at rest, and its mass is greater than its mass when at rest. Moreover, observers in different reference frames may disagree about whether two events are simultaneous and about the speeds of different objects.

Do the effects predicted by relativity really occur?

Experiments with light confirm that its speed is always the same. Experiments with subatomic particles in particle accelerators confirm the predictions of time dilation and mass increase at speeds dose to the speed of light, and time dilation has been verified at relatively low speeds in aircraft and spacecraft. Nuclear power plants and nuclear bombs release energy in accordance with the formula $E = mc^2$, which is also a prediction of special relativity.

2.4 TOWARD A NEW COMMON SENSE

How can we make sense of relativity?

A person moving by you at high speed will see exactly the same effects on you as you see on her; for example, she'll see your time running slowly while you see her time running slowly. Although this might sound contradictory, it simply tells us that time and space must be relative in much the same way that up and down are relative on Earth.

How does special relativity offer us a ticket to the stars?

Although the theory tells us that journeys to the stars will always take many years from the point of view of Earth, it also tells us that the time for the passengers will be much shorter if they travel at speeds close enough to the speed of light. Thus, the passengers may be able to make very distant journeys within their lifetimes, even though their friends back on Earth will not be there to greet them when they return.

EXERCISES AND PROBLEMS: MASTERING ASTRONOMY

For instructor-assigned homework go to www.masteringastronomy.com

REVIEW QUESTIONS

Short-Answer Questions Based on the Reading

1. What is the *theory of relativity*? How does *special relativity* differ from *general relativity*?
2. List five major predictions of the special theory of relativity.
3. What is a *paradox*? How can a paradox lead us to a deeper understanding of an issue?
4. According to the theory of relativity, what are the two absolutes in the universe? Which one is more surprising, and why?
5. What do we mean by a *frame of reference*? What is a *free-float frame*?
6. Suppose you see a friend moving by you at some constant speed. Explain why your friend can equally well say that she is stationary and you are moving by her.
7. Construct your own variation on Thought Experiment 6 to prove that no material object can be observed to travel at or above the speed of light.
8. What is *time dilation*? Explain how and why your measurements of time will differ from those of someone moving by you.
9. Explain why observers in different reference frames will not necessarily agree about the order of two events that occur in different places.
10. What is *length contraction*? How will your measurements of the size of a spaceship differ if the spaceship is moving by you compared to when it is at rest in your reference frame?
11. What is *mass increase*? How does the mass of an object moving by you compare to its rest mass?

SPACE AND GRAVITY

The eternal mystery of the world is its comprehensibility. The fact that it is comprehensible is a miracle.
—Albert Einstein

What is gravity? Newton considered gravity to be a mysterious force that somehow reaches across vast distances of space to hold the Moon in orbit around Earth and the planets in orbit around the Sun. His law of gravity explained the actions and consequences of this mysterious force but said nothing about how the force is transmitted through space.

Einstein removed the mystery of how gravity acts at a distance. As he extended his theory of relativity, Einstein found that he could explain gravity in terms of the structure of space and time. With his new view, the orbits of the Moon and the planets become as natural as motion in a straight line.

In this chapter, we will investigate Einstein's revolutionary view of gravity. As we will see, Einstein's theory is crucial to understanding many astronomical phenomena, ranging from the peculiar orbit of Mercury to the bizarre properties of black holes.

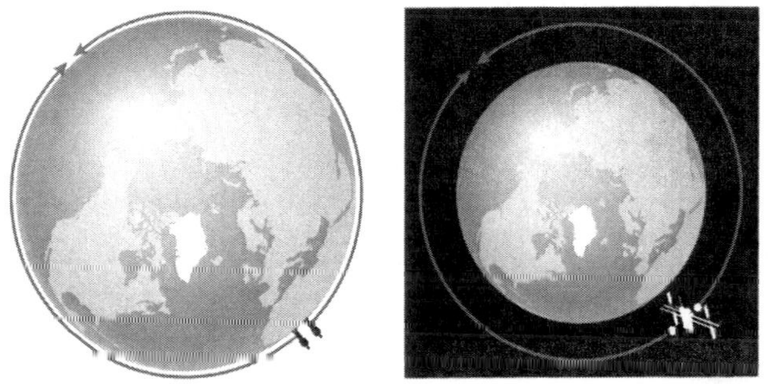

a Travelers going in opposite directions along paths that are as straight as possible will meet as they go around the Earth, a fact that we attribute to the curvature of Earth's surface.

b Two space probes launched in opposite directions in Earth orbit will meet as they orbit Earth, a fact that we usually attribute to the mysterious force of gravity.

Figure 3.1 Travelers on Earth's surface and orbiting objects follow similar-shaped paths, but we usually explain these paths in very different ways.

3.1 EINSTEIN'S SECOND REVOLUTION

Imagine that you and everyone around you believe the Earth to be flat. As a wealthy patron of the sciences, you decide to sponsor an expedition to the far reaches of the world. You select two fearless explorers and give them careful instructions. Each is to journey along a perfectly straight path, but they are to travel in opposite directions. You provide each with a caravan for land-based travel and boats for water crossings, and you tell each to turn back only after discovering "something extraordinary."

Some time later, the two explorers return. You ask, "Did you discover something extraordinary?" To your surprise, they answer in unison, "Yes, but we both discovered the same thing: We ran into each other, despite having traveled

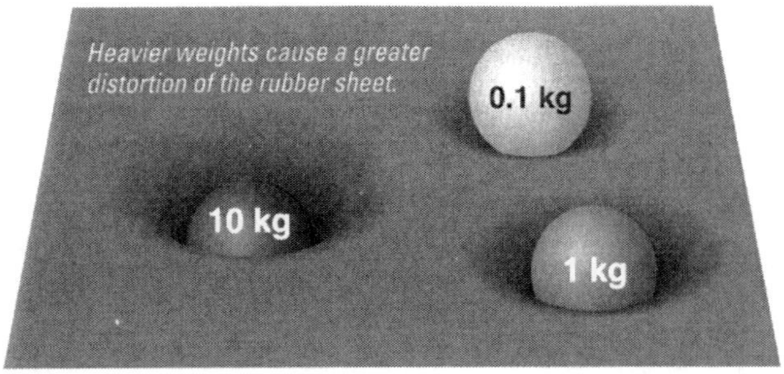

Figure 3.2 A rubber sheet analogy to spacetime: Matter distorts the "fabric" of four-dimensional spacetime in a manner analogous to the way heavy weights distort a taut, two-dimensional rubber sheet. The greater the mass, the greater the distortion of spacetime.

in opposite directions along perfectly straight paths."

Although this outcome would be extraordinarily surprising if you truly believed the Earth to be flat, we are not really surprised because we know that Earth is round (Figure 3.Ia). In a sense, the explorers followed the *straightest possible paths*, but these "straight" lines follow the curved surface of the Earth.

Now let's consider a somewhat more modern scenario. You are floating freely in a spaceship somewhere out in space. Hoping to learn more about space in your vicinity, you launch two small probes along straight paths in opposite directions. Each probe is equipped with a camera that transmits pictures back to your spaceship. Imagine that, to your astonishment, the probes one day transmit pictures of each other! That is, although you launched them in opposite directions and neither has ever fired its engines, the probes have somehow met. This might at first sound surprising, but in fact this situation arises quite naturally with orbiting objects. If you launch two probes in opposite directions from a space station, they will meet as they orbit Earth (Figure 3.1b).

Since the time of Newton, we've generally explained the curved paths of the two probes as an effect caused by the force of gravity. However, by analogy with the explorers journeying in opposite directions on Earth, might we instead conclude that the probes meet because *space* is somehow curved? The idea that space could be curved certainly sounds strange at first. While it's easy to visualize a surface curving *through* space, our minds cannot visualize three-dimensional space itself as being curved. The idea that space can be curved lies at the heart of Einstein's second revolution—a revolutionary view of gravity contained in his *general theory of relativity*, published in 1915.

What are the major ideas of general relativity?

Recall that the special theory of relativity applies only to situations in which we are not concerned with the effects of gravity. From this theory, we already know that space and time are inextricably linked. More specifically, special relativity tells us that the three dimensions of space and the one dimension of time together form an inseparable, *four-dimensional* combination called **spacetime**. When Einstein extended the theory of relativity to the general case that includes gravity, he discovered that matter shapes the "fabric" of spacetime in a manner analogous to the way heavy weights distort a taut rubber sheet or trampoline (figure 3.2). Of course, we cannot place weights "upon" spacetime because all matter exists *within* spacetime, and we cannot visualize distortions of spacetime. However, using a rubber sheet as an analogy, we can begin to appreciate the principles of general relativity.

It is difficult to overstate the significance of general relativity to our understanding of the universe. For example, the following ideas all come directly from Einstein's general theory of relativity:

- Gravity arises from distortions of spacetime. It is *not* a mysterious force that acts at a distance. The presence of mass causes the distortions, and the resulting distortions determine how other objects move through spacetime.
- Time runs slowly in gravitational fields. The stronger the gravity, the more slowly time runs.
- *Black holes* can exist in spacetime, and falling into a black hole means leaving the observable universe.
- The universe has no boundaries and no center, yet it might still have a finite volume.
- Large masses that undergo rapid changes in motion or structure emit *gravitational waves* that travel at the speed of light.

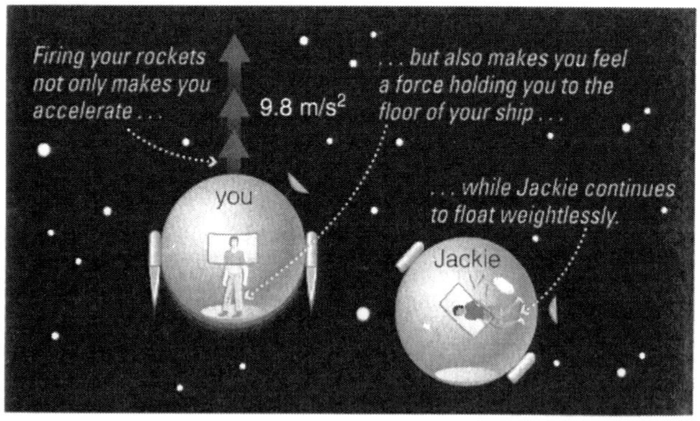

Figure 3.3 Out in deep space with no engines firing, you and Jackie would both float freely in your spaceships. But as soon as you start firing your engines, you feel a force that presses you against the floor of your spaceship. Jackie sees you accelerate away. You might wish to claim that she is accelerating away from you, but how, then, do you explain the force that you feel and the fact that Jackie remains weightless?

Is all motion relative?

Special relativity tells us that there is no single, absolute answer to the question "Who is moving?" when two people pass each other at a constant Velocity in free-float frames. Each individual can claim to be at rest, and each claim is equally valid. However, the situation seems quite different if one of the reference frames is accelerating rather than traveling at constant velocity. Understanding this apparent difference is a good way to begin our study of the general theory of relativity.

A Thought Experiment with Acceleration

Imagine that you and your friend Jackie are both floating freely in space when you decide to fire your rocket engines. Before you fire the engines, you and Jackie are both in free-float reference frames, which means you are both floating weightlessly in your spaceships. Thus, as we saw in Chapter 2, you cannot determine who is really moving no matter what your relative velocity, because you will both experience the laws of nature in exactly the same way. Now, suppose your rocket engines start to fire with enough thrust to give you an acceleration of 1*g*, or 9.8 m/s^2—which is the acceleration of gravity on Earth [Section 4.1). Jackie sees you accelerating away, with your speed growing ever faster, so she sends you a radio message saying, "Good-bye, have a nice trip!"

THINK ABOUT IT

Suppose you start from rest in Jackie's reference frame and she sees you accelerate at 1g (= 9.8 m/s^2). Approximately how fast will Jackie see you going after 1 second? After 10 seconds? After a minute? (*Hint:* See Section 4.1 to review the meaning of acceleration.)

*Technically, this equivalence holds only within small regions of space. Over larger regions, the gravity of a massive object varies in ways that would not occur due to acceleration; for example, such variation gives rise to tidal forces that do not arise from acceleration.

The Equivalence Principle

Figure 3.4 The equivalence principle states that the effects of gravity are exactly equivalent to the effects of acceleration. Therefore, you cannot tell the difference between being in a closed room on Earth and being in a closed room accelerating through space at 1g.

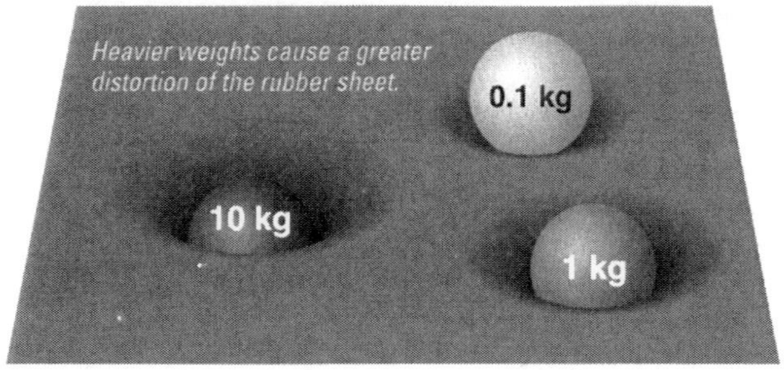

Figure 3.5 According to the equivalence principle, you can claim to be stationary in a gravitational field, using your engines to prevent you from falling. You feel weight due to gravity, while Jackie is weightless because she is in free-fall.

If all motion is relative, you should be free to claim that you are still stationary and that it is Jackie who is receding into the distance at ever-faster speeds. You might therefore wish to reply: "Thanks, but I'm not going anywhere. You're the one accelerating into the distance." However, this situation has a new element that was not present when we dealt with constant velocities: Because your rocket engines are firing, you feel a force inside your spaceship (Figure 3.3). In fact, because the engines are giving you an acceleration of 19, you'll feel a force holding you to the floor of your spaceship that is the same as the force you feel when you stand on Earth's surface. In other words, you'll no longer be weightless but instead will be able to walk on the spaceship floor with your normal Earth weight. Thus, Jackie may respond back: "Oh, yeah? If you're not going anywhere, why are you stuck to the floor of your spaceship, and why do you have your engines turned on? Furthermore, if I'm accelerating, why am I weightless?"

You must admit that Jackie is asking very good questions. It certainly *looks* as if you really are the one who is accelercating, in which case you *cannot* legitimately claim to be stationary. In other words, it seems that motion is no longer relative when we introduce acceleration. This idea did not sit well with Einstein, because he believed that *all* motion should be relative, regardless of whether the motion was at constant velocity or included an acceleration. Einstein therefore needed some way to explain the force you feel due to firing your rocket engines without necessarily assuming that you are accelerating through space. (Einstein did not really do his thought experiments with spaceships, but the basic ideas of his thought experiments were the same.)

The Equivalence Principle

In 1907, Einstein hit upon what he later called "the happiest thought of my life." His revelation consisted of the idea that, whenever you feel weight (as opposed to weight-

Figure 3.6 An object's number of dimensions is the number of independent directions in which movement is possible within the object. It is zero for a point, one for a line, two for a plane, and three for space.

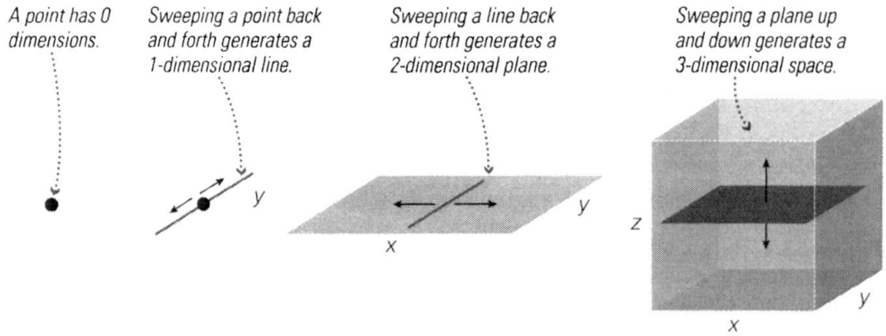

lessness), you can equally well attribute it to effects of either acceleration or gravity. This idea is called the **equivalence principle**. Stated more precisely, it says:

> **The effects of gravity are exactly equivalent to the effects of acceleration.***

To clarify the meaning of the equivalence principle, imagine that you are sitting inside with doors closed and window shades pulled down when your room is magically removed from Earth and sent hurtling through space with an acceleration of 1*g* (Figure 3.4). According to the equivalence principle, you would have no way of knowing that you've left Earth. Any experiment you performed, such as dropping balls of different weights, would yield the same results you'd get on Earth.

Now let's go back to Jackie's questions. She claims that *you* must be accelerating because you feel weight as you stand in your spaceship. However, the equivalence principle tells us that, with equal validity, you can claim to feel weight because of gravity. From this perspective, space is filled with a gravitational field pointing "downward" toward the floor of your spaceship. You are stationary only because your rocket engine *prevents* you from falling (thereby keeping you stationary), and you feel weight just as you would if you were hovering in a helicopter on Earth. And how do you explain Jackie's weightlessness? Easy: Because she is not using her engines, she and her spaceship are falling through the gravitational field that fills space around you—and anyone in free-fall feels weightless [Section 4.1]. In essence, you can claim that the situation is much as it would be if you were hovering over a cliff while Jackie had fallen over the edge (Figure 3.5). Thus, you can respond: "Sorry, Jackie, but I still say that you have it backward. I'm using my engines to prevent my spaceship from falling, and I feel weight because of *gravity*. You're weightless because you're in free-fall. I hope you won't be hurt

SPECIAL TOPIC: EINSTEIN'S LEAP

Given that the similarities in the effects of gravity and of acceleration were well known to scientists as far back as the time of Newton, you may be wondering why the equivalence principle is so surprising. The answer is that the similarities were generally attributed to coincidence—although a very puzzling coincidence. It was as if other scientists imagined that nature was showing them two boxes, one labeled "effects of gravity" and the other labeled "effects of acceleration." They shook, weighed, and kicked the boxes, but could never find any obvious differences between them. They concluded: "What a strange coincidence! The boxes seem the same from the outside even though they contain different things." Einstein's revelation was, in essence, to look at the boxes and say that it is not a coincidence at all. The boxes appear the same from the outside because they contain the same thing.

In many ways, Einstein's assertion of the equivalence principle represented a leap of faith, although it was a faith he would willingly test through scientific experiment. He proposed the equivalence principle because he thought the universe would make more sense if it were true, not because of any compelling observational or experimental evidence for it at the time. This leap of faith sent him on a path far ahead of his scientific colleagues.

From a historical viewpoint, special relativity was a "theory waiting to happen" because it was needed to explain two significant problems left over from the 19th century: the perplexing constancy of the speed of light, demonstrated in the Michelson-Morley experiment [Section 2.3], and some seeming peculiarities of the laws of electromagnetism. Indeed, several other scientists were very close to discovering the ideas of special relativity when Einstein published the theory in 1905, and *someone* was bound to come up with special relativity around that time.

General relativity, in contrast, was a *tour de force* by Einstein. He recognized that unsolved problems remained after completing the theory of special relativity, and he alone took the leap of faith required to accept the equivalence principle. Without Einstein, general relativity probably would have remained undiscovered for at least a couple of decades beyond 1915, the year he completed and published the theory.

by hitting whatever lies at the bottom of this gravitational field!"

Because the equivalence principle says that acceleration affects the laws of physics in exactly the same way as gravity, it allows us to treat all motion as relative. Thus, it is the starting point for general relativity. Just as the consequences of special relativity follow directly from the ideas that the laws of nature are the same for everyone and that everyone always measures the same speed of light, the astounding predictions of general relativity follow from the equivalence principle. Moreover, just as special relativity led us to recognize some underlying truths about nature—such as that space and time are different for observers in different reference frames—general relativity also leads us to a new and deeper understanding of the universe.

3.2 UNDERSTANDING SPACETIME

It's easy to *say* that you can equally well attribute your weight to effects of gravity or of acceleration, but the two effects tend to *look* very different. A person standing on the surface of the Earth appears to be motionless, while an astronaut accelerating through space continually gains speed. How can gravity and acceleration produce such similar effects when they look so different? According to the general theory of relativity, the answer is that they look different only because we're not seeing the whole picture. Instead of looking just at the three dimensions of space, we must learn to "look" at the *four* dimensions of spacetime.

What is spacetime?

The first step in understanding spacetime is understanding what we mean when we say, for example, that something is two-dimensional or three-dimensional. The concept of **dimension** describes the number of independent directions in which movement is possible (Figure 3.6). A **point** has zero dimensions. If you were a "geometric prisoner" confined to a point, you'd have no place to go. Sweeping a point back and forth along one direction generates a **line**. The line is one-dimensional because only one direction of motion is possible (going backward is considered the same as going forward by a negative distance). Sweeping a line back and forth generates a two-dimensional **plane**. The two directions of possible motion are, say, lengthwise and width-wise. Any other direction is just a combination of these two. If we sweep a plane up and down, it fills **three-dimensional space**, with the three independent directions of length, width, and depth.

We live in three-dimensional space and thus cannot visualize any direction that is distinct from length, width, and depth (and combinations thereof). However, just because we cannot *see* "other" directions doesn't mean they don't exist. Thus, we can imagine sweeping space back and forth in some "other" direction to generate a **four-dimensional space**.

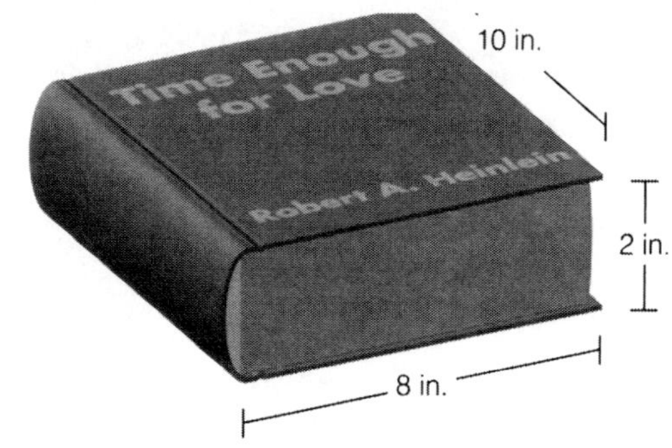

a A book has an unambiguous three-dimensional shape.

b Two-dimensional pictures of the book can look very different.

Figure 3.7 Two-dimensional views of a three-dimensional object (like a book) can appear different, even though the object has only a single real shape. In a similar way, observers in different reference frames may measure space and time differently (because they perceive only three dimensions at once) even though they are all observing the same four-dimensional space-time reality.

Although we have no hope of visualizing a four-dimensional space, we can describe it mathematically. In algebra, we do one-dimensional problems with the single variable x, two-dimensional problems with the variables x and y, and three-dimensional problems with the variables x, y, and z. A four-dimensional problem simply requires adding a fourth variable, as in x, y, z, and w. We could continue to five dimensions, six dimensions, and so forth. Any space with more than three dimensions is called a *hyperspace*, which means "beyond space."

Spacetime

Spacetime is a four-dimensional space in which the four directions of possible motion are length, width, depth, and *time*. Note that time is not "the" fourth dimension. It is simply one of the four. (However, time differs in an important way from the other three dimensions. See Mathematical Insight 3.1.)

We cannot picture all four dimensions of spacetime at once, but we can imagine what things would look like if we could. In addition to the three spatial dimensions of spacetime that we ordinarily see, every object would be stretched out through time. Objects that we see as three-dimensional in our ordinary lives would appear as four-dimensional objects in spacetime. If we could see in four dimensions, we could look through time just as easily as we look to our left or right. If we looked at a person, we could see every event in that person's life. If we wondered what really happened during some historical event, we'd simply look to find the answer.

THINK ABOUT IT

Try to imagine how you would look in four dimensions. How would your body, stretched through time, appear? Imagine that you bumped into someone on the bus yesterday. What would this event look like in spacetime?

This spacetime view of objects provides a new way of understanding why different observers can disagree about measurements of time and distance. Because we can't visualize four dimensions, we'll use a three-dimensional analogy. Suppose you give the same book to many different people and ask each person to measure the book's dimensions. Everyone will get the same results, agreeing on the three-dimensional structure of the book (Figure 3.7a). Now, suppose instead that you show each person only a two-dimensional picture of the book rather than the book itself. The pictures may look very different, even though they show the same book in all cases (Figure 3.7b). If the people believed that the two-dimensional pictures reflected reality, they might argue endlessly about what the book really looks like.

In our ordinary lives, we perceive only three dimensions, and we assume that this perception reflects reality. But space-time is actually four-dimensional. Just as different people can see different two-dimensional pictures of the same three-dimensional book, different observers can see different three-dimensional "pictures" of the same spacetime reality. These different "pictures" are the differing perceptions of time and space of observers in different reference frames. Thus, different observers will get different results when they measure time, length, or mass, even though they are all actually looking at the same spacetime reality. In the words of a famous textbook on relativity:

Space is different for different observers.
Time is different for different observers.
*Spacetime is the same for everyone.**

3.3 A NEW VIEW OF GRAVITY

Newton's law of gravity claims that every mass exerts a gravitational attraction on every other mass, no matter how far away they are from each other. However, on close examination, this idea of "action at a distance" is rather mysterious. For example, how does Earth feel the Sun's attraction and know to orbit it? Newton himself was troubled by this idea. A few years after publishing his law of gravity in 1687, Newton wrote:

*That one body may act upon another at a distance through a vacuum, ... and force may be conveyed from one to another, is to me so great an absurdity, that I believe no man, who has a competent faculty in thinking, can ever fall into it.**

Nevertheless, for more than 230 years after Newton pub-

*From E. F. Taylor and J. A. Wheeler, *Spacetime Physics*, 2nd ed. (Freeman, 1992)

*Letter from Newton, 1692–1693, as quoted in J. A. Wheeler, *A Journey into Gravity and Spacetime* (Scientific American Library, 1990, p. 2).

a On a flat rubber sheet, evenly spaced circles all have circumference $2\pi r$.

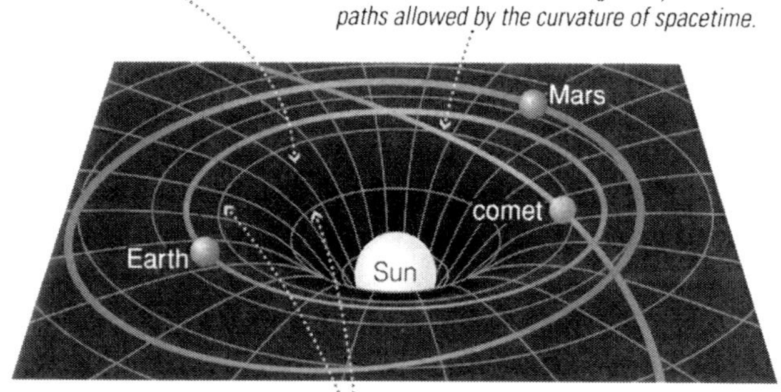

b The Sun curves spacetime much like a heavy weight curves a rubber sheet.

Figure 3.13 According to general relativity, planets orbit the Sun for much the same reason that you can make a marble go around in a salad bowl: Each planet is going as straight as it can, but the curvature of spacetime causes its path through space to go round and round.

lished his gravitational law, no one found any better way to explain gravity's mysterious "action at a distance." Einstein changed all that, when he realized that the equivalence principle allowed him to explain the action of gravity without requiring any long-distance force.

What is gravity?

Einstein's general theory of relativity removes the idea of "action at a distance" by stating that Earth feels *no* force tugging on it in its orbit, and therefore follows the straightest possible path through spacetime. Thus, the fact that Earth goes around the Sun tells us that spacetime itself is curved near the Sun. In other words:

What we perceive as gravity arises from the curvature of spacetime.

Rubber Sheet Analogy

We cannot actually picture the curvature of spacetime, but a two-dimensional analogy can help us understand the idea. We represent spacetime in the analogy with a stretched rubber sheet. To make the analogy work, we have to ignore any effects of friction on the rubber sheet, because there is no friction in space.

Figure 3.13a shows a flat rubber sheet representing spacetime in a region where it has a flat geometry. Notice that the radial distances between each of the circles shown on the sheet are the same, and all the circles have circumferences that follow the flat geometry formula of 27πr. If you rolled a marble across this frictionless sheet, it would roll in a straight line at constant speed. This fact essentially illustrates Newton's first law of motion, in which objects move at constant velocity when they are not affected by gravity or any other forces.

Figure 3.13b shows what happens to spacetime around the Sun. We represent the Sun with a heavy mass on the rubber sheet, which causes the sheet to curve and form a bowl-like depression. The circles that were evenly spaced on the flat sheet now become more widely separated (with circumferences increasingly less than 2πr) near the bottom of the bowl, showing that gravity becomes stronger and the curvature of spacetime becomes greater as we approach the Sun's surface. (Notice that the curvature does *not* continue to increase with depth inside the Sun; in fact, the strength of gravity actually weakens near the Sun's center.) If you rolled marbles on this rubber sheet, they could not go in straight lines because the sheet itself is curved. Instead, the marbles would follow the *straightest possible* paths given the curvature of the sheet. A particular marble's path would depend on the speed and direction with which you rolled it. You'd find that marbles rolled relatively slowly and close to the center would follow circular or elliptical "orbits" around the center of the bowl, while marbles rolled from farther away or at higher speeds could loop around the center on unbound parabolic or hyperbolic paths.

By analogy, general relativity tells us that, depending on their speed and direction, planets or other objects moving freely in space can follow circular, elliptical, or unbound parabolic or hyperbolic orbits—the same orbital shapes that Newton's universal law of gravitation allows [Section 4.4J. However, the explanation for these orbits is now quite different from that in Newton's view of gravity. Rather than orbiting because of a mysterious force exerted on them by the distant Sun, the planets orbit because they follow the *straightest possible paths* allowed by the shape of spacetime around them. The central mass of the Sun is not grabbing them, communicating with them, or doing anything else to influence their motion. Instead, it is simply dictating the shape of spacetime around it. In other words:

A mass like the Sun causes spacetime to curve, and the curvature of spacetime determines the paths of freely moving masses like the planets.

Weightlessness in Space

This idea gives us a new way to explain the weightlessness of astronauts in space. Just as the Sun curves spacetime into a "bowl shape" (but in four dimensions) that makes the straightest possible paths of the planets go round and round, Earth also curves spacetime in a way that makes orbiting spacecraft go round and round. In other words, spacecraft orbit Earth because, as long as their engines are off and they are unaffected by atmospheric drag, circular or elliptical orbits are the straightest possible paths they can follow through spacetime in Earth's vicinity. Thus, instead of having to invoke the idea of free-fall caused by a gravitational attraction to Earth, we can explain the weightlessness of astronauts in the Space Station simply by recognizing that they are following the straightest possible paths through spacetime.

The same idea holds true for any other orbital trajectory. For example, if we launched a human mission to Mars, we would need to give the spaceship escape velocity from Earth. In the rubber sheet analogy, this means launching it with enough speed so that it can escape from the bowl-shaped region around Earth, like a marble shot fast enough to roll out of the bowl and onto the flatter region far away from it. Except when their rockets are firing, the astronauts would still be weightless throughout the trip because they would be following the straightest possible path. Firing the engines, either to accelerate away from Earth or to decelerate near Mars, would make the spaceship deviate from the straightest possible path, so the astronauts would feel weight during those portions of their journey.

Limitations of the Analogy

The rubber sheet analogy is useful for understanding how

mass affects spacetime, but it also has limitations because it is a two-dimensional representation of a four-dimensional reality. In particular, the analogy has three important limitations that you should keep in mind whenever you use it:

- The rubber sheet is supposed to represent the universe, but it makes no sense to think of placing a mass like the Sun "upon" the universe. Instead, we should think of the masses as being *within* the rubber sheet.
- The rubber sheet allows us to picture only two of the three dimensions of space. For example, it allows us to show that different planets orbit at different distances from the Sun and that some have more highly elliptical orbits than others, but it does not allow us to show the fact that the planets do not all orbit the Sun in precisely the same plane.
- The rubber sheet analogy does not show the *time* part of spacetime. Bound orbits on the sheet or in space appear to return to the same point with each circuit of the Sun. However, objects cannot return to the same point in spacetime, because they always move forward through time. For example, with each orbit of the Sun, Earth returns to the same place in space (relative to the Sun) but to a time that is a year later (Figure 3.14).

What is a black hole?

Greater curvature of spacetime means stronger gravity, and the rubber sheet analogy suggests two basic ways to increase the strength of gravity. First, a larger mass causes greater curvature at any particular distance away from it. For example, the Sun curves spacetime more than any planet, and Earth curves spacetime more than the Moon. Note that this idea is consistent with Newton's law of gravity, in which increasing the mass of an object increases the gravitational attraction at all distances.

The second way to increase the curvature of spacetime around an object is to leave its mass alone but increase its density by making it smaller in size. For example, suppose we could compress the Sun into a type of "dead" star called a *white dwarf* [Section 18.1]. Because its total mass would still be the same, there would be no effect on the curvature of spacetime far from the Sun. However, spacetime would be much more curved near the compressed Sun's surface, reflecting the fact that gravity is much stronger on the surface of a compressed white dwarf than on the Sun. Again, the idea that the surface gravity on an object of a particular mass grows stronger as the object shrinks in radius is consistent with Newton's law of gravity.

Now, imagine that we could continue to compress the Sun to smaller and smaller sizes. Far from the Sun, this compression would have no effect at all, because the same total mass would still be causing the curvature of spacetime. Near the Sun's surface, however, spacetime would become increasingly curved as we shrank the Sun in size. In fact, if we shrank the Sun enough, we could eventually curve spacetime so much that it would become a bottomless pit—a hole in the observable universe. This is what we call a **black hole** (Figure 3.15). Note that Newton's view of gravity does not really have any analog to a black hole, because it does not envision the possibility of holes in the universe.

To summarize, a black hole is a place where spacetime is so curved that nothing that falls into it can ever escape. The boundary that marks the "point of no return" is called the **event horizon**, because events that occur within this boundary can have no influence on our observable universe. The idea of black holes is so bizarre that for decades after Einstein published his general theory of relativity, most scientists did not think they could really exist. However, we now have very strong evidence suggesting that black holes are in fact quite common.

How does gravity affect time?

Given that gravity arises from the curvature of spacetime, you should not be surprised to learn that gravity affects time as well as space. We can learn about the effects of gravity on time by considering the effects of accelerated motion and then invoking the equivalence principle.

Imagine that you and Jackie are floating weightlessly at opposite ends of a spaceship. You both have watches that flash brightly each second, which you synchronized beforehand. Because you are both floating freely with no relative motion between you, you are both in the same reference frame. Therefore, you will see each other's watches flashing at the same rate.

Now suppose you fire the spaceship engines so that the spaceship begins to accelerate, with you at the front and Jackie at the back. When the ship begins accelerating, you and Jackie will no longer be weightless. The acceleration introduces an even more important change into the situation, which we can understand by imagining the view of someone floating weightlessly outside the spaceship. Remember that observers moving at different relative speeds are in different reference frames. When the spaceship is accelerating, its speed is constantly increasing relative to the outside observer, which means that both you and Jackie are constantly changing reference frames. Moreover, the flashes from your watches take a bit of time to travel the length of the spaceship. Thus, by the time a particular flash from Jackie's watch reaches you (or a flash from your watch reaches Jackie), both your reference frames are different from what they were at the time the flash was emitted.

Because you are in the *front* of the accelerating spaceship, your changing reference frames are always carrying you *away* from the point at which each of Jackie's flashes is emitted. Thus, the light from each of her flashes will take a little *longer* to reach you than it would if the ship were not accelerating. As a result, instead of seeing Jackie's flashes 1 second apart, you'll see them coming a little *more* than 1 second apart. That is, you'll see Jackie's watch flashing more slowly than yours (Figure 3.16a). You will therefore conclude that time is running more slowly at the back end of the spaceship.

From Jackie's point of view at the *back* of the accelerating spaceship, her changing reference frames are always carrying her *toward* the point at which each of your flashes is emitted. Thus, the light from each of your flashes will take a little *less* time to reach her than it would if the ship were not accelerating, so she'll see them coming a little less than 1 second apart. She will see your watch flashing *faster* than hers and conclude that time is running *fast* at the front end of the spaceship. Note that you and Jackie agree: Time is running more slowly at the back end of the spaceship and faster at the front end. The greater the acceleration of the spaceship, the greater the difference in the rate at which time passes at the two ends of the spaceship.

Now we apply the equivalence principle, which tells us that we should get the same results for a spaceship at rest in a gravitational field as we do for a spaceship accelerating through space. Thus, if the spaceship were at rest on a planet, time would also have to be running more slowly at the bottom of the spaceship than at the top (Figure 3.16b). That is, time must run more slowly at lower altitudes than at higher altitudes in a gravitational field. This effect is known as **gravitational time dilation.**

The stronger the gravity—and hence the greater the curvature of spacetime—the larger the factor by which time runs slowly. On an object with relatively weak gravity, like Earth, the slowing of time is barely detectable compared to the rate at which time passes in deep space. However, time runs noticeably more slowly on the surface of the Sun than on Earth, and more slowly on the surface of a white dwarf star than on the Sun. Perhaps you've already guessed that the extreme case is a black hole: To anyone watching from a distance, time comes to a stop at the event horizon. If you could observe clocks placed at varying distances from the black hole, you'd see that clocks nearer the event horizon run more slowly and clocks at the event horizon would show time to be frozen.

THINK ABOUT IT

Where would you age more slowly, on Earth or on the Moon? Would you expect the difference to be significant? Explain.

3.4 TESTING GENERAL RELATIVITY

Starting from the principle of equivalence, we've used logic and analogies to develop the ideas of general relativity. However, as always, we should not accept these logical conclusions unless they withstand observational and experimental tests.

How do we test the predictions of the general theory of relativity?

Like the predictions of special relativity, those of general relativity have faced many tests and have passed with flying colors. Let's examine some of the most important tests of general relativity.

Mercury's Peculiar Orbit

The first observational test passed by the theory of general relativity concerned the orbit of the planet Mercury. Newton's law of gravity predicts that Mercury's orbit should precess slowly around the Sun because of the gravitational influences of other planets (Figure 3.17). Careful observations of Mercury's orbit during the 1800s showed that it does indeed precess, but careful calculations made with Newton's law of gravity could not completely account for the observed precession. Although the discrepancy was small, further observations verified that it was real.

Einstein was aware of this discrepancy and, from the time he first thought of the equivalence principle in 1907, he

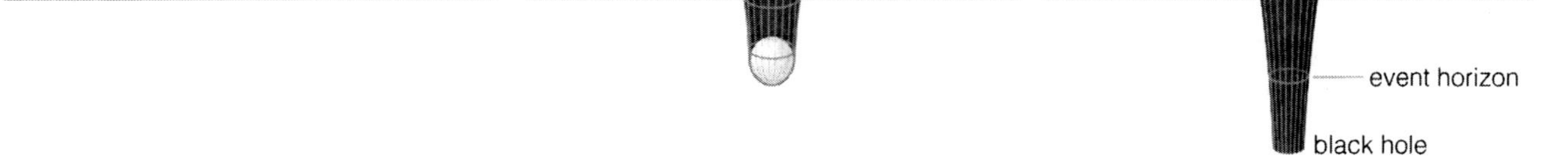

Figure 3.15: Interactive figure. According to general relativity, a black hole is like a bottomless pit in spacetime. Once an object crosses the event horizon, it has left our observable universe.

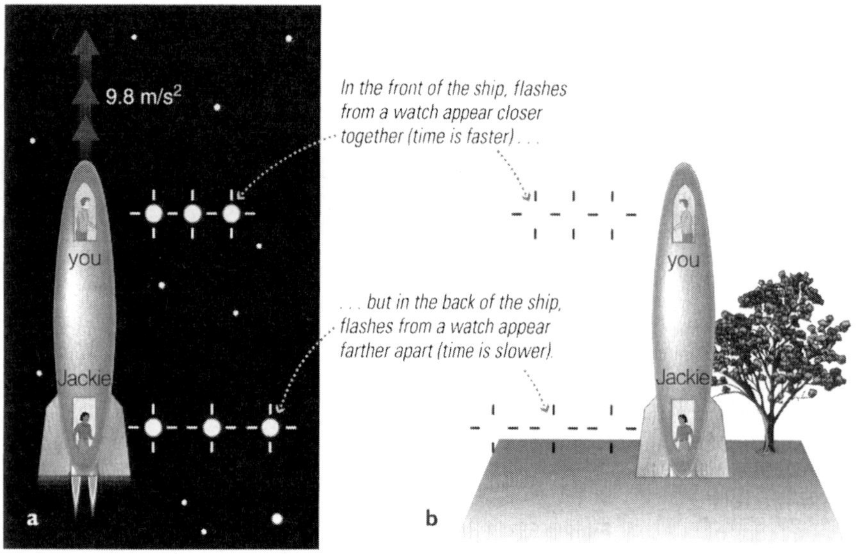

a In an *accelerating* spaceship (but not in one at constant velocity), time must run faster at the front end and more slowly at the back end. The yellow dots represent the flashes from the watches, and the spacing between the dots represents the time between the flashes.

b By the equivalence principle, time must also run more slowly at lower altitudes in a gravitational field.

Figure 3.16 Gravity causes time to run more slowly at lower altitudes than at higher altitudes, an effect called *gravitational time dilation.* (Note that the effect occurs even in a uniform gravitational field; that is, it does not depend on the additional fact that gravity tends to weaken at higher altitudes.)

SPECIAL TOPIC: THE TWIN PARADOX

Imagine two twins, one of whom stays on Earth while the other takes a high-speed trip to a distant star and back. In Chapter 2, we said that the twin who takes the trip will age less than the twin who stays home on Earth. Shouldn't the traveling twin be allowed to claim that she stayed stationary while Earth made a trip away from her and back? And in that case, shouldn't the twin on Earth be the one who ages less? This question underlies the so-called *twin paradox*, which can be analyzed in several different ways. We will take an approach that offers some insights into the nature of spacetime.

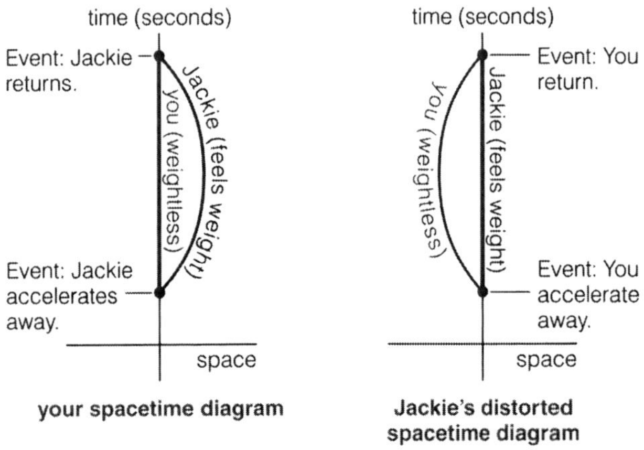

Figure 1 A person floating weightlessly must be following the straightest possible path through spacetime (left). Because this is not the case in Jackie's diagram (right), her diagram must be distorted.

Suppose you and Jackie are floating weightlessly next to each other with synchronized watches. While you remain weightless, Jackie uses her engines to accelerate a short distance away from you, decelerate to a stop a bit farther away, and then turn around and return. From your point of view, Jackie's motion means that you'll see her watch ticking more slowly than yours. Thus, upon her return, you expect to find that less time has passed for her than for you. But how does Jackie view the situation?

The two of you can argue endlessly about who is really moving, but one fact is obvious to both of you: During the trip, you remained weightless while Jackie felt *weight* holding her to the floor of her spaceship. Jackie can account for her weight in either of two ways. First, she can agree with you that she was the one who accelerated. However, because we know that time runs more slowly in an accelerating spaceship, she'll therefore agree that her watch ran more slowly than yours. Alternatively, she can claim that she felt weight because her engines counteracted a gravitational field in which she was stationary while you fell freely, but we also know that time runs slowly in gravitational fields. Therefore, she'll still agree that her watch ran more slowly than yours. Thus, no matter how you or Jackie looks at it, the result is the same: less time passes for Jackie.

The left side of Figure 1 shows a spacetime diagram for this experiment. You and Jackie both moved between the same two events in spacetime: the start and end points of Jackie's trip. However, your path between the two events is shorter than Jackie's. Because we have already concluded hoped he would be able to explain it. When he finally succeeded in November 1915, he was so excited that he was unable to work for the next 3 days. He later called the moment of this success the high point of his scientific life.

In essence, Einstein showed that the discrepancy arose because Newton's law of gravity assumes that time is absolute and space is flat. In reality, time runs more slowly and space is more curved on the part of Mercury's orbit that is nearer the Sun. The equations of general relativity take this distortion of spacetime into account, providing a predicted orbit for Mercury that precisely matches its observed orbit.

THINK ABOUT IT

Suppose the perihelion of Mercury's orbit were even closer to the Sun than it actually is. Would you expect the discrepancy between the actual orbit and the orbit predicted by Newton's laws to be greater than or less than it actually is? Explain.

Gravitational Lensing

We can also test Einstein's claim that space is curved by observing the trajectories of light rays moving through the universe. Because light always travels at the same speed, which means it never accelerates or decelerates, light must always follow the straightest possible path. If space itself is curved, then light paths will appear curved as well.

Suppose we could carefully measure the angular separation between two stars during the daytime just when the light from one of the stars passes near the Sun. The curvature of space near the Sun should cause the light beam passing closer to the Sun to curve more than the light beam from the other star (Figure 3.18). Therefore, the angular separation of the two stars should appear smaller than their true angular separation (which we would know from nighttime measurements). This effect was first observed in 1919, when astronomers traveled far and wide to measure stellar positions near the Sun during a total eclipse. The results agreed with Einstein's predictions, and the media attention drawn by the eclipse expeditions brought Einstein worldwide fame.

Even more dramatic effects occur when a distant star or galaxy, as seen from Earth, lies directly behind another object with a strong gravitational field (Figure 3.19). The mass of the intervening object curves spacetime in its vicin-

that less time passes for Jackie, we are led to a remarkable insight about the passage of time:

Between any two events in spacetime, more time passes on the shorter (and hence straighter) path.

The maximum amount of time you can record between two events in spacetime occurs if you follow the straightest possible path—that is, the path on which you are weightless.

The subtlety arises if Jackie chooses to claim that she is at rest and attributes her weight to gravity. In that case, she might be tempted to draw the spacetime diagram on the right in Figure 1, on which she appears to have the shorter path through spacetime. The rule that more time passes on shorter paths would then seem to imply that *your* watch should have recorded less time than Jackie's, contradicting our earlier claims. The contradiction is an illusion. If Jackie wishes to assert that she felt gravity, she must also claim that the gravity she felt implies that spacetime is curved in her vicinity. Therefore, she should not have drawn a spacetime diagram on a flat piece of paper.

Jackie's problem is analogous to that of a pilot who plans a trip from Philadelphia to Beijing on a flat map of Earth (Figure 2). On the flat map, it appears that the pilot has plotted the straightest possible path. However, this appearance is an illusion: The shortest and straightest path really is the great-circle route shown in Figure 3.11b, which appears curved on the flat map of Earth. A flat map of Earth distorts reality because the actual geometry of Earth's surface is spherical.

Just as the distortions in a map of the world do not change the actual distances between cities, the way we choose to draw a spacetime diagram does not alter the reality of spacetime. The solution to the twin paradox is that the two twins do not share identical situations. The twin who turns around at the distant star must have a more strongly curved worldline than the stay-at-home twin. Thus, more time must pass for the stay-at-home twin, and the traveler does indeed age less during the journey.

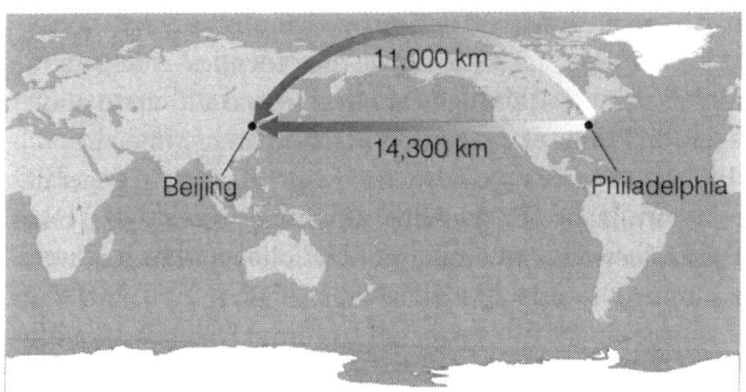

Figure 2 This flat map shows the same two paths on Earth shown in Figure 3.11 b. However, the distortion involved in making the map flat means that what looks like a straight line is not really as straight or as short as possible.

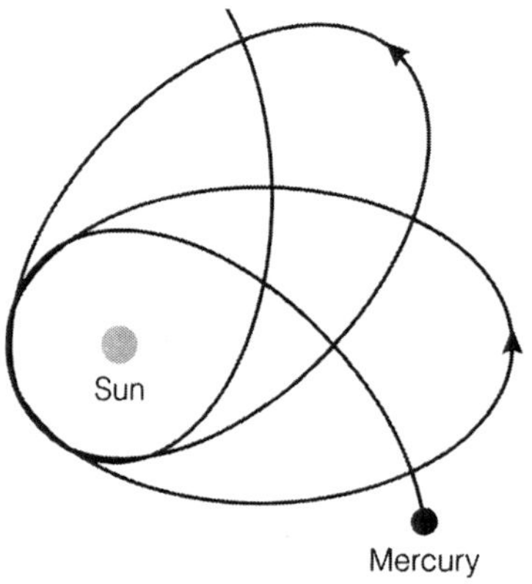

Note: The amount of progression with each orbit is highly exaggerated in this picture.

Figure 3.17 Mercury's orbit slowly precesses around the Sun.

ity, altering the trajectories of light beams passing nearby. Different light paths can curve so much that they end up converging at Earth, grossly distorting the appearance of the star or galaxy. Depending on the precise four-dimensional geometry of spacetime between us and the observed star or galaxy, the image we see may be magnified or distorted into arcs, rings, or multiple images of the same object (Figure 3.20). This type of distortion is called **gravitational lensing**, analogous to the lensing of light when it is bent by a glass lens.

Gravitational Time Dilation

We can test the prediction of gravitational time dilation by comparing clocks located in places with different gravitational field strengths. Even in Earth's weak gravity, experiments demonstrate that clocks at low altitude tick more slowly than identical clocks at higher altitude. Although the effect would add up to only a few billionths of a second over a human lifetime, the differences agree precisely with the predictions of general relativity. In fact, the global positioning system (GPS) takes these effects into account; if it didn't, it would be far less accurate in locating positions on Earth.

Surprisingly, it's even easier to compare the passage of time on Earth with the passage of time on the surface of the Sun and other stars. Because stellar gases emit and absorb *spectral lines* with particular frequencies [Section 5.4], they serve as natural atomic clocks. Suppose that, in a laboratory on Earth, we find that a particular type of gas emits a spectral line with a frequency of 500 trillion cycles per second. If this same gas is present on the Sun, it will also

However, general relativity claims that time should be running slightly more slowly on the Sun than on Earth. That is, 1 second on the Sun lasts *longer* than 1 second on Earth, or, equivalently, a second on Earth is shorter than a second on the Sun. Thus, during 1 second on Earth, we will see *fewer* than 500 trillion cycles from the gas on the Sun. Because lower frequency means longer or redder wavelengths, the spectral lines from the Sun ought to be *redshifted*. This redshift has nothing to do with the Doppler shifts that we see from moving objects [Section 5.5]. Instead, it is a **gravitational redshift**, caused by the fact that time runs slowly in gravitational fields. Gravitational redshifts have been measured for spectral lines from the Sun and from many other stars. The results agree with the predictions of general relativity, confirming that time slows down in stronger gravitational fields.

Measuring Earth's Effect on Spacetime

Earth has a relatively weak gravitational field, which means that it causes a fairly small amount of spacetime curvature. Nevertheless, Earth's effects on spacetime should be measurable in principle. In 2004, after decades of development, a satellite designed to measure these effects was launched into space. Known by the name "Gravity Probe B," the satellite used precision gyroscopes—which consist of the most perfectly round objects ever made—to look for subtle effects of spacetime curvature and the effect of Earth's rotation on spacetime. Unfortunately, early results from Gravity Probe B show too much noise for firm conclusions to be drawn, but scientists are continuing to analyze the data. Meanwhile, similar tests have been conducted by careful analysis of satellite motions. These tests have confirmed that Earth does indeed affect spacetime as predicted by general relativity, at least to within an accuracy of about 10%.

In the more than 100 years since Einstein first published his special theory of relativity, both the special and general theories have been subjected to extensive and precise testing. So far, both theories have passed every test. While it remains possible that future tests will force modifications to the theories of relativity, it seems inconceivable that new discoveries could change their underlying message about the intertwined nature of spacetime, with its implications to our understanding of everything from the passage of time to black holes to the overall geometry of the universe.

The astonishing transformation that relativity brings to

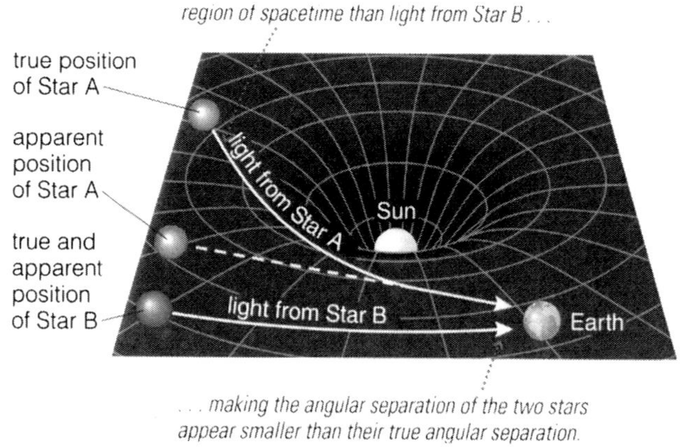

Figure 3.18 When we see starlight that passes near the Sun during a total eclipse, the curvature of spacetime causes a shift in the star's apparent position.

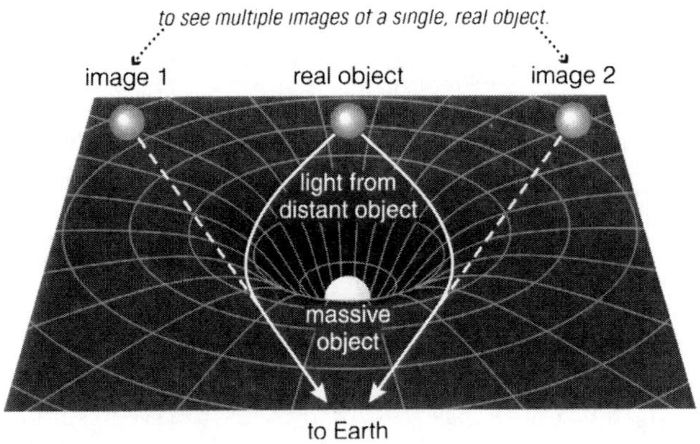

Figure 3.19 Gravitational lensing can create distorted or multiple images of a distant object whose light passes by a massive object on its way to Earth.

our view of space and time has also provided grist for philosophers, who debate its implications to concepts such as fate and free will. These debates are fascinating and well worth exploring, but remember that they do not affect the data from which the scientific theories are built. In other words, while there are many different viewpoints on the philosophical implications of relativity, only the scientific aspects of the theory are confirmed by data.

With that in mind, for our last word on relativity let's turn to Einstein himself. Here is what he said about a month before his death, on April 18, 1955:

*"Death signifies nothing ... the distinction between past, present, and future is only a stubbornly persistent illusion."**

THE BIG PICTURE

Putting Chapter 3 into Context

Just as the Copernican revolution overthrew the ancient belief in an Earth-centered universe, Einstein's revolution overthrew the common belief that space and time are distinct and absolute. We have explored Einstein's revolution in some detail in the past two chapters. Keep in mind the following "big picture" ideas:

- We live in four-dimensional spacetime. Disagreements among different observers about measurements of time and space occur because the different observers are looking at a single four-dimensional reality from different three-dimensional perspectives.
- Gravity arises from curvature of spacetime. Once we recognize this fact, the orbits of planets, moons, and all other objects are perfectly natural consequences of the curvature, rather than results of a mysterious "force" acting over great distances.
- Although the predictions of relativity may seem quite bizarre, they have been well verified by many observations and experiments.
- Some questions remain well beyond our current understanding. In particular, we do not yet know whether travel through hyperspace might be possible, allowing some of the imaginative ideas of science fiction to be come reality.

*This quotation was found with the aid of Alice Calaprice, author of *The Quotable Einstein* (Princeton University Press, 1996).

BLACK HOLES: THE ULTIMATE ENDS

If you throw a ball upward, it loses kinetic energy under the action of gravity, slows, and falls back to Earth. If you could throw it upward at 11.2 km/s, the terrestrial escape velocity, the strength of the Earth's gravitational field is insufficient ever to bring it to a stop. Although the ball would perpetually slow, it would go on forever into space, never returning.

In 1916 Albert Einstein showed, in his general theory of relativity, that mass—the source of gravity—curves spacetime. One of the fundamental precepts of relativity is that the speed of light in a vacuum is a constant, independent of the speed of the source or observer. Since spacetime is curved, a ray of light, although massless, must be affected by gravity in a way similar to the ball. Shine a flashlight upward; the beam must lose energy, but its speed, unlike the ball's, will stay constant. Since the energy of a photon is $E = h\nu$, the energy will instead be lost via a decrease in frequency. The wavelength gets longer, and the light reddens.

Now, in your mind, move into space and look back at the Earth, from which a companion shines a green light directly at you. Because the Earth's gravitational field is weak, the light will be reddened only imperceptibly. At the same time, however, your friend begins to compress the Earth. The gravitational field at the surface goes up, as do the escape velocity and the degree of reddening. By the time the

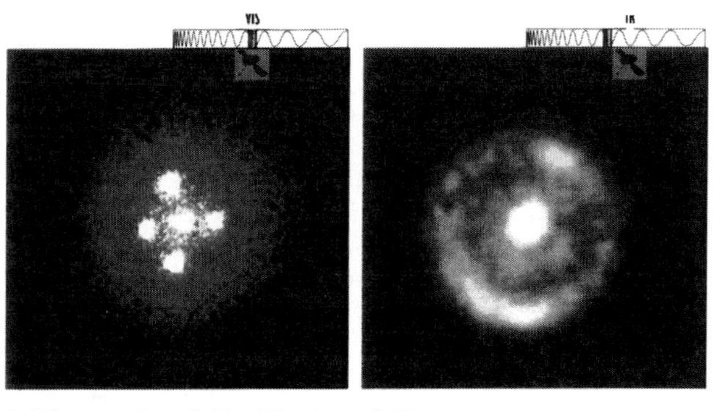

a In this case of gravitational lensing, called Einstein's Cross, the gravity of a foreground galaxy (center) bends light from a single bright background object so that it reaches us along four different paths—creating four distinct images of a single object.

b When one galaxy lies directly behind another, the foreground galaxy can bend light on all sides so that the light converges on Earth, forming an Einstein Ring like that pictured here.

Figure 3.20 Examples of gravitational lensing.

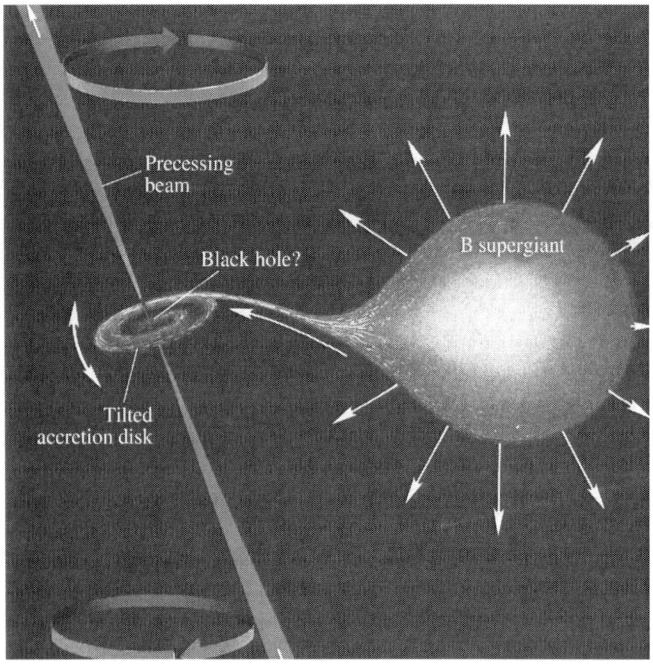

The collapsed object—a black hole or neutron star—in SS 433 accretes matter from a class B companion, then sprays it out from its poles. The accretion disk precesses with a 164-day period, causing the twin beams to oscillate wildly.

Earth is compressed to a little less than 2 centimeters across, the escape velocity reaches *c*: Light loses all its energy, the wavelengths are shifted to infinity, and both the flashlight beam and the Earth itself disappear from view. Our planet has fallen into a black hole, a puncture in spacetime, from which nothing can ever escape. Anything in that condition must collapse forever, the mass eventually reaching infinite density at the center. The laws of physics within the black hole have lost their meaning, because no information can ever exit. However, as the black hole collapses, it leaves behind an apparent surface called the event horizon, the limiting radius at which the strength of gravity just barely stops the light from escaping. The size of the event horizon, which is taken as the size of the black hole, depends only upon the black hole's mass and not upon its degree of compression, so it stays constant over time.

Although the theory of black holes was first worked out in 1916 by Karl Schwarzschild, it was not until the last 20 years or so that astronomers began to find objects in the sky whose explanations seemed to demand it. The phenomenon seems so bizarre as to be impossible; but the conditions are not that far from those of neutron stars, which we *know* exist. Support by degenerate neutrons has a limit just like the Chandrasekhar limit of a white dwarf. Above about 3 solar masses, a neutron star can no longer support itself and must collapse to a black hole. The event horizon is a few kilometers wide, not much smaller than the neutron star.

But do black holes really exist? If they do, where are they? And if no radiation can escape, how can we ever hope to find them? The key lies in the same phenomenon that produces novae, symbiotic stars, and (possibly) Type I supernovae: tidal interactions. In the constellation Cygnus is an otherwise ordinary O (or possibly B) dwarf that radiates powerfully in the X-ray spectrum. The star, called Cygnus X-l, displays absorption lines that Doppler-shift back and forth, showing it to be a member of a binary system—yet there is no trace of its companion.

If we know the velocities of both members of a spectroscopic binary as well as the inclination of the orbit to the line of sight (as observed from an eclipse), then we can determine the stellar masses from Kepler's third law. Velocities from a double-line spectroscopic binary without the inclination yield the mass ratio and lower limits on the individual masses. A single-line version gives even less information, only a complex function of masses and the inclination. But if we can estimate the mass of the O star in CygX-1, then we can at least place a lower limit on the mass of the invisible companion. It turns out to be over 3 solar masses, and it may be close to 16. Any normal star of that mass should be visible. It is too massive for a white dwarf or a neutron star, and therefore we expect it to be a black hole. The X-rays are caused by material that overflows the tidal surface of the O dwarf and enters an accretion disk that is heated to a high temperature before disappearing forever down the celestial drain. Without the X-rays we never would have noticed the star in the first place. There are two similar candidates: LMC X-3 in the Large Magellanic Cloud and the obscure A0620-00 in Monoceros, both of which seem to be closer to 10 M_o.

But in spite of their enormous X-ray luminosities, these three black hole candidates are relatively normal compared with one called SS 433, which has one of the strangest spectra known anywhere. Superimposed on an ordinary set of absorption lines are twin sets of Doppler-shifted hydrogen and helium emissions that wander across the spectrum, criss-crossing back and forth with a period of 164 days. They imply material that is alternately approaching and receding with a speed *at least one-sixth that of light*. From the small Doppler shifts in the absorptions, we see a B star in orbit about an invisible companion. Apparently, it is again sending mass onto a collapsed star of some kind.

Not all matter that flows into SS 433's accretion disk falls into the central gravitational source. Some of it leaves in violent twin jets from the rotation poles. Subtle Doppler motions show that the accretion plane, and the poles, are precessing at the 164-day period, the jets spraying wildly around like a garden hose gone berserk. Radio telescopes even show us the corkscrews of matter spurting away. The mass of the source of this extraordinary behavior is close to the black hole limit; moreover, it is inside a known supernova remnant called W 50, which is lit in the X-ray by the ferocity within. SS 433 may well be a neutron star. Nevertheless the evidence for the existence of black holes is quite strong. In addition, at the centers of galaxies, we see huge masses packed into probable black holes, sometimes with jets of matter shooting out in opposite directions, a phenomenon to be explored in the next chapter.

So far we have seen an escalation of stellar strangeness, from the apparently simple Sun, through the evolution of the giants and the super-giants, to the end products—the dense collapsed white dwarfs, neutron stars, and black holes. Now we come to perhaps the most remarkable part of the story: the beginning.

SUMMARY OF KEY CONCEPTS

3.1 EINSTEIN'S SECOND REVOLUTION

What are the major ideas of general relativity?

General relativity tells us that gravity arises from curvature of spacetime and that the curvature arises from the presence of masses. This idea leads us to a view of gravity in which time runs more slowly in gravitational fields, black holes can exist in spacetime, and the universe has no center or edges. It also predicts the existence of gravitational waves propagating through space.

Is all motion relative?

Special relativity shows that motion at constant velocity is always relative, but the relative nature of motion is less evident when gravity and acceleration enter the picture. However, the **equivalence principle** allows us to continue treating all motion as relative because it tells us that the effects of gravity are exactly equivalent to the effects of acceleration. The physical effects on someone in an accelerating reference frame are identical to those on someone who is stationary in a gravitational field.

3.2 UNDERSTANDING SPACETIME

What is spacetime?

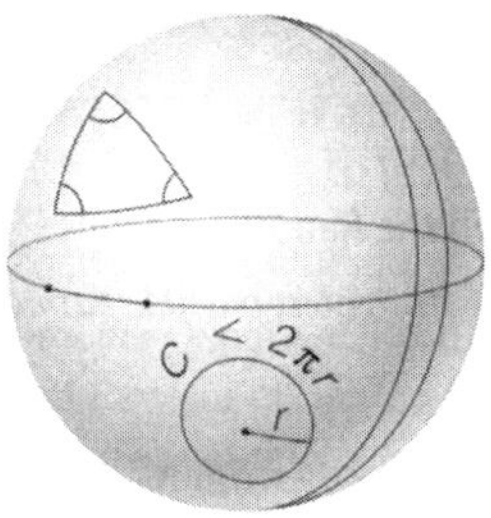

Spacetime is the four-dimensional combination of space and time that forms the "fabric" of our universe.

What is curved spacetime?

Spacetime can be curved much as a sheet of paper can be curved, but in more dimensions. We can recognize spacetime curvature from the rules of geometry. The three possible geometries are a **flat geometry**, in which the ordinary laws of flat (Euclidean) geometry apply; a **spherical geometry**, in which lines that start out parallel tend to converge; and a **saddle-shaped geometry**, in which lines that start out parallel tend to diverge. Spacetime may have different geometries in different places.

3.3 A NEW VIEW OF GRAVITY

What is gravity?

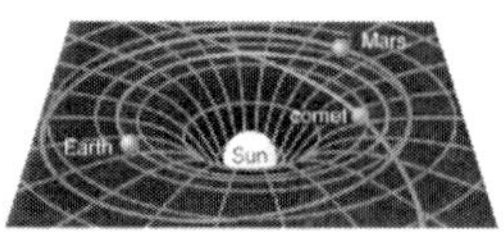

Gravity arises from curvature of spacetime. Mass causes spacetime to curve, and the curvature of spacetime determines the paths of freely moving masses.

What is a black hole?

A **black hole** is a place where spacetime is curved so much that it essentially forms a bottomless pit, making it a true hole in spacetime.

How does gravity affect time?

Time runs more slowly in places where gravity is stronger, an effect called **gravitational time dilation**.

3.4 TESTING GENERAL RELATIVITY

How do we test the predictions of the general theory of relativity?

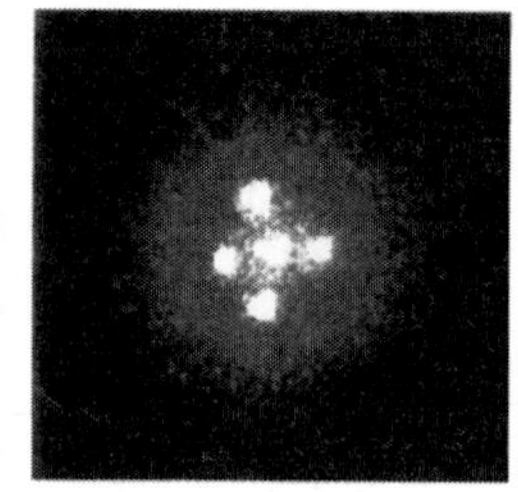

Observations of the precession of Mercury's orbit match the precession predicted by Einstein's theory, Observations of stars during eclipses and photos of **gravitational lensing** provide spectacular confirmation of the idea that light can travel curved paths through space. **Gravitational redshifts** observed in the light of objects with strong gravity confirm the slowing of time predicted by general relativity, a prediction that has also been confirmed with docks at different altitudes on Earth.

What are gravitational waves?

General relativity predicts that accelerating masses produce gravitational waves that travel at the speed of light. Observations of binary neutron stars provide solid indirect evidence that gravitational waves really exist.

3.5 HYPERSPACE, WORMHOLES, AND WARP DRIVE

Where does science end and science fiction begin?

No known physical laws prevent hyperspace, **wormholes**, or warp drive from offering "loopholes" that could allow us to get from one place to another in less time than we could by traveling through ordinary space. However, if any one of them proves to be real, then cause and effect might not be absolute, a proposition troubling to many scientists.

3.6 THE LAST WORD

How has relativity changed our view of space and time?

Prior to Einstein, space and time were viewed as separate and distinct. We now know that they are deeply intertwined as *spacetime* and that understanding spacetime is crucial to understanding many aspects of astronomy, including black holes and the overall geometry of the universe.

CHAPTER 12
HOW ARE STARS BORN?

THE INTERSTELLAR MEDIUM

Stars come from the matter that lies in the spaces between them, its existence made vividly clear by the diffuse nebulae and by the structure of the Milky Way itself, with its clouds of obscuring dust. These phenomena, however, represent but the simplest features of an enormously complex system only recently revealed to us by an advancing technology that allows observations by radio telescopes operating at a variety of frequencies and by ground- and space-based infrared detectors. Only when the intricate details of the interstellar medium are understood can we comprehend the complicated mechanisms by which stars are created.

The existence of a general interstellar gas pervading the galactic disk was first revealed in 1904, when astronomers noted the K line of Ca II in absorption in the spectrum of Delta Orionis. This star is a spectroscopic binary whose stellar absorption features move back and forth over a 5.7-day period as a result of Doppler shifts. But the Ca II line just sits there unmoving, revealing that it is not a part of the star but must be produced in the space along the line of sight. Today we know of hundreds of these interstellar lines (particularly in the part of the ultraviolet recently made accessible by spacecraft), formed by a wide range of elements that includes neutral and ionized carbon, sodium, silicon, magnesium, zinc, nickel, and iron. Even the traces of some simple interstellar molecules—CH and CN, for example—are found in stellar spectra, a portent of the great discoveries of interstellar molecular chemistry to come. Quite clearly, interstellar gas is not confined to diffuse nebulae. The lines are commonly split by small Doppler shifts, showing that the matter is arranged in discrete clouds with densities of 10 or so atoms per cubic centimeter that are slowly moving (at a few km/s) relative to one another.

The insidious nature of the dust delayed its discovery. Before 1930, astronomers thought that except for the few obvious dark clouds, interstellar space was clear and transparent. That notion was shattered by Robert Trumpler of Lick Observatory in his research on open clusters, the kind found in the galactic disk. Distances determined on the basis of stellar brightness through main sequence fitting were invariably larger than those found from the clusters' angular diameters based upon the assumption of common physical size. Moreover, the stars of the more distant clusters appeared redder than expected for their spectral classes. Suddenly it was clear that dust was everywhere in the galactic plane, dimming the light of the stars by about a magnitude for every thousand parsecs of distance and making the clusters appear farther away than they were. At about the same time, Edwin Hubble finally showed that the fuzzy "spiral nebulae" are actually distant galaxies, stellar systems similar to our own. It had long been known that they did not appear in the plane of the Milky Way. The obvious explanation of this "zone of avoidance" was the dust, which almost completely prevents us from seeing through the plane of the disk to the outside. The dust is distributed very irregularly, in great clumps and patches, the thickest of which, the Bok globules (after the Dutch-American astronomer Bart Bok, 1906-1983) are easily seen against the background of diffuse nebulae and the Milky Way.

The dimming or extinction of background starlight is caused by the scattering of photons rather than by absorption. The scattering efficiency increases in inverse proportion to wavelength, causing distant stars to appear anomalously red and demonstrating that the particles are typically well under a micron (a thousandth of a millimeter) in diameter. The dust is optically illuminated only when it lies in the neighborhood of a bright star, where it produces a reflection nebula. It is always found mixed in with the gas, the particles making about 1 percent of the total mass. The dust has a powerful impact on studies of the universe because any distances based on inferred absolute magnitudes (or vice versa) require correction; fortunately, total extinction is related to the degree of reddening, which is usually measurable. The dust is so thick, however, that regions like the galactic center and sites of active star formation (the interiors of Bok globules, for example) are forever hidden from optical view.

Clearly, any picture of the universe confined to the optical band of the electromagnetic spectrum must be very limited. A true understanding of the nature of interstellar space had to wait for developments in radio and infrared astronomy. These longer waves can penetrate the dust and provide information on low-temperature systems that have no optical effects.

The great initial step was taken in 1951 with the discovery by Edward Purcell and Harold Ewen of Harvard (based on a 1944 prediction by the Dutch astronomer Hendrik van de Hulst) that interstellar neutral hydrogen produces a powerful spectrum line at a wavelength of 21 centimeters. A magnetic interaction between the electron and the proton splits the ground energy state of the hydrogen atom into two very closely spaced hyperfine levels. The energy is a little greater when the proton and electron are spinning in the same direction (parallel) than when they are spinning oppositely (antiparallel). The electrons are knocked into the parallel state by collisions with neighbors. When they spontaneously reverse direction, they emit photons at 21 cm. If the hydrogen happens to be silhouetted against a bright radio source, the 21-cm line can also be seen in absorption. Doppler shifts and radial velocities of discrete radiating or absorbing clouds are easily found, allowing the determination of the speed of galactic rotation as a function of distance from the center, as well as the estimation of the distances of the individual clouds. As a result, we see that the neutral hydrogen outlines spiral arms, and we can map them across the Galaxy.

But the 21-cm observations show only a small part of the picture. Beginning in the 1960s, radio astronomers began to find an abundance of interstellar molecules far in excess of that inferred from optical data. Molecules are free to rotate and vibrate in quantized energy states that split the electronic energy levels into fine structures. These are again further split into hyperfine levels. The transitions between a variety of closely spaced states produce low-energy photons in the radio spectrum. The first to be discovered were simple structures—the OH radical, ammonia, water, and carbon monoxide—several of them observable in circumstellar shells. Then came real astonishment, as rarer and more complex species were uncovered: formaldehyde, acetylene, methyl and ethyl alcohol, and (including some ions) nearly 80 more. The most intriguing—long organic chains of up to 13 atoms that cannot be produced in terrestrial laboratories—can be identified only by predicting their complex spectra theoretically.

The vast majority of the gas is simple molecular hydrogen, H_2, which is quite difficult to observe as it does not produce, radio lines. However, we theoretically expect, and indeed observe, that the different kinds of molecules are linked together as one is created from another. Therefore, the location of the H_2 is traced out by powerful emissions of much less abundant carbon monoxide. Some molecular gas resides in globules, but most of it is found within giant molecular clouds with which the globules are frequently associated and whose dusty natures protect the fragile molecules from destructive stellar radiation. Some 6000 giant molecular clouds are known. Up to 100 pc across, they are the most massive discrete structures in the Galaxy, typically containing enough matter to create over 200,000 Suns. They are closely confined to the spiral arms of the Galaxy; indeed, it seems to be the spiral arms that make the clouds in the first place.

These graceful features are not permanent streams of stars, which would wrap up as the galaxy rotates. Instead, they are waves of density propagating through the interstellar material. They are initiated by gravitational perturbations possibly produced by a rotating bar or a nearby companion like the Large Magellanic Cloud. The gravitational disturbances pile up matter, producing more disturbances, and the waves then spread throughout the galaxy, developing a spiral pattern as the result of galactic rotation. Stars and clouds move in and out of them as they go by. The waves compress the interstellar gas, creating regions of higher density, and the clouds develop with the aid of their own gravity. These vast molecular assemblies are not the regions that produce the 21-cm emissions, which tend to lie along the outside edges of the true spiral arms. There is an evolutionary relation between the two that probably involves the way interstellar gas is consumed and transformed in star formation, but its nature is far from clear.

The last step involves a more difficult spectral domain, the infrared. Advances in understanding the interstellar medium had to wait for improved detector technology and the ability to observe from space. Once observation became possible, the view was stunning. At longer wavelengths the cool, dusty clouds are no longer dark but glow brightly. Spectral absorption bands in heavily reddened starlight show that the grains of dust come in two basic varieties: silicates, and carbon in a form akin to graphite (recall the same division in circumstellar shells. Optical chemical analysis of the interstellar gas shows it is strongly depleted in heavier elements, such as iron, which appear to have condensed from the gas onto the grains. The grains grow over time, accumulating atoms from interstellar space and becoming coated with ices. Much of the molecular chemistry takes place on the grain surfaces. After the molecules are formed, they are kicked back into space.

The infrared gives an even more intriguing view of interstellar chemistry at work. Several objects, including planetary, reflection, and diffuse nebulae, exhibit infrared emission lines that have been identified with polycyclic aromatic hydrocarbons (PAHs), components of soot built up from benzene rings. Alternatively, the spectral features may be produced by C^{60}, a soccer-ball shaped molecule named buckminsterfullerene after its structural resemblance to a geodesic dome and known familiarly as "bucky-ball." The deeper we look, the more we see. What are the most complex molecules that exist? Can we find the seeds of life itself?

Big clouds, little clouds, some dense, some tenuous, all netted within a thinner medium of neutral hydrogen and partially ionized gas. Occasionally, some of this gas comes in contact with a hot star, and the ghostly glow of a diffuse of reflection nebula marks its secret existence. Threading its way through this chaos, like tunnels in a giant sponge, is a hot gas, its temperatures in the hundreds of thousands of degrees, the result of overlapping bubbles created by supernovae.

INFANT STARS

We cannot formulate the processes by which stars develop without further evidence. Can we find newborn stars? If so, what are their characteristics?

Look at the dark clouds of the Milky Way in optical radiation and you see nothing, only a blank wall of dust. But infrared radiation can penetrate the obscuring haze, and we see that the clouds, from the small Bok globules to the larger complexes that sprawl over constellation boundaries, are filled with stars. We immediately surmise that these stars are very young and that the clouds are their birthplaces. That supposition is powerfully supported by the massive but short-lived O stars that are always associated with interstellar clouds, which they illuminate in great diffuse nebulae. Somehow, the gas and dust conspire to produce the thousands of mature lights that fill the evening sky—and the one that fills the day.

Proximity alone, however, does not prove that the clouds are stellar nurseries, nor does it tell us anything about the birth process itself. We need an observational progression of formation from the very beginning right onto the main sequence. And we have it. Sprinkled across the faces of the vast, dark clouds that sprawl through Orion, Scorpius-Ophiuchus, and Taurus-Auriga are myriad odd variables named after their prototype, T Tauri, which has been known for 50 years. Rather like the O stars, they are loosely organized into gravitationally unbound systems called T associations, showing that they had to be created together. They are typically G and K stars that exhibit considerable instability; T Tauri itself, for example, is normally about magnitude 11, but can erratically brighten to 10th or fade to 14th. Their youth is revealed by their spectra, as they exhibit strong absorption lines of lithium. This element is easily destroyed by nuclear reactions at temperatures far below those required for fusion. The lithium lines in the 5-billion-year-old Sun, for example, are quite weak. As convection circulates the solar gases downward to higher temperatures, the lithium in the envelope gradually diminishes. But the T Tauri stars have a full complement, comparable to that found in interstellar space.

Emission lines in T Tauri spectra disclose that the stars are involved with circumstellar gas. Many have P Cygni lines that reveal gas outflows. but we occasionally also see *inverse* P Cyg features that demonstrate matter flowing *onto* the star. These stars are so young that they are still growing, accreting mass from their surroundings. One, DR Tauri, can show both infall and outflow at the same time! As in their evolved cousins, accretion seems to take place from a disk that surrounds the new star. The highly unstable nature of the accretion process accounts for at least some of the erratic variations that these stars undergo. Occasionally, the infall rate will suddenly increase, and the star will brighten by several magnitudes.

Long and short-wave spectra reveal more surprises. The stars are much brighter in the infrared than they should be for their spectral types, the enhancement coming from clouds (or a disk) of circumstellar dust warmed by the stellar radiation. The ultraviolet is anomalously bright as well, a result of rapidly infalling gas and intense chromospheric activity that can also generate powerful X-ray flares. Since stellar activity steadily declines as a result of magnetic braking, the strong chromospheres again demonstrate how truly young these stars are.

The roles of the T Tauri stars in the flow of stellar evolution are clear when they are placed on the HR diagram, where they lie well above and to the right of the main sequence. Since they, are demonstrably young, they must be evolving onto it rather than away from it. Further proof of the direction of evolution is that the' so-called naked T Tauri stars, those that display only a vestige of the activity of this class, are closer to the zero-age main sequence than are the others. They have shed or used up much of their circumstellar disks and shrouds, enabling us to see through to the stars themselves with little interference. The whole collection falls right into the region predicted by theory for masses between about a third and double solar. Farther up the main sequence we see hotter-emission-line stars up to about 8 solar masses behaving the same way. Given time, they will all move to the left along their evolutionary tracks and become good, quiet citizens of the main sequence like our own Sun—which some 5 billion years ago must surely have been one of these unruly infants.

DISKS AND JETS

In the 1950s, George Herbig of Lick Observatory and Guilliermo Haro of the University of Mexico independently found a few small blobs of bright matter among the dark clouds of Orion. These Herbig-Haro (HH) objects show considerable structure, consisting of several individual knots. They are everywhere that we find dark clouds and T Tauri stars. The knots display emission lines but have no stellar cores nor obvious means of illumination. Even odder, their structures are changeable over a few years, a new knot appearing where none had been before. For a time, astronomers thought the objects were new stars undergoing gravitational collapse. Instead, these by-products of star formation provide powerful clues to the process.

HH objects are commonly paired, as seen in the photograph of HH 34. When we look closely enough, with deep optical or infrared imaging, we invariably find a star between the two. Some of the central sources are in fact T Tauri stars, so the phenomenon is clearly linked with stellar youth. Emerging from the star are oppositely directed twin jets of gas that are aimed right at the HH objects. These blobs are not new stars at all, but condensations of the interstellar medium that have been rammed and compressed by a bipolar flow of matter speeding along at several hundred kilometers per second. Over time, we can even see the HH objects being pushed radially outward under the force of the blow. Their illumination comes from the pounding shock wave that is set up as the jet comes to a screaming halt.

The narrowness of the twin beams is remarkable. Something within is tightly collimating them. Most likely, the central stars are surrounded by thick circumstellar disks, the same ones suspected of enclosing and feeding matter to the T Tauri stars, which display their own evidence for mass outflow. The two groups of objects are thus bound even more tightly. Evidently mass loss is a natural part of mass accretion. As matter falls into the star from the thick surrounding ring, some of it is hurled away, but it can escape only through the poles of the disk, where there is little or no mass. Exactly the same phenomenon was encountered for active galaxies with black hole cores and for the one-time supernova and black hole candidate SS 433. In those cases the matter enters the disk from an evolving companion or from disrupted stars; with HH objects, it is acquired directly from the interstellar medium. The evidence that it occurs is overwhelming.

If we pull back from HH 34 to look at its environs, our view encompasses more and more of these fascinating objects. Several stars are being formed at the same time from the parental cloud, generating what will eventually be a T association. More remarkably, the jets are nearly all aliened! [?] Something, most likely a broad magnetic field, has choreographed the spins of the interstellar clouds that collapse to create first the HH objects and then the T Tauri stars. The existence of such magnetic fields within the Galaxy is demonstrated by the partial polarization of obscured starlight (perfectly polarized light waves oscillate in a single plane), showing that the elongated grains that make up the clouds and filter the light are magnetically constrained to lie in a common direction.

Evidence for bipolar flows is not confined to the obvious HH objects. Extending far beyond these optical shocks can be long streams of matter traceable with radio radiation from the CO molecule. Doppler shifts show that the CO is streaming away from a central source.

Despite this ample evidence, however, the collimating disks have so far in the discussion only been hypothesized. They are indeed there, and they can be seen. The ammonia molecule is a good tracer of dense matter. Radio observations show disks of ammonia radiation set perpendicular to the streams of CO emission. *Hubble* has even shown one directly.

FORMATION

Star formation is as complex as stellar evolution. There is no single scenario, no one simple sequence that takes stars from their conception to their births as hydrogen-burning dwarfs. Some clouds generate high-mass stars, associations, even tight clumps of O stars, whereas others specialize in the lower main sequence. Whole clusters can be born with all stellar masses aboard, and so can isolated multiples, pairs, perhaps even singles. We do not yet understand why some clouds behave differently from others, nor are the paths of star formation all mapped out in detail. But the broad outline is becoming abundantly clear. So now we gather the evidence and place it in a context of theory.

The first step in star formation is the production of these clouds. The chief triggering agents seem to be the density waves that create, the Galaxy's spiral structure. The clouds, once formed, are held together by their own gravity but are kept from collapsing by their internal turbulence, rotation, and quite likely by weak but extensive magnetic fields. However, here and there, self-gravity will win out, and the cold gas and dust will begin to contract into more compact zones called dense cores. At least some of these are likely to be triggered by blast waves from supernova explosions that provide additional compression, star death initiating star birth.

The big problem for a developing blob is its angular momentum. As the blob contracts, it must spin faster and faster. If nothing intervenes, the nascent star will tear itself apart long before it has a chance to be born. There are two chief ways of disposing of the excess angular momentum. The magnetic fields that thread through the clump of matter are tied to the surrounding gas. They can transfer the angular momentum outward, and thereby apply a brake, or the condensing mass can divide into two or more parts, the angular momentum going into orbital rather than rotational motion. Presumably, cloud fragmentation is the reason we see so many double and multiple stars: They are a natural by-product of the star-forming process.

Follow along now the development of stars like the Sun. The dense cores begin to collapse from the center outward, accreting more matter as time proceeds. At their centers they are developing protostars that over a few hundred thousand years will build to real stellar proportions. Growing protostars will become hot and very bright as a result of the conversion of gravitational energy into radiation. Once they become hot enough at their centers (about a million degrees Kelvin) the deuterium invested by the Big Bang begins to fuse into helium, so the protostars swell. The stars become fully convective, so that the deuterium falling in from the surrounding dense cores can be swept into the protostars and burned, which maintains their sizes. At the same time, the accreting matter is spinning out into disks as a result of the conservation, of angular momentum. With radio telescopes, we now detect bipolar flows that emanate from the poles in the disks. The protostars then cross the "birthline" and now become visible as producers of Herbig-Haro objects and finally as T Tauri stars.

The winds of the protostars eventually develop winds, which terminate the infall and begin to clear away the surrounding opaque hazes. As the stars slowly move along their evolutionary tracks the deuterium fuel supply and thus deuterium burning diminishes and then finally stops. The stars then move toward the zero age main sequence. Eventually, the intense surface activity generated from the accretion disks diminishes and the naked T Tauri stars emerge. These pre-main sequence stars finally quiet down, become hot enough to fuse common hydrogen in their cores by the proton-proton chain, and settle into life as ordinary dwarfs. The whole process has taken a few tens of millions of years. The newly minted stars will now leave their birthplaces to wander free in space and to live serenely for billions of years, their origins long lost in time.

The higher-mass stars—up to about 8 solar masses—behave similarly. Some of the phenomena associated with star birth reach their acmes among the high-mass O stars, which not only produce bipolar flows but also radiate with such intensity that they can sweep great bubbles in the interstellar medium, aiding in its compression. The brilliant O and B stars live such short lives that they never move very far before they expire. They therefore act as markers of the spiral arms and show us where stellar birth takes place. Without them, the arms, which are only enough to get the birth process under way, would effectively disappear from optical view.

CHAPTER 13
"STAR STUFF"

from Carl Sagan's Cosmos, pp. 231–243

THE LIVES OF STARS

Hydrogen fusion cannot continue forever: in the Sun or any other star, there is only so much hydrogen fuel in its hot interior. The fate of a star, the end of its life cycle, depends very much on its initial mass. If, after whatever matter it has lost to space, a star retains two or three times the mass of the Sun, it ends its life cycle in a startlingly different mode than the Sun. But the Sun's fate is spectacular enough. When the central hydrogen has all reacted to form helium, five or six billion years from now, the zone of hydrogen fusion will slowly migrate outward, an expanding shell of thermonuclear reactions, until it reaches the place where the temperatures are less than about ten million degrees. Then hydrogen fusion will shut itself off. Meanwhile the self-gravity of the Sun will force a renewed contraction of its helium-rich core and a further increase in its interior temperatures and pressures. The helium nuclei will be jammed together still more tightly, so much so that *they* begin to stick together, the hooks of their short-range nuclear forces becoming engaged despite the mutual electrical repulsion. The ash will become fuel, and the Sun will be triggered into a second round of fusion reactions.

This process will generate the elements carbon and oxygen and provide additional energy for the Sun to continue shining for a limited time. A star is a phoenix, destined to rise for a time from its own ashes.* Under the combined influence of hydrogen fusion in a thin shell far from the solar interior and the high temperature helium fusion in the core, the Sun will undergo a major change: its exterior will expand and cool. The Sun will become a red giant star, its visible surface so far from its interior that the gravity at its surface grows feeble, its atmosphere expanding into space in a kind of stellar gale. When the Sun, ruddy and bloated, becomes a red giant, it will envelop and devour the planets Mercury and Venus—and probably the Earth as well. The inner solar system will then reside within the Sun.

Billions of years from now, there will be a last perfect day on Earth. Thereafter the Sun will slowly become red and distended, presiding over an Earth sweltering even at the poles. The Arctic and Antarctic icecaps will melt, flooding the coasts of the world. The high oceanic temperatures will release more water vapor into the air, increasing cloudiness, shielding the Earth from sunlight and delaying the end a little. But solar evolution is inexorable. Eventually the oceans will boil, the atmosphere will evaporate away to space and a catastrophe of the most immense proportions imaginable will overtake our planet.† In the meantime, human beings will almost certainly have evolved into something quite different. Perhaps our descendants will be able to control or moderate stellar evolution. Or perhaps they will merely pick up and leave for Mars or Europa or Titan or, at last, as Robert Goddard envisioned, seek out an uninhabited planet in some young and promising planetary system.

The Sun's stellar ash can be reused for fuel only up to a point. Eventually the time will come when the solar interior is all carbon and oxygen, when at the prevailing temperatures and pressures no further nuclear reactions can occur. After the central helium is almost all used up, the interior of the Sun will continue its postponed collapse, the temperatures will rise again, triggering a last round of nuclear reactions and expanding the solar atmosphere a little. In its death throes, the Sun will slowly pulsate, expanding and contracting once every few millennia, eventually spewing its atmosphere into space in one or more concentric shells of gas. The hot exposed solar interior will flood the shell with ultraviolet light, inducing a lovely red and blue fluorescence extending beyond the orbit of Pluto. Perhaps half the mass of the Sun will be lost in this way. The solar system will then be filled with an eerie radiance, the ghost of the Sun, outward bound.

When we look around us in our little corner of the Milky Way, we see many stars surrounded by spherical shells of glowing gas, the planetary nebulae. (They have nothing to do with planets, but some of them seemed reminiscent in inferior telescopes of the blue-green discs of Uranus and Neptune.) They appear as rings, but only because, as with soap bubbles, we see more of them at the periphery than at the center. Every planetary nebula is a token of a star *in extremis*. Near the central star there may be a retinue of dead worlds, the remnants of planets once full of life and now airless and ocean-free, bathed in a wraithlike luminance. The

*Stars more massive than the Sun achieve higher central temperatures and pressures in their late evolutionary stages. They are able to rise more than once from their ashes, using carbon and oxygen as fuel for synthesizing still heavier elements.

†The Aztecs foretold a time "when the Earth has become tired … when the seed of Earth has ended." On that day, they believed, the Sun will fall from the sky and the stars will be shaken from the heavens.

remains of the Sun, the exposed solar core at first enveloped in its planetary nebula, will be a small hot star, cooling to space, collapsed to a density unheard of on Earth, more than a ton per teaspoonful. Billions of years hence, the Sun will become a degenerate white dwarf, cooling like all those points of light we see at the centers of planetary nebulae from high surface temperatures to its ultimate state, a dark and dead black dwarf.

Two stars of roughly the same mass will evolve roughly in parallel. But a more massive star will spend its nuclear fuel faster, become a red giant sooner, and be first to enter the final white dwarf decline. There should therefore be, as there are, many cases of binary stars, one component a red giant, the other a white dwarf. Some such pairs are so close together that they touch, and the glowing stellar atmosphere flows from the distended red giant to the compact white dwarf, tending to fall on a particular province of the surface of the white dwarf. The hydrogen accumulates, compressed to higher and higher pressures and temperatures by the intense gravity of the white dwarf, until the stolen atmosphere of the red giant undergoes thermonuclear reactions, and the white dwarf briefly flares into brilliance. Such a binary is called a nova and has quite a different origin from a supernova. Novae occur only in binary systems and are powered by hydrogen fusion; supernovae occur in single stars and are powered by silicon fusion.

Atoms synthesized in the interiors of stars are commonly returned to the interstellar gas. Red giants find their outer atmospheres blowing away into space; planetary nebulae are the final stages of Sun-like stars blowing their tops. Supernovae violently eject much of their stellar mass into space. The atoms returned are, naturally, those most readily made in the thermonuclear reactions in stellar interiors: Hydrogen fuses into helium, helium into carbon, carbon into oxygen and thereafter, in massive stars, by the successive addition of further helium nuclei, neon, magnesium, silicon, sulfur, and so on are built—additions by stages, two protons and two neutrons per stage, all the way to iron. Direct fusion of silicon also generates iron, a pair of silicon atoms, each with twenty-eight protons and neutrons, joining, at a temperature of billions of degrees, to make an atom of iron with fifty-six protons and neutrons.

These are all familiar chemical elements. We recognize their names. Such stellar nuclear reactions do not readily generate erbium, hafnium, dyprosium, praseodymium or yttrium, but rather the elements we know in everyday life, elements returned to the interstellar gas, where they are swept up in a subsequent generation of cloud collapse and star and planet formation. All the elements of the Earth except hydrogen and some helium have been cooked by a kind of stellar alchemy billions of years ago in stars, some of which are today inconspicuous white dwarfs on the other side of the Milky-Way Galaxy. The nitrogen in our DNA, the calcium in our teeth, the iron in our blood, the carbon in our apple pies were made in the interiors of collapsing stars. We are made of starstuff.

Some of the rarer elements are generated in the supernova explosion itself. We have relatively abundant gold and uranium on Earth only because many supernova explosions had occurred just before the solar system formed. Other planetary systems may have somewhat different amounts of our rare elements. Are there planets where the inhabitants proudly display pendants of niobium and bracelets of protactinium, while gold is a laboratory curiosity? Would our lives be improved if gold and uranium were as obscure and unimportant on Earth as praseodymium?

The origin and evolution of life are connected in the most intimate way with the origin and evolution of the stars. First: The very matter of which we are composed, the atoms that make life possible, were generated long ago and far away in giant red stars.

The relative abundance of the chemical elements found in the Cosmos matches the relative abundance of atoms generated in stars so well as to leave little doubt that red giants and supernovae are the ovens and crucibles in which matter has been forged. The Sun is a second- or third-generation star. All the matter in it, all the matter you see around you, has been through one or two previous cycles of stellar alchemy. Second: The existence of certain varieties of heavy atoms on the Earth suggests that there was a nearby supernova explosion shortly before the solar system was formed. But this is unlikely to be a mere coincidence; more likely, the shock wave produced by the supernova compressed interstellar gas and dust and triggered the condensation of the solar system. Third: When the Sun turned on, its ultraviolet radiation poured into the atmosphere of the Earth; its warmth generated lightning; and these energy sources sparked the complex organic molecules that led to the origin of life. Fourth: Life on Earth runs almost exclusively on sunlight. Plants gather the photons and convert solar to chemical energy. Animals parasitize the plants. Farming is simply the methodical harvesting of sunlight, using plants as grudging intermediaries. We are, almost all of us, solar-powered. Finally, the hereditary changes called mutations provide the raw material for evolution. Mutations, from which nature selects its new inventory of life forms, are produced in part by cosmic rays—high-energy particles ejected almost at the speed of light in supernova explosions. The evolution of life on Earth is driven in part by the spectacular deaths of distant, massive suns.

Imagine carrying a Geiger counter and a piece of uranium ore to some place deep beneath the Earth—a gold mine, say, or a lava tube, a cave carved through the Earth by a river of molten rock. The sensitive counter clicks when exposed to gamma rays or to such high-energy charged particles as protons and helium nuclei. If we bring it close to the uranium ore, which is emitting helium nuclei in a spontaneous nuclear decay, the count rate, the number of clicks per minute, increases dramatically. If we drop the uranium ore into a heavy lead canister, the count rate declines substantially; the lead has absorbed the uranium radiation. Bur some clicks can still be heard. Of the remaining counts, a fraction come from natural radioactivity in the walls of the cave. But there are more clicks than can be accounted for by radioactivity. Some of them are caused by high-energy charged particles penetrating the roof. We are listening to cosmic rays, produced in another age in the depths of space. Cosmic rays, mainly electrons and protons, have bombarded the Earth for the entire history of life on our planet. A star destroys itself thousands of light-years away and produces cosmic rays that spiral through the Milky Way Galaxy for millions of years, until, quite by accident, some of them strike the Earth, and our hereditary material. Perhaps some key steps in the development of the genetic code, or the Cambrian explosion, or bipedal stature among our ancestors were initiated by cosmic rays.

On July 4, in the year 1054, Chinese astronomers recorded what they called a "guest star" in the constellation of Taurus, the Bull. A star never before seen became brighter than any star in the sky. Halfway around the world, in the American Southwest, there was then a high culture, rich in astronomical tradition, that also witnessed this brilliant new star.* From carbon 14 dating of the remains of a charcoal fire, we know that in the middle eleventh century some Anasazi, the antecedents of the Hopi of today, were living under an overhanging ledge in what is today New Mexico. One of them seems to have drawn on the cliff overhang, protected from the weather, a picture of the new star. Its position relative to the crescent moon would have been just as was depicted. There is also a handprint, perhaps the artist's signature.

This remarkable star, 5,000 light-years distant, is now called the Crab Supernova, because an astronomer centuries later was unaccountably reminded of a crab when looking at the explosion remnant through his telescope. The Crab Nebula is the remains of a massive star that blew itself up. The explosion was seen on Earth with the naked eye for three months. Easily visible in broad daylight, you could read by it at night. On the average, a supernova occurs in a given galaxy about once every century. During the lifetime of a typical galaxy, about ten billion years, a hundred million stars will have exploded—a great many, but still only about one star in a thousand. In the Milky Way, after the event of 1054, there was a supernova observed in 1572, and described by Tycho Brahe, and another, just after, in 1604, described by Johannes Kepler.† Unhappily, no supernova explosions have been observed in our Galaxy since the invention of the telescope, and astronomers have been chafing at the bit for some centuries.

Supernovae are now routinely observed in other galaxies. Among my candidates for the sentence that would most thoroughly astonish an astronomer of the early 1900's is the following, from a paper by David Helfand and Knox Long in the December 6, 1979, issue of the British journal *Nature:* "On 5 March, 1979, an extremely intense burst of hard X-rays and gamma rays was recorded by the nine interplanetary spacecraft of the burst sensor network, and localized by time-of-flight determinations to a position coincident with the supernova remnant N49 in the Large Magellanic Cloud." (The Large Magellanic Cloud, so-called because the first inhabitant of the Northern Hemisphere to notice it was Magellan, is a small satellite galaxy of the Milky Way, 180,000 light-years distant. There is also, as you might expect, a Small Magellanic Cloud.) However, in the same issue of Nature, E. P. Mazets and colleagues of the Ioffe Institute, Leningrad—who observed this source with the gamma-ray burst detector aboard the Venera 11 and 12 spacecraft on their way to land on Venus—argue that what is being seen is a flaring pulsar only a few hundred light-years away. But despite the close agreement in position Helfand and Long do not insist that the gamma-ray outburst is associated with the supernova remnant. They charitably consider many alternatives, including the surprising possibility that the source lies within the solar system. Perhaps it is the exhaust of an alien starship on its long voyage home. But a rousing of the stellar fires in N49 is a simpler hypothesis: we are sure there are such things as supernovae.

The fate of the inner solar system as the Sun becomes a red giant is grim enough. But at least the planets will never be melted and frizzled by an erupting supernova. That is a fate reserved for planets near stars more massive than the Sun. Since such stars with higher temperatures and pressures run rapidly through their store of nuclear fuel, their lifetimes are much shorter than the Sun's. A star tens of times more massive than the Sun can stably convert hydrogen to helium for only a few million years before moving briefly on to more exotic nuclear reactions. Thus there is almost certainly not enough time for the evolution of advanced forms of life on any accompanying planets; and it will be rare that beings

*Moslem observers noted it as well. But there is not a word about it in all the chronicles of Europe.

†Kepler published in 1606 a book called *De Stella Nova*, "On the New Star," in which he wonders if a supernova is the result of some random concatenation of atoms in the heavens. He presents what he says is "… not my own opinion, but my wife's: Yesterday, when weary with writing, I was called to supper, and a salad I had asked for was set before me. 'It seems then,' I said, 'if pewter dishes, leaves of lettuce, grains of salt, drops of water, vinegar, oil and slices of eggs had been flying about in the air for all eternity, it might at last happen by chance that there would come a salad.' 'Yes,' responded my lovely, 'but not so nice as this one of mine.'"

elsewhere can ever know that their star will become a supernova; if they live long enough to understand supernovae, their star is unlikly to become one.

The essential preliminary to a supernova explosion is the generation by silicon fusion of a massive iron core. Under enormous pressure, the free electrons in the stellar interior are forceably melded with the protons of the iron nuclei, the equal and opposite electrical charges canceling each other out; the inside of the star is turned into a single giant atomic nucleus, occupying a much smaller volume than the precursor electrons and iron nuclei. The core implodes violently, the exterior rebounds and a supernova explosion results. A supernova can be brighter than the combined radiance of all the other stars in the galaxy within which it is embedded. All those recently hatched massive blue-white supergiant stars in Orion are destined in the next few million years to become supernovae, a continuing cosmic fireworks in the constellation of the hunter.

The awesome supernova explosion ejects into space most of the matter of the precursor star—a little residual hydrogen and helium and significant amounts of other atoms, carbon and silicon, iron and uranium. Remaining is a core of hot neutrons, bound together by nuclear forces, a single, massive atomic nucleus with an atomic weight about 10^{56}, a Sun thirty kilometers across; a tiny, shrunken, dense, withered stellar fragment, a rapidly rotating neutron star. As the core of a massive red giant collapses to form such a neutron star, it spins faster. The neutron star at the center of the Crab Nebula is an immense atomic nucleus, about the size of Manhattan, spinning thirty rimes a second. Its powerful magnetic field, amplified during the collapse, traps charged particles rather as the much tinier magnetic field of Jupiter does. Electrons in the rotating magnetic field emit beamed radiation not only at radio frequencies but in visible light as well. If the Earth happens to lie in the beam of this cosmic lighthouse, we see it flash once each rotation. This is the reason it is called a pulsar. Blinking and ticking like a cosmic metronome, pulsars keep far better time than the most accurate ordinary clock. Long-term timing of the radio pulse rate of some pulsars, for instance, one called PSR 0329 + 54, suggests that these objects may have one or more small planetary companions. It is perhaps conceivable that a planet could survive the evolution of a star into a pulsar; or a planet could be captured at a later time. I wonder how the sky would look from the surface of such a planet.

Neutron star matter weighs about the same as an ordinary mountain per teaspoonful—so much that if you had a piece of it and let it go (you could hardly do otherwise), it might pass effortlessly through the Earth like a falling stone through air, carving a hole for itself completely through our planet and emerging out the other side—perhaps in China. People there might be out for a stroll, minding their own business, when a tiny lump of neutron star plummets out of the ground, hovers for a moment, and then returns beneath the Earth, providing at least a diversion from the routine of the day. If a piece of neutron star matter were dropped from nearby space, with the Earth rotating beneath it as it fell, it would plunge repeatedly through the rotating Earth, punching hundreds of thousands of holes before friction with the interior of our planet stopped the motion. Before it comes to rest at the center of the Earth, the inside of our planet might look briefly like a Swiss cheese until the subterranean flow of rock and metal healed the wounds. It is just as well that large lumps of neutron star matter are unknown on Earth. But small lumps are everywhere. The awsome power of the neutron star is lurking in the nucleus of every atom, hidden in every teacup and dormouse, every breath of air, every apple pie. The neutron star teaches us respect for the commonplace.

A star like the Sun will end its days, as we have seen, as a red giant and then a white dwarf. A collapsing star twice as massive as the Sun will become a supernova and then a neutron star. But a more massive star, left, after its supernova phase, with, say, five times the Sun's mass, has an even more remarkable fate reserved for it—its gravity will turn it into a black hole. Suppose we had a magic gravity machine—a device with which we could control the Earth's gravity, perhaps by turning a dial. Initially the dial is set at 1 g* and everything behaves as we have grown up to expect. The animals and plants on Earth and the structures of our buildings are all evolved or designed for 1 g. If the gravity were much less, there might be tall, spindly shapes that would not be tumbled or crushed by their own weight. If the gravity were much more, plants and animals and architecture would have to be short and squat and sturdy in order not to collapse. But even in a fairly strong gravity field, light would travel in a straight line, as it does, of course, in everyday life.

Consider (page 236) a possibly typical group of Earth beings. As we lower the gravity, things weigh less. Near 0 g the slightest motion sends our friends floating and tumbling up in the air. Spilled tea—or any other liquid—forms throbbing spherical globs in the air: the surface tension of

*1 g is the acceleration experienced by falling objects on the Earth, almost 10 meters per second every second. A falling rock will reach a speed of 10 meters per second after one second of fall, 20 meters per second after two seconds, and so on until it strikes the ground or is slowed by friction with the air. On a world where the gravitational acceleration was much greater, falling bodies would increase their speed by correspondingly greater amounts. On a world with 10 g acceleration, a rock would travel 10x10 m/sec or almost 100 m/sec after the first second, 200 m/sec after the next second, and so on. A slight stumble could be fatal. The acceleration due to gravity should always be written with a lowercase g, to distinguish it from the Newtonian gravitational constant, G, which is a measure of the strength of gravity everywhere in the universe, not merely on whatever world or Sun we are discussing. (The Newtonian relationship of the two quantities is F = mg = GMm/r2; g = GM/r2, where F is the gravitational force, M is the mass of the planet or star, m is the mass of the falling object, and r is the distance from the falling object to the center of the planet or star.)

the liquid overwhelms gravity. Balls of tea are everywhere. If now we dial 1 g again, we make a rain of tea. When we increase the gravity a little—from 1 g to, say, 3 or 4 g's—everyone becomes immobilized: even moving a paw requires enormous effort. As a kindness we remove our friends from the domain of the gravity machine before we dial higher gravities still. The beam from a lantern travels in a perfectly straight line (as nearly as we can see) at a few g's, as it does at 0 g. At 1000 g's, the beam is still straight, but trees have become squashed and flattened; at 100,000 g's, rocks are crushed by their own weight. Eventually, nothing at all survives except, through a special dispensation, the Cheshire cat. When the gravity approaches a billion g's, something still more strange happens. The beam of light, which has until now been heading straight up into the sky, is beginning to bend. Under extremely strong gravitational accelerations, even light is affected. If we increase the gravity still more, the light is pulled back to the ground near us. Now the cosmic Cheshire cat has vanished; only its gravitational grin remains.

When the gravity is sufficiently high, nothing, not even light, can get out. Such a place is called a black hole. Enigmatically indifferent to its surroundings, it is a kind of cosmic Cheshire cat. When the density and gravity become sufficiently high, the black hole winks out and disappears from our universe. That is why it is called black: no light can escape from it. On the inside, because the light is trapped down there, things may be attractively well-lit. Even if a black hole is invisible from the outside, its gravitational presence can be palpable. If, on an interstellar voyage, you are not paying attention, you can find yourself drawn into it irrevocably, your body stretched unpleasantly into a long, thin thread. But the matter accreting into a disk surrounding the black hole would be a sight worth remembering, in the unlikely case that you survived the trip.

Thermonuclear reactions in the solar interior support the outer layers of the Sun and postpone for billions of years a catastrophic gravitational collapse. For white dwarfs, the pressure of the electrons, stripped from their nuclei, holds the star up. For neutron stars, the pressure of the neutrons staves off gravity. But for an elderly star left after supernova explosions and other impetuosities with more than several times the Sun's mass, there are no forces known that can prevent collapse. The star shrinks incredibly, spins, reddens and disappears. A star twenty times the mass of the Sun will shrink until it is the size of Greater Los Angeles; the crushing gravity becomes 10^{10} g's, and the star slips through a self-generated crack in the space-time continuum and vanishes from our universe.

Black holes were first thought of by the English astronomer John Michell in 1783. But the idea seemed so bizarre that it was generally ignored until quite recently. Then, to the astonishment of many, including many astronomers, evidence was actually found for the existence of black holes in space. The Earth's atmosphere is opaque to X-rays. To determine whether astronomical objects emit such short wavelengths of light, an X-ray telescope must be carried aloft. The first X-ray observatory was an admirably international effort, orbited by the United States from an Italian launch platform in the Indian Ocean off the coast of Kenya and named Uhuru, the Swahili word for "freedom." In 1971, Uhuru discovered a remarkably bright X-ray source in the constellation of Cygnus, the Swan, flickering on and off a thousand times a second. The source, called Cygnus X-l, must therefore be very small. Whatever the reason for the flicker, information on when to turn on and off can cross Cyg X-1 no faster than the speed of light, 300,000 km/sec. Thus Cyg X-l can be no larger than [300,000 km/sec] x[(1/1000) sec] = 300 kilometers across. Something the size of an asteroid is a brilliant, blinking source of X-rays, visible over interstellar distances. What could it possibly be? Cyg X-l is in precisely the same place in the sky as a hot blue supergiant star, which reveals itself in visible light to have a massive close but unseen companion that gravitationally tugs it first in one direction and then in another. The companion's mass is about ten times that of the Sun. The supergiant is an unlikely source of X-rays, and it is tempting to identify the companion inferred in visible light with the source detected in X-ray light. But an invisible object weighing ten times more than the Sun and collapsed into a volume the size of an asteroid can only be a black hole. The X-rays are plausibly generated by friction in the disk of gas and dust accreted around Cyg X-l from its supergiant companion. Other stars called V861 Scorpii, GX339-4, SS433, and Circinus X-2 are also candidate black holes. Cassiopeia A is the remnant of a supernova whose light should have reached the Earth in the seventeenth century, when there were a fair number of astronomers. Yet no one reported the explosion. Perhaps, as I. S. Shklovskii has suggested, there is a black hole hiding there, which ate the exploding stellar core and damped the fires of the supernova. Telescopes in space are the means for checking these shards and fragments of data that may be the spoor, the trail, of the legendary black hole.

A helpful way to understand black holes is to think about the curvature of space. Consider a fiat, flexible, lined two-dimensional surface, like a piece of graph paper made of rubber. If we drop a small mass, the surface is deformed or puckered. A marble rolls around the pucker in an orbit like that of a planet around the Sun. In this interpretation, which we owe to Einstein, gravity is a distortion in the fabric of space. In our example, we see two-dimensional space warped by mass into a third physical dimension. Imagine we live in a three-dimensional universe, locally distorted by matter into a fourth physical dimension that we cannot perceive directly. The greater the local mass, the more intense the local gravity, and the more severe the pucker, distortion or warp of space. In this analogy, a black hole is a kind of bottomless pit. What happens if you fall in? As seen from the outside, you would take an infinite amount of time to fall in, because all your clocks—mechanical and biological—would be perceived as having stopped. But from *your* point of view, all your clocks would be ticking away normally. If you could somehow survive the gravitational tides and radiation flux, and (a likely assumption) if the black hole were rotating, it is just possible that you might emerge in another part of spacetime—somewhere else in space, somewhen else in time. Such worm holes in space, a little like those in an apple, have been seriously suggested, although they have by no means been proved to exist. Might gravity tunnels provide a kind of interstellar or intergalactic subway, permitting us to travel to inaccessible places much more rapidly than we could in the ordinary way? Can black holes serve as time machines, carrying us to the remote past or the distant future? The fact that such ideas are being discussed even semi-seriously shows how surreal the universe may be.

We are, in the most profound sense, children of the Cosmos. Think of the Sun's heat on your upturned face on a cloudless summer's day; think how dangerous it is to gaze at the Sun directly. From 150 million kilometers away, we recognize its power. What would we feel on its seething self-luminous surface, or immersed in its heart of nuclear fire? The Sun warms us and feeds us and permits us to see. It fecundated the Earth. It is powerful beyond human experience. Birds greet the sunrise with an audible ecstasy. Even some one-celled organisms know to swim to the light. Our ancestors worshiped the Sun,* and they were far from foolish. And yet the Sun is an ordinary, even a mediocre star. If we must worship a power greater than ourselves, does it not make sense to revere the Sun and stars? Hidden within every astronomical investigation, sometimes so deeply buried that the researcher himself is unaware of its presence, lies a kernel of awe.

The Galaxy is an unexplored continent filled with exotic beings of stellar dimensions. We have made a preliminary reconnaissance and have encountered some of the inhabitants. A few of them resemble beings we know. Others are bizarre beyond our most unconstrained fantasies. But we are at the very beginning of our exploration. Past voyages of discovery suggest that many of the most interesting inhabitants of the galactic continent remain as yet unknown and unanticipated. Not far outside the Galaxy there are almost certainly planets, orbiting stars in the Magellanic Clouds and in the globular clusters that surround the Milky Way. Such worlds would offer a breathtaking view of the Galaxy rising—an enormous spiral form comprising 400 billion stellar inhabitants, with collapsing gas clouds, condensing planetary systems, luminous supergiants, stable middle-aged stars, red giants, white dwarfs, planetary nebulae, novae, supernovae, neutron stars and black holes. It would be clear from such a world, as it is beginning to be clear from ours, how our matter, our form and much of our character is determined by the deep connection between life and the Cosmos.

*The early Sumerian pictograph for god was an asterisk, the symbol of the stars. The Aztec word for god was Teotl, and its glyph was a representation of the Sun. The heavens were called the Teoatl, the godsea, the cosmic ocean.

APPENDICES

APPENDIX 1: THE ANCIENT CONSTELLATIONS

NAME	MEANING	α(h)	δ°	GENITIVE	ABBREVIATION
The Zodiac					
Aries	ram	3	+20	Arietis	Ari
Taurus	bull	5	+20	Tauri	Tau
Gemini	twins	7	+20	Geminorum	Gem
Cancer	crab	8.5	+15	Cancri	Cnc
Leo	lion	11	+15	Leonis	Leo
Virgo	virgin	13	0	Virginis	Vir
Libra[1]	scales	15	-15	Librae	Lib
Scorpius	scorpion	17	-30	Scorpii	Sco
Sagittarius[2]	archer	19	-25	Sagittarii	Sgo
Capricornus[3]	water-goat	21	-20	Capricornii	Cap
Aquarius[3]	waterman	22	-10	Aquarii	Aqr
Pisces[3]	fishes	1	+10	Piscium	Psc
Ursa Major					
Ursa Major	larger bear	11	+60	Ursae Majoris	UMa
Ursa Minor[4]	smaller bear	16	+80	Ursae Minoris	Umi
Boötes	proper name; herdsman, wagoner	15	+30	Bootis	Boo
Orion					
Orion	proper name; hunter, giant	6	0	Orionis	Ori
Canis Major	larger dog	7	-20	Canis Majoris	CMa
Canis Minor	smaller dog	8	+5	Canis Minoris	CMi
Lepus	hare	6	-20	Leporis	Lep
Perseus					
Persus	proper name; hero	3	+45	Persei	Per
Andromeda	proper name; princess	1	+40	Andromedae	And
Cassiopeia	proper name; queen	1	+60	Cassiopeiae	Cas
Cepheus	proper name; king	22	+65	Cephei	Cep
Cetus	whale	2	-10	Ceti	Cet
Pegasus	proper name; winged horse	23	+20	Pegasi	Peg
Argo					
Carina[5]	keel	9	-60	Carinae	Car
Puppis[5]	stern	8	-30	Puppis	Pup
Vela[5]	sails	10	-45	Velorum	Vel
Hercules[6]	proper name; hero	17	+30	Herculis	Her

APPENDIX 1: THE ANCIENT CONSTELLATIONS (CONTINUED)

NAME	MEANING	α(h)	δ°	GENITIVE	ABBREVIATION
Argo (continued)					
Hydra	water serpent	12	-25	Hydrae	Hya
Aries[7]	ram; here, the golden fleece	3	+20	Arietis	Ari
Centaurus					
Centaurus[8]	centaur	13	-45	Centauri	Cen
Lupus	wolf	15	-45	Lupi	Lup
Ara	altar	17	-55	Arae	Ara
Ophiuchus[9]					
Ophiuchus[9]	serpent-bearer	17	0	Ophiuchi	Oph
Serpens[9]	serpent	17	0	Serpentis	Ser
Single Constellations					
Aquila	eagle	20	+15	Aquilae	Aql
Auriga	charioteer	6	+40	Aquilae	Aur
Corona Australis[10]	southern crown	19	+40	Coronae Australis	CrA
Corona Borealis[11]	northern crown	16	+30	Coronae Borealis	CrB
Corvus[12]	crow, raven	12	-20	Corvi	Crv
Crater	cup	11	-15	Crateris	Crt
Cygnus	swan	21	+40	Cygni	Cyg
Delphinus[3]	dolphin	21	+10	Delphini	Del
Draco[13]	dragon	15	+60	Draconis	Dra
Equuleus	little horse	21	+10	Equulei	Equ
Eridanus	proper name; river	4	-30	Eridani	Eri
Lyra[12]	lyre	19	+35	Lyrae	Lyr
Piscis Austrinus[3]	southern fish	22	-30	Piscis Austrini	PsA
Sagitta	arrow	20	+20	Sagittae	Sge
Triangulum	triangle	2	+30	Trianguli	Tri

[1]Originally the claws of Scorpius.
[2]Contains the galactic center.
[3]Constellations of the area of the sky known as the wet quarter for its many watery images.
[4]Contains the north celestial pole.
[5]Carina, Puppis, and Vela are modern subdivisions of the old constellation Argo and together make one of the ancient 48.
[6]One of the oldest constellations known.
[7]Also included in the Zodiac.
[8]Sometimes included in the Argo myth; the centaur Chiron was Jason's foster father.
[9]Ophiucus is identified with the physician Aesculapius, and Serpens with the caduceus.
[10]Sometimes considered as Sagittarius' crown.
[11]Ariadne's crown.
[12]Corvus was Orpheus' companion, Lyra his harp.
[13]Contains the north ecliptic pole.

APPENDIX 2: THE MODERN CONSTELLATIONS

NAME	MEANING	α(h)	δ°	GENITIVE	ABBREVIATION
Antlia	air pump	10	-35	Antliae	Ant
Apus	bee	16	-75	Apodis	Aps
Caelum	graving tool	5	-40	Caeli	Cae
Camelopardalis	giraffe	6	+70	Camelopardalis	Cam
Canes Venatici	hunting dogs	13	+40	Canum Venaticorum	CVn
Chamaeleon	chameleon	10	-80	Chamaeleontis	Cha
Circinus	compasses	15	-65	Circini	Cir
Columba	dove	6	-35	Columbae	Col
Coma Berenices[1]	Berenice's hair	13	+20	Comae Berenices	Com
Crux[2]	southern cross	12	-60	Crucis	Cru
Dorado[3]	swordfish	6	-55	Doradus	Dor
Fornax	furnace	3	-30	Fornacis	For
Grus	crane	22	-45	Gruis	Gru
Horologium	clock	3	-55	Horologii	Hor
Hydrus	water snake	2	-70	Hydri	Hyi
Indus	Indian	22	-70	Indi	Ind
Lacerta	lizard	22	+45	Lacertae	Lac
Leo Minor	smaller lion	10	+35	Leonis Minoris	LMi
Lynx	lynx	8	+45	Lyncis	Lyn
Microscopium	microscope	21	-40	Microscopii	Mic
Monoceros	unicorn	7	0	Monocerotis	Mon
Mensa	table	6	-75	Mensae	Men
Musca (Australis)[4]	(southern) fly	12	-70	Muscae	Mus
Norma	square	16	-50	Normae	Nor
Octans[5]	octant	—	-90	Octantis	Oct
Pavo	peacock	20	-70	Pavonis	Pav
Phoenix	phoenix	1	-50	Phoenicis	Phe
Pictor	easel	6	-55	Pictoris	Pic
Pyxis[6]	compass	9	-30	Pyxidis	Pyx
Reticulum	net	4	-60	Reticuli	Ret
Sculptor[7]	sculptor's workshop	1	-30	Sculptoris	Scl
Scutum[8]	shield	19	-10	Scuti	Sct
Sextans	sextant	10	0	Sextantis	Sex
Telescopium	telescope	19	-50	Telescopii	Tel
Triangulum Australe	southern triangle	16	-65	Trianguli Australis	TrA
Tucana[9]	toucan	0	-65	Tucanae	Tuc
Volans	flying fish	8	-70	Volantis	Vol
Vulpecula	fox	20	+25	Vulpeculae	Vul

[1] Star cluster with many old references, but not considered one of the ancient 48. Referred to by Eratosthenes as Ariadne's Hair; contains the north galactic pole.
[2] Originally a part of Centaurus.
[3] Contains the Large Magellanic Cloud and the south ecliptic pole.
[4] Originally named Musca Australis to distinguish it from Musca Borealis, the northern fly, which is now defunct; "Australis" is now dropped.
[5] Contains the south celestial pole.
[6] Grouped with the ancient constellation Argo.
[7] Originally named by Lacaille l'Atelier du Sculpteur (in Latin, Apparatus Sculptoris); now known simply as Sculptor. Contains the south galactic pole.
[8] Shield of the Polish hero John Sobieski.
[9] Contains the Small Magellanic Cloud.

APPENDIX 3: THE 40 BRIGHTEST STARS

NAME	MEANING[1]	GREEK LETTER NAME	MAGNITUDE[2]	DISTANCE[3] (PC)	ABSOLUTE[3] MAGNITUDE	SPECTRAL[3] TYPE
Sirius	scorching (Gr)	aCMa	-1.46	2.64	1.43	A0 V
Canopus	proper name; pilot (Gr)	a Car	-0.72	96	-5.6	F0II
Rigil Kentaurus	the Centaur's foot	a Cen A	-0.01	1.35	4.34	G2 V
		a Cen B	1.33	1.29	5.78	Kl V
Arcturus	bear-watcher (Gr)	a BOO	-0.04	11.3	-0.31	Kl III
Vega	swooping eagle	a Lyr	0.03	7.76	0.58	A0 V
Capella[4]	she-goat (Lat)	a Aur A	0.08	12.9	-0.47	G5 III
		a Aur B				GO III
Rigel	foot of the Central One	b Ori	0.12	240	-7	B8 Ia
Procyon	before the dog (Gr)	b CMi	0.38	3.49	2.67	F5 IV
Achernar	end of the river	a Eri	0.46	44	-2.8	B3 V
Betelgeuse	armpit of the Central One (corr)	a Ori	0.50v	130	-5.1	M2 Ia
		b Cen	0.61	160	-5.4	Bl III
Altair	flying eagle	a Aql	0.77	5.14	2.22	A7 V
Aldebaran	follower	a Tau	0.85	20.0	-0.66	K5 III
Antares	like Mars (Gr)	a Sco	0.96	190	-5.4	Ml.5Ib
Spica[5]	ear of wheat (Lat)	aVir	0.98	80	-3.5	B2 V
		a Cru A	1.58	98	-3.8	B0.5 IV
Pollux	proper name; twin (Lat)	b Gem	1.14	10.3	1.08	K0 III
Fomalhaut	fish's mouth	a PsA	1.16	7.69	1.73	A3 V
Deneb	tail	a Cyg	1.25	500	-7	A2 Ia
		b Cru	1.25	110	-4.0	B0.5 III
Regulus	little king (Lat)	a Leo	1.35	24	-0.55	B7 V
Adhara	virgins	Œ CMa	1.50	130	-4.1	B2 II
Castor[6]	proper name; twin (Lat)	a Gem	1.58	15.8	0.59	Al V
		_ Cru	1.63	36	-1.14	M3.5 III
Shaula	stinger	_ SCO	1.63	200	-5	B2 IV
Bellatrix	female warrior (Lat)	_ Ori	1.64	75	-2.7	B2 III
Elnath	butting one	b Tau	1.65	40	-1.4	B7 IV
Miaplacidus	uncertain; placidus = calm (Lat)	b Car	1.68	34	-1.0	A2 IV
Alnilam	string of pearls; refers to Orion's belt	Œ Ori	1.70	460	-6.6	B0 la
Al Nair	bright one in the fish's tail	a Gru	1.74	31	-0.7	B7 IV
Alioth	the bull; (corr)	eUMa	1.77	25	-0.2	AO V
Regor	uncertain derivation	_ 2 Vel	1.78	260	-5	WC8 + 09 I
Dubhe	bear	a UMa	1.79	38	-1.1	K0 III
Mirfak	elbow	a Per	1.79	180	- 4	F5 Ib
Wezen	weight	d CMa	1.84	740	-7.5	F8 Ia

NAME	MEANING[1]	GREEK LETTER NAME	MAGNITUDE[2]	DISTANCE[3] (PC)	ABSOLUTE[3] MAGNITUDE	SPECTRAL[3] TYPE
Alkaid	chief of mourners	_ UMa	1.86	31	-0.6	B3 V
Avior	uncertain derivation	Œ Car	1.86	190	-5	K3 III + B2 V
Menkalinen	shoulder of the rein holder	b Aur	1.90	25	-0.1	A2 IV

[1]Names are of Arabic derivation unless otherwise stated. "Lat" and "Gr" refer to Latin and Greek names; serious corruption is indicated by "corr." Principal source is P. Kunitzsch and T. Smart, *Short Guide to Modern Star Names and Their Derivations.* Weisbaden: Harrasowitz, 1986.

[2]Visual magnitudes, or V; "V"denotes variable magnitude.

[3]Distance in parsecs, absolute visual magnitude (Mv), and spectral types. Distances under 400 pc are from Hipparcos parallax measurements; the others are derived from spectral class. These new values are not reflected in the HR diagram on p. 89.

[4]Visually unresolved binary; magnitudes are combined values.

[5]Binary with roughly equal components.

[6]Multiple star.

APPENDIX 4

STAR AND CONSTELLATION MAPS

The following six maps locate most of the constellations and brighter stars. The first shows the north polar region down to about 50° declination. The next four are seasonal equatorial maps that display stars between 60°N and 60°S declinations, and the last shows the south polar region. Declinations are shown along a central hour circle. Right ascensions are noted around the peripheries of the polar maps and along the celestial equator for the seasonal maps.

Stars, generally selected to indicate constellation positions and outlines, are shown through fourth magnitude, although their census is not complete. Fifth magnitude stars are included where they are important parts of their constellations. A few nonstellar objects—clusters, nebulae, and galaxies—are also noted.

The maps show the broad outline of the Milky Way; however, much of its intricate detail is omitted. The galactic equator (the mid-line of the Galaxy) is marked with galactic longitude starting at the galactic center in Sagittarius.

The dates along the edges of the polar maps and along the tops and bottoms of the equatorial maps indicate the appearance of the sky at approximately 8:30 P.M. (20^h30^m) local time. To use the north polar map, face north and rotate the map so that the current month appears at the top. In the southern hemisphere, use the south polar map similarly. The celestial pole will have an elevation in degrees equal to your latitude. To use the equatorial maps in the northern hemisphere, face south and line up the current month with the celestial meridian. To use them in the southern hemisphere, face north and turn them upside down. The equator point (the intersection between the celestial equator and the meridian) will have an elevation in degrees equal to 90° minus the latitude.

For each hour past 20^h 30^m, shift or rotate the map one hour to the west (that is, align an hour circle that is one additional hour to the east). For every 2 hours past 20^h 30^m, add one month to your current month. For example, if it is March 15 at 20^h 30^m, you would align "March" on Map 4 with the celestial meridian. If it is 2^h 30^m, set "June" (Map 5) on the meridian.

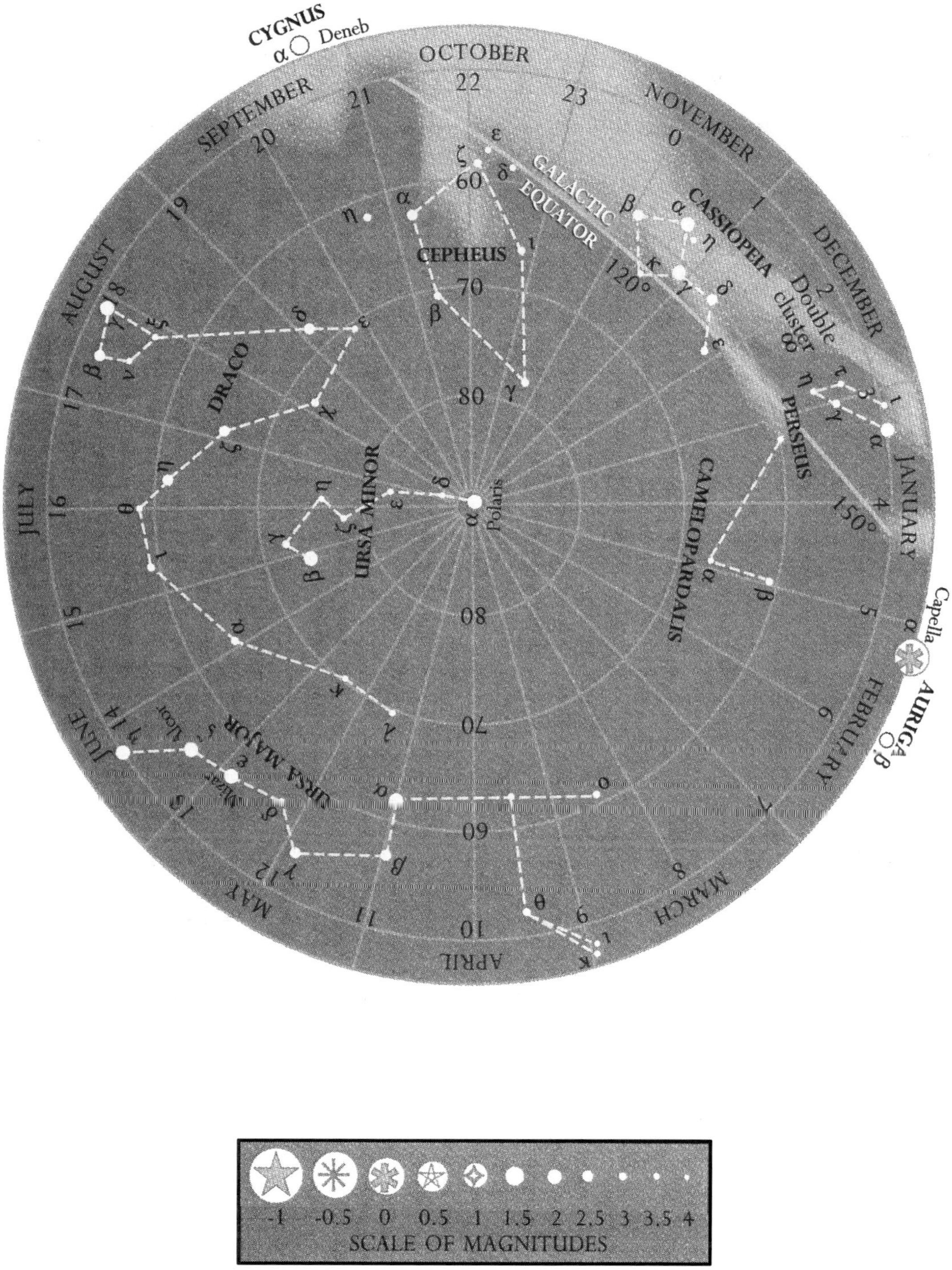

Map 1. The North Polar Constellations

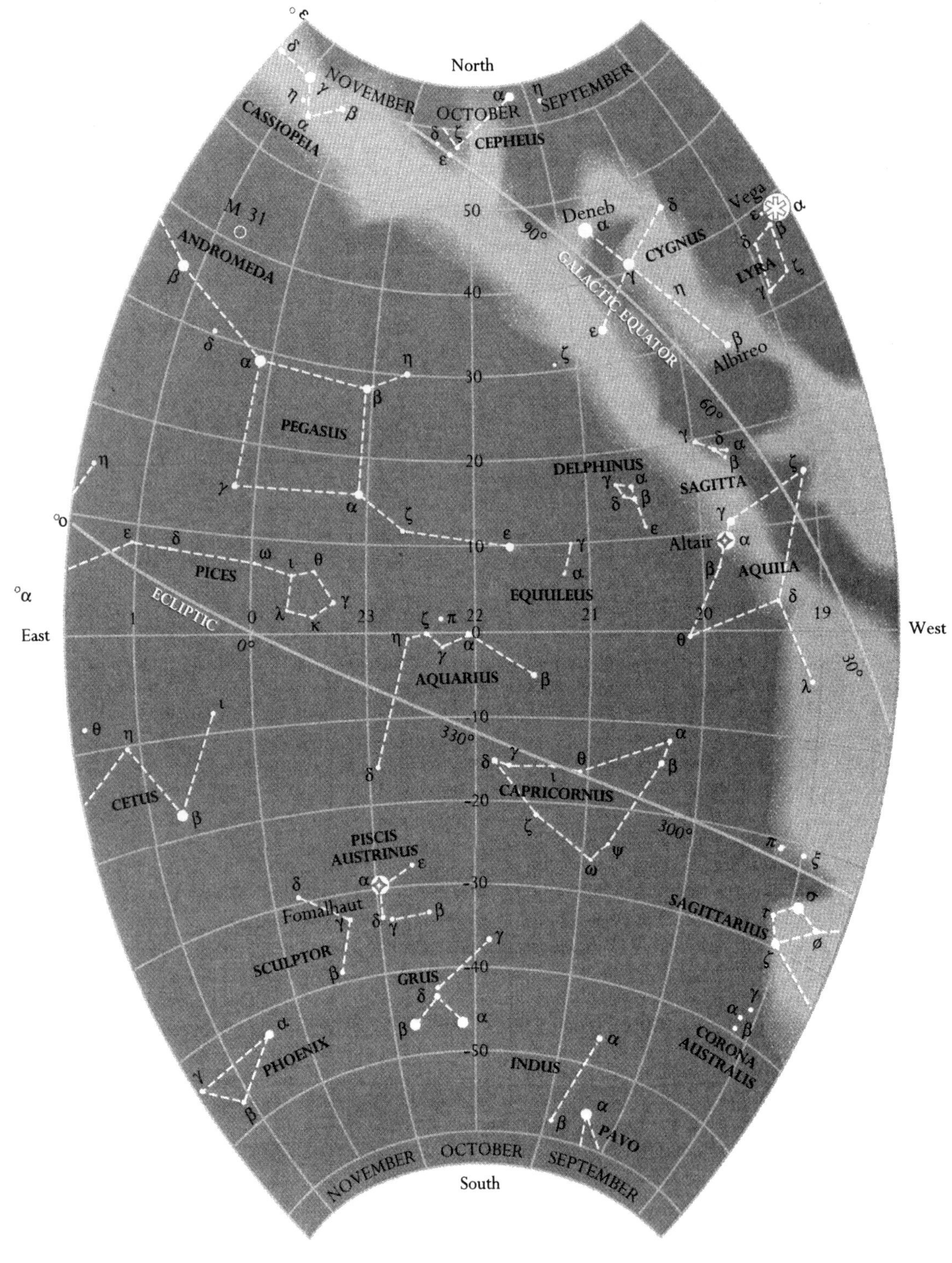

Map 2. The Constellations of Northern Autumn, Southern Spring

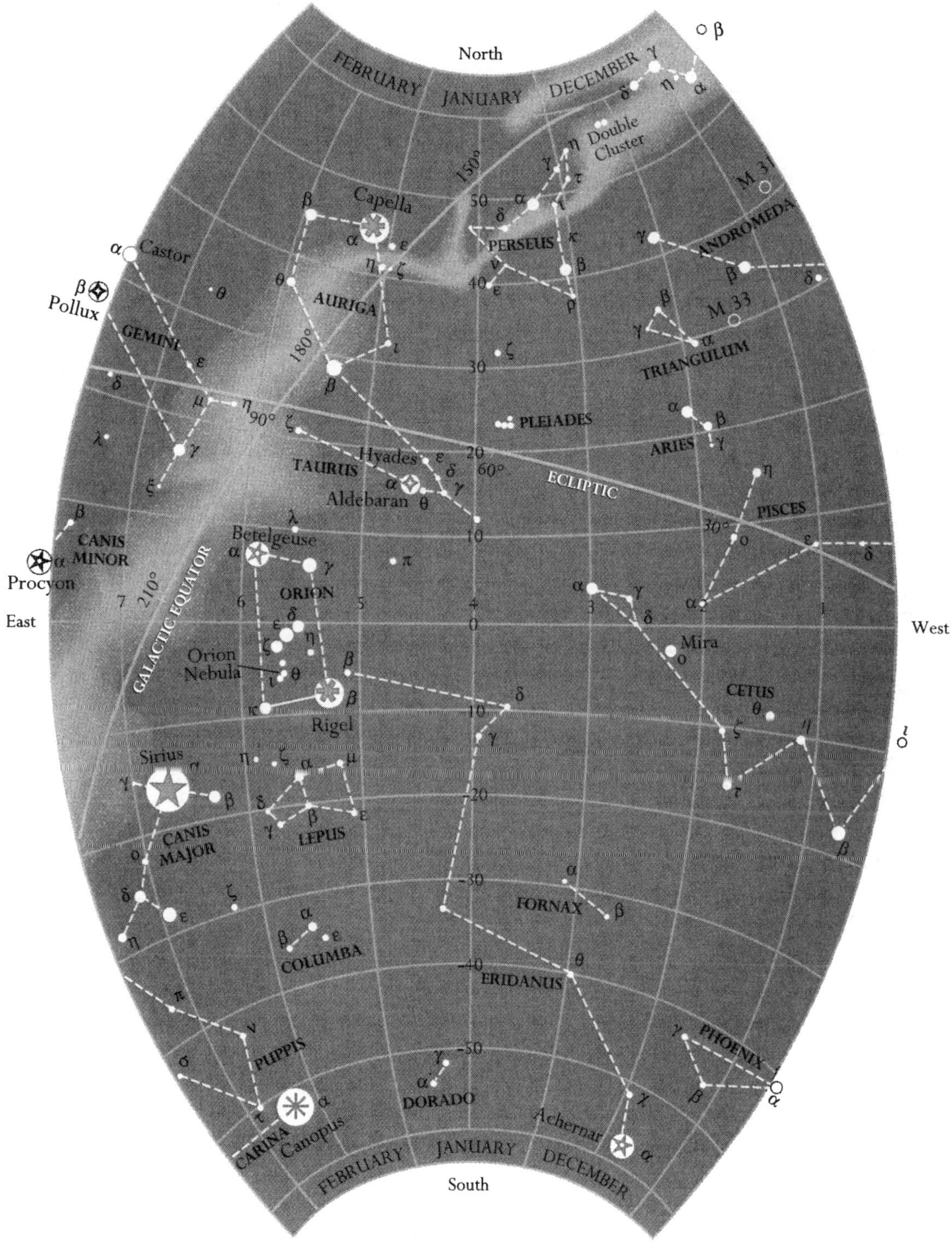

Map 3. The Constellationws of Northern Winter, Southern Summer

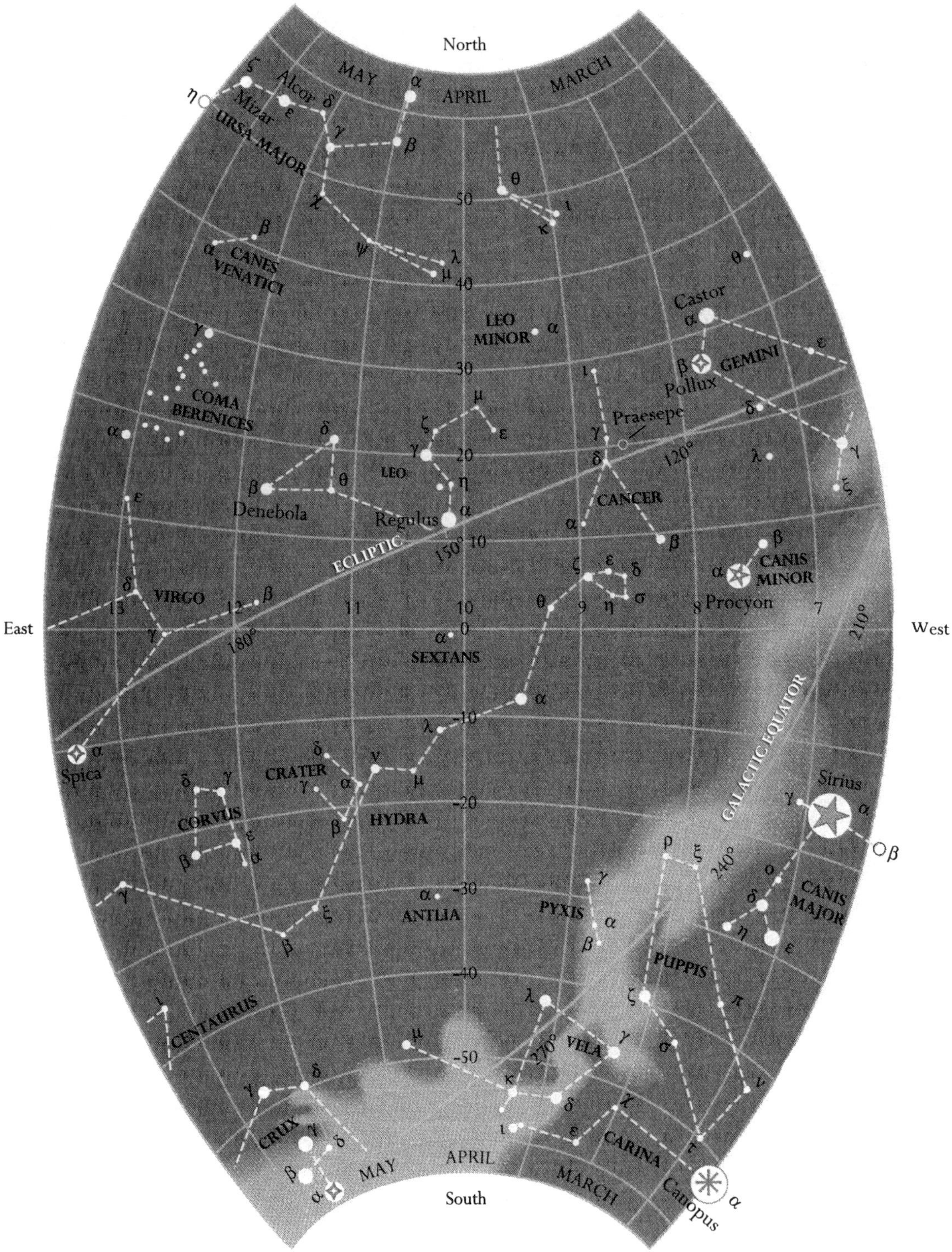

Map 4. The Constellations of Northern Spring, Southern Autumn

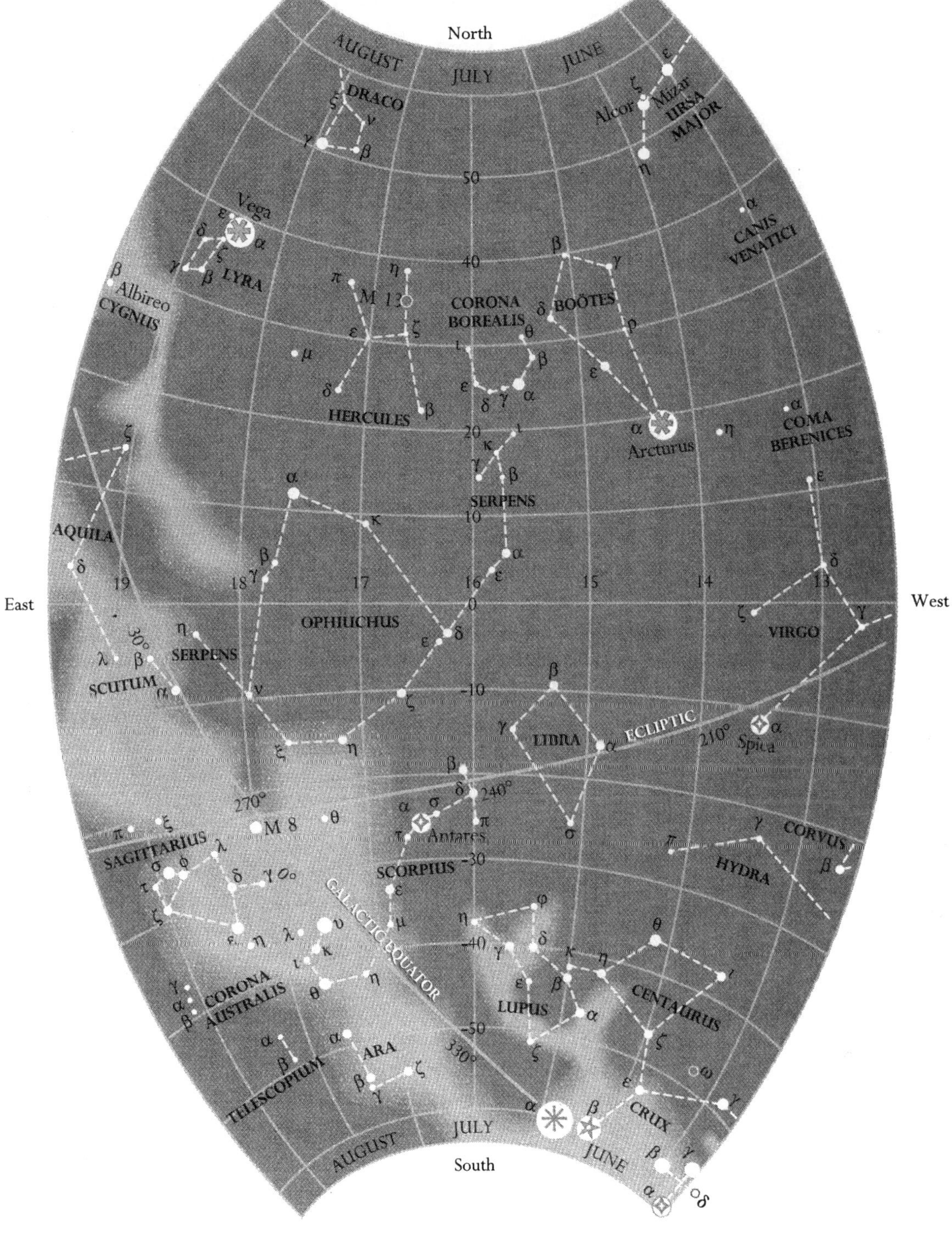

Map 5. The Constellations of Northern Summer, Southern Winter

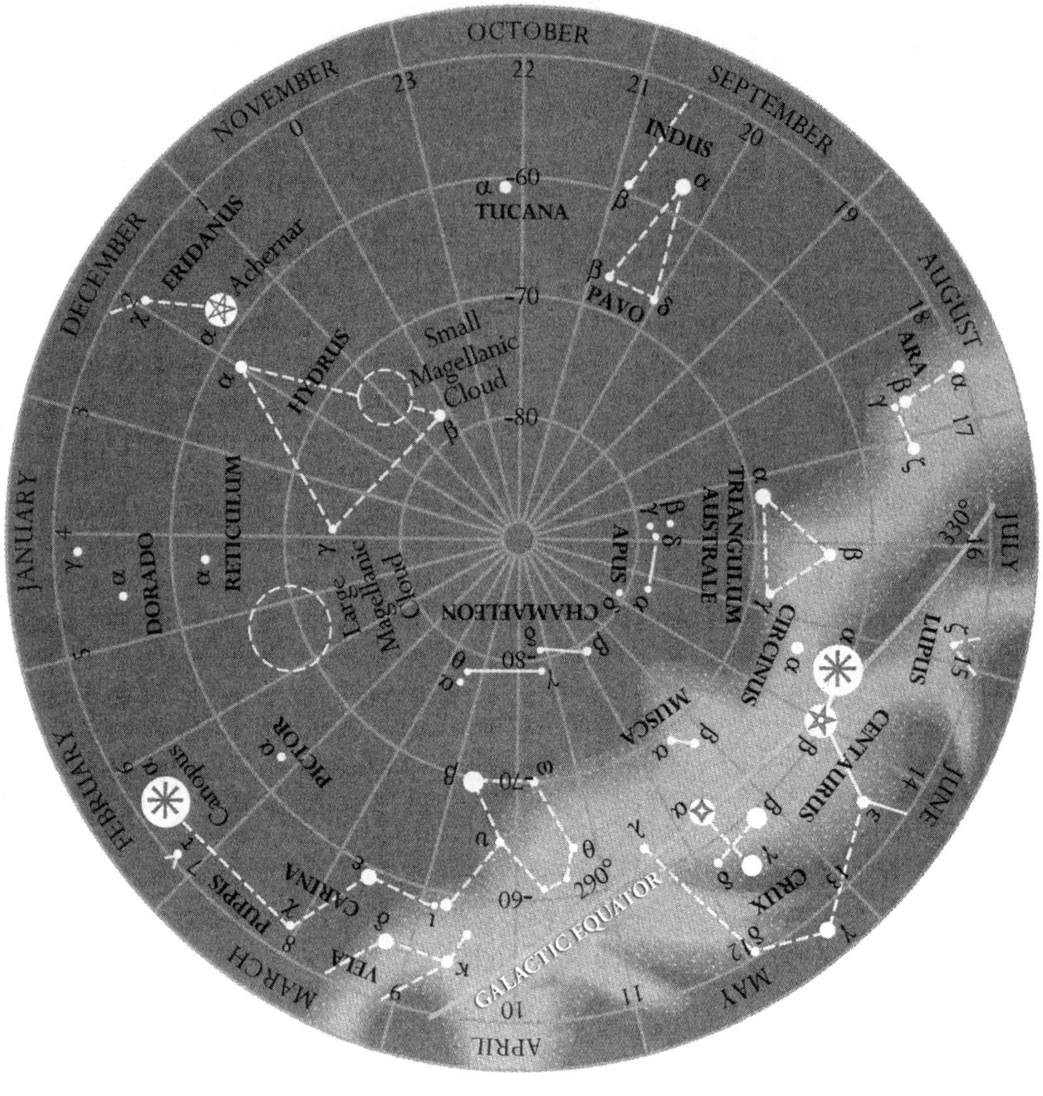

Map 6. The South Polar Constellations